# Sustainable Industrial Wastewater Treatment and Pollution Control

Maulin P. Shah
Editor

# Sustainable Industrial Wastewater Treatment and Pollution Control

Springer

*Editor*
Maulin P. Shah
Environmental Microbiology Lab
Ankleshwar, Gujarat, India

ISBN 978-981-99-2559-9 ISBN 978-981-99-2560-5 (eBook)
https://doi.org/10.1007/978-981-99-2560-5



This Springer imprint is published by the registered company Springer Nature Singapore Pte Ltd.
The registered company address is: 152 Beach Road, #21-01/04 Gateway East, Singapore 189721, Singapore

# Contents

# Microalgae for Treating Wastewater

**C. Marimuthu, J. Arun, M. Subathra, P. Priyadharsini, N. Nirmala, and S. Sarojadevi**

## 1 Introduction

India is the world's second most populated country, with a population growth rate of 1.2% yearly. Water is becoming insufficient in India as a result of the country's rapid population growth. Water shortages will worsen as the world's population is anticipated to reach 1.6 billion by 2050 (Krishnamoorthy and Manickam 2021). In addition to the water shortage, wastewater discharge is increasing as a result of increased industrial, agricultural, and urban activities (Al-Jabri et al. 2021). The largest sources of heavy metal waste in the environment include industries connected to pulp and paper mills, aircraft plating, petroleum refineries, steelworks, tanneries, and textiles. The content of heavy metals varies depending upon the industry. Individual nations establish heavy metal discharge restrictions to safeguard natural water bodies and aquatic life (Chia et al. 2021). Furthermore, numerous governmental entities have recently adopted the circular economy idea in order to decrease and control wastewater pollution (Al-Jabri et al. 2021).

Over the last decade, there has been a revived interest in utilizing fast-growing microalgae for biofuels. These microalgae can contain up to 90% carbohydrates, lipids, and proteins, making them an excellent choice for biofuel production. Microalgae also contain pigments, vitamins, and bio-active molecules, which are

C. Marimuthu
Department of Chemical Engineering, Higher College of Technology, Muscat, Oman

C. Marimuthu · J. Arun (✉) · P. Priyadharsini · N. Nirmala
Centre for Waste Management—International Research Centre, Sathyabama Institute of Science and Technology, Jeppiaar Nagar (OMR), Chennai, Tamil Nadu 600119, India
e-mail: arunjayaseelan93@gmail.com

M. Subathra · S. Sarojadevi
Department of Biotechnology, Aarupadai Veedu Institute of Technology, Vinayaka Nagar, Paiyanoor (OMR), Chennai, Tamil Nadu 603104, India



M. P. Shah (ed.), *Sustainable Industrial Wastewater Treatment and Pollution Control*,
https://doi.org/10.1007/978-981-99-2560-5_1

employed in the pharmaceutical, cosmetics, food and feed sectors (Shandilya and Pattarkine 2019). Due to microalgae's high nutrient absorption capacity and high carbon dioxide ($CO_2$) fixation potential through photosynthesis, microalgae-based wastewater treatment offers a sustainable and cost-effective alternative to traditional wastewater treatment systems. Different microalgae species have been evaluated for nutrient removal efficiency and biomass generation on a variety of wastewater types, including municipal, agricultural, aquacultural, and industrial discharges (Su 2020).

Microalgal genera were identified and attempted for cultivation in a variety of waste stabilization ponds. *Scenedesmus, Euglena, Chlorella, Chlamydomonas, Ankistrodesmus, Golenkinia, Micractinium*, and *Oscillatoria* were the algae discovered in order of quantity and frequency of occurrence (Abdel-Raouf et al. 2012; Shah 2020). Due to the significant land-space requirements of microalgal wastewater treatment systems, efforts are being undertaken to develop wastewater treatment systems based on the utilization of hyper-concentrated algal culture. This was shown to be extremely effective at removing N and P in very short timespan, such as less than 1 h (Lavoie and la Noüe 1985).

Microalgae-based bioremediation, also known as phytoremediation, is a unified idea that may aid the bioeconomy and rising civilizations by delivering value-added goods in addition to solving problems such as pollution (Batista et al. 2015; Shah 2021a, b). Zero-draft wastewater treatment systems with microalgal culture were developed as a way to reduce the cost of microalgae production, but they are now used as a substitute for conventional treatment of wastewater. Due to the enormous potential of microalgae in terms of nutrient intake and biomass production, a significant amount of research is being conducted to assess their economic viability (Kalra, Gaur and Goel 2021; Grobbelaar 2010).

## 2 Industrial Wastewater

Wastewater from varied sectors comprises a variety of complicated chemical compositions. Industrial wastewater often contains low P and N concentrations as well as high levels of heavy metals and toxins, all of which can interfere with microalgae production and photosynthesis. Industrial effluent contains significant levels of metals such as Cadmium, Arsenic, Mercury, Selenium, Chromium, Lead as well as nitrogenous and Sulphur chemicals. Because of their huge surface area and strong binding affinity, microalgal cells have been shown to effectively sequester heavy metals. Industrial wastewaters are frequently contaminated with organic chemicals such as hydrocarbons and aromatic compounds (Gupta et al. 2016; Kumar et al. 2015; Shah 2021a, b). Pharmaceuticals and personal care compounds pollute wastewater, which causes the severe health and environmental risk. It contains diclofenac, benzothiazole, and OH-Benzothiazole, etc. (Al-Jabri et al. 2021).

Two techniques are widely utilized in the treatment of industrial wastewater: primary treatment and secondary treatment. Coarse solids/particles are removed in primary treatment, whereas organic materials are bioremediated using microorganisms in subsequent treatment. There are certain disadvantages to conventional methods. They often need a significant quantity of energy, a significant area, and substantial operational and maintenance expenses. Because microalgae can extract nutrients and convert them to biomass, they provide an alternate technique in biological treatment (Udaiyappan et al. 2017; Chinnasamy et al. 2010). Some microalgal species have the tendency to absorb pollutants from wastewater, such as heavy metals, nitrogen, and other toxic chemicals (Udaiyappan et al. 2017; El-Kassas and Mohamed 2014).

In comparison to plants, microalgae remove 10–50 times more $CO_2$. Additionally, these photosynthetic microbes convert solar energy to biomass three times faster than plants, giving them a remarkable feedstock. Microalgae, like plants, have minimal nutritional needs and may flourish in a variety of water environments, including freshwater, seawater, and even toxic wastewaters. The use of wastewater is deemed sustainable due to the simultaneous benefit of nutrient/toxic removal and satisfying vast quantities of water requirements for the growth of microalgae (Debnath et al. 2021).

## 3 Microalgae on Wastewater Treatment

Microalgae being a sustainable resource can be used as a promising remedy for treating the wastewater. Wastewater treatment is always a challenge in the globe, since it involves many different types of contaminants. Each contaminant requires different technology for purification process (Muñoz et al. 2009). Existing conventional wastewater treatment techniques focus on the removal of solids by using activated sludge and reduce the biological oxygen demand. They also perform degradation process to break down the organic and inorganic constituents. But this technology holds up with a limitation regards with the removals of heavy metals, heavy loads of nutrients and xenobiotic (Sousa et al. 2018; Eerkes-Medrano et al. 2019). This limitation can be swiped off by using microalgae wastewater treating technique since microalgae are flexible metabolically and can be able to perform any kind of metabolism such as phototrophic, autotrophic, heterotrophic according to the source available in the wastewater (Hu et al. 2018). Microalgae usage in treating wastewater also has added advantage since they involve in direct transformation of contaminants present in the wastewater and offers valuable sustainable bio-products as shown in Fig. 1. Microalgae also aggregate with bacteria to improve the purification process by supplying required oxygen for the photosynthesis process. This activity reduces both the direct and indirect cost-like oxygen supply and stirring performance (Soto-Sierra et al. 2018; Quijano et al. 2017). The activity of the microalgae also depends on many parameters which directly and indirectly influence the microalgae growth, thereby influencing its performance against treating wastewater.

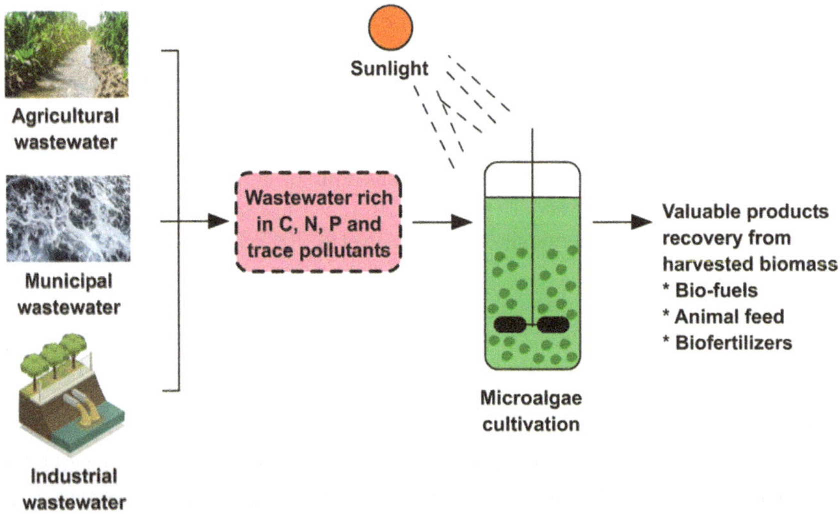


**Fig. 1** Algae biotechnology in wastewater treatment and recovery of valuable products

## 4 Parameters Influencing Microalgae-based Wastewater Treatment

The important factors influencing algal growth are

- Light
- Temperature
- pH and salinity
- Carbon dioxide
- Nutrients
- Mixing
- Reactor design.

### *4.1 Light*

Light availability plays a major role in carrying out photosynthesis process which directly impacts on the growth and production of microalgae. Light remains an important factor in microalgae growth. The amount of light supply can be abundant which increases the productivity. On the other side, excess light or higher oxygen level may damage the photosynthetic apparatus present inside the microalgae and pull down the growth of the microalgal species (Tredici and Zittelli 1998). Light source can be in any form such as, sunlight, lamps, LEDs, etc., which produce irradiance. But photosynthetically active radiation whose range is between 400 and 700 nm is used by the microalgae species (Scott et al. 2010). The intensity of the light also impacts on the biomass production. If the light intensity is less, the cells on the outer layer get exposed to the light and photosynthesis is being performed. But the cells present in the inner part of the species remain in dark condition and fail to perform photosynthesis which pulls down the biomass concentration. To rule out this, a light source with optimum intensity is a must (Ota et al. 2015).

### 4.2 Temperature

Temperature is one of the important factors that influence the microalgae growth in many means like rate of growth, requirement of nutrient by the cell, size of the cell, biochemical composition, etc. Temperature to the microalgae is obtained from the light source which is supplied to the reactor (Deb et al. 2017). Optimal temperature (20–35 °C) increases the microalgae growth. But low temperature as well as high temperature reduce the yield and damage the cells (Bernard and Rémond 2012). The natural source of light—sunlight—also tends to increase the reactor's temperature based on the season. Thus seasonal changes must be considered in order to control and maintain the optimal temperature (Sánchez et al. 2008). To prevent the microalgae culture from overheating, many methods have been used such as,

i. Heat exchangers (Watanabe et al. 1995)
ii. Placing the reactor under shade (Vonshak 1997)
iii. Submerged cultures (Becker 1994)
iv. Spraying cold water over the reactor surface (Becker 1994).

### 4.3 pH and Salinity

pH plays a significant role in microalgae culture because it not only affects the growth of microalgae but also influences the concentration of minerals present in the reactor, thereby affecting the microalgae's growth efficiency (Qiu et al. 2017). The optimum pH for the microalgae species is from 6 to 8. There are few other species which can withstand higher and lower pH range apart from the specified optimum level (Rai and Gupta 2017). The hydroxide ion accumulation in the medium leads to the increase in pH during the photosynthesis $CO_2$ fixation process. This increased pH results in the carbonate formation which cannot be used by the microalgae as a carbon source (Lower 1999). But the decrease in pH results in carbon dioxide formation which is considered as the important factor for the microalgae growth. The nature of substrate with nitrogen source also influences the pH of the medium. If the nitrogen source supplied is in the form of ammonia, the pH gradually increases and thereby decreases the amount of nitrogen availability for the growing cell. Phosphorus concentration also increases the pH by precipitating the phosphate salt in the form of iron phosphate, aluminium phosphate, calcium phosphate, etc. due to the precipitation action, the amount of phosphorus availability to the medium gets reduced (Cai et al. 2013).

pH also directly affects the microalgae growth by inactivating the enzyme at low pH and causing cell disruption at high pH (Chiranjeevi and Mohan 2016). Salinity causes growth inhibition since most of the microalgae are grown in freshwater. On considering the wastewater, according to the salinity present, microalgae species have to be selected.

### 4.4 Carbon Dioxide

Carbon dioxide is another vital source for the microalgae growth. Almost double the volume of carbon dioxide is in need to focus the production of single volume of biomass. Since the carbon dioxide concentration needed is very high, the amount of carbon dioxide present in the environment is insufficient to carry out photosynthesis by microalgae cells. So, it is necessary to supply carbon in various other forms like bicarbonate, salts, carbon dioxide air injection (Durán et al. 2018).

### 4.5 Nutrients

The major macro nutrients required for the microalgae growth are Carbon, Nitrogen, and Phosphorus (Becker 1994). Among the above three, carbon occupies 50% of the microalgae composition. Carbon source is supplied in the form of $CO_2$, bicarbonates, sugar, alcohols, and acids. This is considered as the vital nutrients since it is an important constituent of all the organic substances produced by the cells which include carbohydrates, lipids, proteins, vitamins, nucleic acids, etc. Nitrogen is the second important nutrient in the microalgae growth, since it involves in the production of structural and operating element. This occupies almost 14% of the biomass composition (Richmond et al. 2004). Nitrogen shortage in the microalgae cultivation results in discoloration of the biomass which has to be optimized to attain good volume of biomass (Goiris et al. 2015). Wastewater will almost have all the nutrients needed for the microalgae growth, still over nutrient supply also inhibits the biomass production which has to be taken into account.

### 4.6 Mixing

Mixing is carried out to ensure the overall $CO_2$ supply and maintenance of pH inside the reactor. Proper mixing increases the movement of cell in dark and light mode, thereby promoting the photosynthesis process. Mixing also reduces the temperature. But vigorous mixing may damage the cells based on its cell wall thickness. So based on the microalgae species, mixing has to be carried out (Barbosa and Wijffels 2004).

## 4.7 *Reactor Design*

The wastewater treatment using bacterial activated sludge is commonly used method. But the disadvantage in this treatment is removal of excess nutrients, and hazardous metals are very less. This can be overcome by using microalgae. Microalgae designed in a photobioreactor can be efficiently used to treat the wastewater efficiently than any other conventional method. In general, photobioreactors are of two kinds. One is open and another is closed system. In addition to photobioreactors, many other culture treatment methods can be used to treat the wastewater as follows.

### 4.7.1 Hyper-concentrated Cultures

Hyper-concentrated cultures are algal biomass with concentrated number of cells. This concentration of cells can be achieved by flocculation process by using flocculants. This method increases the removal of nitrogen, ammonia, and phosphorous on comparing with other methods. The efficiency of the system is directly proportional to the concentration of the culture present in the reactor. This culture can be used in pond area; when it is used in large scale, this also reduces the residence time (Hashimoto and Furukawa 1989).

### 4.7.2 Stabilization Pond

Stabilization ponds can be used as best alternative to existing conventional method, since the cost and energy consumption are very less since it can be operated with natural process. The stabilization ponds are constructed like a shallow basin, where both the algae and bacteria can be used. This method is a cost-effective method, operation mode is user-friendly and reliable. This method can be used to treat industrial and domestic wastewater efficiently. This entirely works on natural light energy—sunlight—and requires very less maintenance care (Mara 1987). There are many types of stabilization ponds which include,

- Facultative pond
- Anaerobic pond
- Maturation pond.

### 4.7.3 Tubular Photobioreactor

The most promising reactor is tubular photobioreactor which is a closed system. The algae are grown inside the pipe which are circulated with the help of pump; a gas exchanger is fixed to supply $CO_2$ and heat exchanger may be fixed either for cooling or heating process based on the climatic condition. These reactors can be constructed either in vertical rows or flat tubes arranged on the ground or tubes that are spirally wounded or helical-shaped tubes (Lee and Bazin 1990). The tubes are built with glass, PVC, or Perspex with the diameter ranging from 24 mm to 24 cm. These reactors are constructed in the way that temperature, circulation speed, $O_2$, and $CO_2$ can be controlled (Borowitzka 1989).

## 5 Advantages of Microalgae-based Wastewater Treatment

For many countries around the world, global warming is a big worry. By 2030, the globe will have a 40% water shortfall, posing a significant barrier to societal and economic progress. Treatment of various types of wastewater using physicochemical or biological approaches can be inefficient and energy-intensive. The need to upgrade the traditional wastewater treatment process is being driven by a lack of water resources and environmental concerns. Microalgae are minute organisms that grow in water bodies that contain the nutrients they require. The microalgae-based wastewater treatment (MBWT) process, which is one of the most promising technologies for improved wastewater treatment and nutrient recovery, has gotten a lot of attention in recent years. Many researchers have demonstrated the viability of employing microalgae in wastewater treatment as a supplement for tertiary wastewater treatment due to its high nutrient removal efficiency in the advanced treatment of urban, agricultural, and industrial wastewaters. Microalgae are oxygen-evolving photosynthetic microorganisms that look like plants and can be found in a range of aquatic habitats, including freshwater and seawater, as well as municipal, agricultural, industrial, and other types of effluents. Some species can even grow on plants, soils, rocks, and other materials if they have enough C (organic or inorganic carbon), N (ammonium, nitrate, urea, yeast extract), and P (ammonium, nitrate, urea, yeast extract, etc.) and other needed trace elements.

Most wastewaters contain high levels of ammonium, nitrate, and phosphorus, which are normally removed during treatment. Carbon is oxidized to $CO_2$, nitrogen (N) is stripped to the environment in the form of $N_2$, and phosphorus (P) is normally precipitated to avoid nutrient valorization in traditional activated sludge treatment plants (Adav et al. 2008). Because microalgae-bacteria consortia may remove nutrients while creating important biomass, they've been explored extensively for wastewater treatment during the last two decades. Microalgae release oxygen into the medium during photosynthesis, which is used by aerobic bacteria to break down organic materials into $CO_2$, soluble phosphorus, and several inorganic nitrogen sources ($NH_4^+$, $NO_3^-$, and $NO_2^-$). Microalgae then take in inorganic carbon ($CO_2$) and solubilized macro- and micronutrients to develop, producing a clean effluent as well as valuable biomass. As a result, when compared to traditional wastewater treatment plants, using microalgae-bacteria consortia for wastewater treatment offers the benefit of nutrient recycling. Microalgae biomass is frequently used as a feedstock for biofuel production or high-value products, and as a result, its energy consumption and production costs are linked to these processes. Other studies have found that producing energy (i.e. biofuels) from microalgal biomass will only be commercially viable if it is combined with an algae-based wastewater treatment system that is cost-effective in terms of input energy. As a result, recycling N and P from wastewater, in conjunction with wastewater clean-up, will offer some nutrients for microalgae development, lowering energy costs (Molinuevo-Salces et al. 2019).

$CO_2$ emissions in a microalgae-bacteria wastewater treatment plant are significantly lower than those produced in traditional aerated activated sludge processes, and when flue gas rich in $CO_2$ is fed into the process, a negative balance can also emerge (Organization 2018). Since $CO_2$ injection lowers the pH of the culture medium, the injection rate in the ponds must be carefully monitored. Many different optimal $CO_2$ concentrations and injection rates for maximizing biomass production have already been published in the scientific literature, but there are significant differences between them due to the wide variety of microalgal species, culture mediums used, photobioreactor types used, and operational conditions tested (light intensity, salinity, and temperature) (Shuey 2001).

The presence of bacteria in wastewater provides various advantages for microalgae nitrogen absorption. Aerobic bacteria oxidize proteins and nucleic acids to $NH_4^+$, which are ingested by microalgae when they are utilized for agro-industrial wastewater bioremediation. When the concentration of $NH_4^+$ in a solution is greater than 100 mg/L and the pH is greater than 8, the proportion of $NH_4^+$ that converts to $NH_3$ becomes poisonous to algae, preventing their growth (Park and Craggs 2010). P is chemically removed from wastewater in traditional wastewater treatment plants by precipitation (Richmond 2008). P removal occurs concurrently with nitrogen assimilation in microalgal-based systems, allowing P to be recovered within the biomass. Furthermore, microalgae can be made to accumulate polyphosphates inside the cell structure regardless of biomass productivity under appropriate operational settings (Brown and Shilton 2014). When microalgae are exposed to particular conditions, such as light and temperature, polyphosphates accumulate inside the cell (luxury uptake), allowing for high P removal efficiency. The photosynthetic efficiency of the MBWT process is a critical component in determining biomass productivity and operating costs. It denotes the efficiency with which solar energy is captured and turned into chemical energy in biomass.

Modelling is thought to be one technique to help optimize the MBWT process with improved efficiencies and sufficient practical experiment data of real wastewater treatment with suitable operational durations. All influence factors (such as microalgae species, wastewater nutrient profile, illumination, temperature, local climate conditions) must be taken into account, as well as the impact of upscaling, which may result in unexpected variations in the performance and cost-effectiveness of the MBWT process. Because there are few standards or principles for the design and operation of the MBWT process, more effort should be put into developing fundamental rules and advanced recommendations to improve the flexibility of the MBWT process to various wastewater sources and application scenarios.

In comparison to the existing conventional wastewater treatment process, the MBWT has shown considerable benefits, and it is also in line with emerging wastewater treatment trends such as nutrient recovery, energy conservation, and low-carbon operation. The high nutrient removal efficiency and biomass productivity of MBWT treating specific types of wastewaters, such as municipal and agricultural wastewater, are enticing researchers to investigate its potential further. However, the successful implementation of the MBWT process is still influenced by the design and operation optimization of the MBWT process based on wastewater parameters and microalgae species preference with lower costs.

## 6 Conclusion

Wastewater treatment in an economical and eco-friendly manner is the ultimate aim of environmentalists and researchers worldwide for reducing water pollution. Industries are the prime source of pollutant discharge onto the atmosphere. The pollutant concentration, type, and nature vary based upon the type of industrial sector. Wastewater treatment by algal biotechnology provides edge over other chemical-based techniques. Chemical-based wastewater treatment regimens result in sufficient amount of pollutant removal; however, at some point they end up in contributing to secondary pollution. Algal-based wastewater treatment provides dual advantage as it generates valuable biomass with treated wastewater. These algal biomasses can be further valorized or processed via appropriate methods for recovery of economically important products (bio-oils, bio-hydrogen, platform chemicals, etc.). Further research is needed towards commercialization and better storage and transport facility for the products recovered from algae biomass.

**Acknowledgements** The authors wish to thank Sathyabama Institute of Science and Technology for the support rendered throughout the research.

**Conflict of Interest** None.

## References

Abdel-Raouf N, Al-Homaidan AA, Ibraheem I (2012) Microalgae and wastewater treatment. Saudi J Biol Sci 19(3):257–275

Adav SS et al (2008) Aerobic granular sludge: recent advances. Biotechnol Adv 26(5):411–423. https://doi.org/10.1016/j.biotechadv.2008.05.002

Al-Jabri H et al (2021) Treatment of wastewaters by microalgae and the potential applications of the produced biomass—a review. Water 13(1):27

Barbosa MJ, Wijffels RH (2004) Overcoming shear stress of microalgae cultures in sparged photobioreactors. Biotechnol Bioeng 85(1):78–85

Batista AP et al (2015) Combining urban wastewater treatment with biohydrogen production—an integrated microalgae-based approach. Biores Technol 184:230–235

Becker EW (1994) Microalgae: biotechnology and microbiology. Cambridge University Press

Bernard O, Rémond B (2012) Validation of a simple model accounting for light and temperature effect on microalgal growth. Biores Technol 123:520–527

Borowitzka LJ (1989) Industrial production: methods and economics. Algal Cyanobact Biotechnol:294–316
Brown N, Shilton A (2014) Luxury uptake of phosphorus by microalgae in waste stabilisation ponds: current understanding and future direction. Rev Environ Sci Bio/technol 13(3):321–328
Cai T, Park SY, Li Y (2013) Nutrient recovery from wastewater streams by microalgae: status and prospects. Renew Sustain Energy Rev 19:360–369
Chia SR et al (2021) $CO_2$ mitigation and phycoremediation of industrial flue gas and wastewater via microalgae-bacteria consortium: possibilities and challenges. Chem Eng J:131436
Chinnasamy S et al (2010) Microalgae cultivation in a wastewater dominated by carpet mill effluents for biofuel applications. Biores Technol 101(9):3097–3105
Chiranjeevi P, Mohan SV (2016) Critical parametric influence on microalgae cultivation towards maximizing biomass growth with simultaneous lipid productivity. Renew Energy 98:64–71
Deb UK et al (2017) The effect of irradiance related temperature on microalgae growth in a tubular photo bioreactor for cleaner energy. Am J Comput Math 7(3):371–384
Debnath C et al (2021) Microalgae: sustainable resource of carbohydrates in third-generation biofuel production. Renew Sustain Energy Rev 150:111464
Durán I, Rubiera F, Pevida C (2018) Microalgae: potential precursors of $CO_2$ adsorbents. J $CO_2$ Util 26:454–464
Eerkes-Medrano D, Leslie HA, Quinn B (2019) Microplastics in drinking water: a review and assessment. Curr Opin Environ Sci Health 7:69–75
El-Kassas HY, Mohamed LA (2014) Bioremediation of the textile waste effluent by *Chlorella vulgaris*. The Egypt J Aquat Res 40(3):301–308
FAO (Food and Agriculture Organization) (2018) FAOSTAT online database. FAO, Rome
Goiris K et al (2015) Impact of nutrient stress on antioxidant production in three species of microalgae. Algal Res 7:51–57
Grobbelaar JU (2010) Microalgal biomass production: challenges and realities. Photosynth Res 106(1):135–144
Gupta PL, Lee S-M, Choi H-J (2016) Integration of microalgal cultivation system for wastewater remediation and sustainable biomass production. World J Microbiol Biotechnol 32(8):1–11
Hashimoto S, Furukawa K (1989) Nutrient removal from secondary effluent by filamentous algae. J Ferment Bioeng 67(1):62–69
Hu J et al (2018) Heterotrophic cultivation of microalgae for pigment production: a review. Biotechnol Adv 36(1):54–67
Kalra R, Gaur S, Goel M (2021) Microalgae bioremediation: a perspective towards wastewater treatment along with industrial carotenoids production. J Water Process Eng 40:101794
Krishnamoorthy S, Manickam P (2021) Phycoremediation of industrial wastewater: challenges and prospects. In: Bioremediation for environmental sustainability. Elsevier, pp 99–123
Kumar KS et al (2015) Microalgae–a promising tool for heavy metal remediation. Ecotoxicol Environ Saf 113:329–352
Lavoie A, la Noüe J (1985) Hyperconcentrated cultures of *Scenedesmus obliquus*: a new approach for wastewater biological tertiary treatment? Water Res 19(11):1437–1442
Lee ET-Y, Bazin MJ (1990) A laboratory scale air-lift helical photobioreactor to increase biomass output rate of photosynthetic algal cultures. New Phytol 116(2):331–335
Lower SK (1999) Carbonate equilibria in natural waters. Simon Fraser University, p 544
Mara D (1987) Waste stabilization ponds: problems and controversies. Water Qual Int (1)
Molinuevo-Salces B et al (2019) Microalgae and wastewater treatment: advantages and disadvantages. Microalgae Biotechnol Dev Biofuel Wastewater Treat:505–533. https://doi.org/10.1007/978-981-13-2264-8_20
Muñoz I et al (2009) Chemical evaluation of contaminants in wastewater effluents and the environmental risk of reusing effluents in agriculture. TrAC Trends Anal Chem 28(6):676–694
Ota M et al (2015) Effects of light intensity and temperature on photoautotrophic growth of a green microalga, *Chlorococcum Littorale*. Biotechnol Rep 7:24–29

Park JBK, Craggs RJ (2010) Wastewater treatment and algal production in high rate algal ponds with carbon dioxide addition. Water Sci Technol 61(3):633–639. https://doi.org/10.2166/wst.2010.951

Qiu R et al (2017) Effects of pH on cell growth, lipid production and $CO_2$ addition of microalgae *Chlorella sorokiniana*. Algal Res 28:192–199

Quijano G, Arcila JS, Buitrón G (2017) Microalgal-bacterial aggregates: applications and perspectives for wastewater treatment. Biotechnol Adv 35(6):772–781

Rai MP, Gupta S (2017) Effect of media composition and light supply on biomass, lipid content and FAME profile for quality biofuel production from *Scenedesmus abundans*. Energy Convers Manage 141:85–92

Richmond A (2008) Handbook of microalgal culture: biotechnology and applied phycology. John Wiley & Sons

Richmond A et al (2004) Biological principles of mass cultivation. In: Handbook of microalgal culture: biotechnology and applied phycology, pp 125–177

Sánchez JF et al (2008) Influence of culture conditions on the productivity and lutein content of the new strain *Scenedesmus almeriensis*. Process Biochem 43(4):398–405

Scott SA et al (2010) Biodiesel from algae: challenges and prospects. Curr Opin Biotechnol 21(3):277–286

Shah MP (2020) Microbial bioremediation & biodegradation. Springer

Shah MP (2021a) Removal of refractory pollutants from wastewater treatment plants. CRC Press

Shah MP (2021b) Removal of emerging contaminants through microbial processes. Springer

Shandilya KK, Pattarkine VM (2019) Using microalgae for treating wastewater. In: Advances in feedstock conversion technologies for alternative fuels and bioproducts. Elsevier, pp 119–136

Shuey R (2001) CRS issue brief for congress. Theater missile defense: issues for congress

Soto-Sierra L, Stoykova P, Nikolov ZL (2018) Extraction and fractionation of microalgae-based protein products. Algal Res 36:175–192

Sousa JCG et al (2018) A review on environmental monitoring of water organic pollutants identified by EU guidelines. J Hazard Mater 344:146–162

Su Y (2020) Revisiting carbon, nitrogen, and phosphorus metabolisms in microalgae for wastewater treatment. Sci Total Environ:144590

Tredici MR, Zittelli GC (1998) Efficiency of sunlight utilization: tubular versus flat photobioreactors. Biotechnol Bioeng 57(2):187–197

Udaiyappan AFM et al (2017) A review of the potentials, challenges and current status of microalgae biomass applications in industrial wastewater treatment. J Water Process Eng 20:8–21

Vonshak A (1997) *Spirulina platensis* arthrospira: physiology, cell-biology and biotechnology. CRC Press

Watanabe Y, de la Noüe J, Hall DO (1995) Photosynthetic performance of a helical tubular photobioreactor incorporating the cyanobacterium *Spirulina platensis*. Biotechnol Bioeng 47(2):261–269

# Application of Membrane Technology Combined with Sequencing Batch Reactor for Treating Milk Wastewater

**Khac-Uan Do and Minh-Hang Tran**

## 1 Introduction

### *1.1 Source of Milk Wastewater*

The demand for milk in Vietnam has been increasing day by day (Chu et al. 2004). Dairy products are increasing. Dairy processing industry has fast developed and contributed for national economy (Ha and Dang 2021; Hoang et al. 2021). However, this industry also generates a large amount of wastewater which contains high pollutants. Wastewater at many factories may not be treated effectively, causing environmental pollution to the receiving source (Ha and Dang 2021). Therefore, it is necessary to choose and apply an appropriate technique for treating mild wastewater. This will help to limit and minimize the adverse impacts of milk wastewater on the surrounding environment. In fact, water plays an important role in milk processing. Water is used in many manufacturing processes, including washing, disinfecting, sanitizing, and cooling. In milk factories, wastewater could be generated from both production stages and domestic wastewater. In particular, wastewater from cleaning and washing equipment could be accounted for about 50–80% of the total water consumption in the dairy factory (Boguniewicz-Zablocka et al. 2019). Wastewater could be estimated to be about 2.5 times of the amount of produced milk. Applying the Good Manufacturing Practice (GMP) standards can reduce the average volume

K.-U. Do (✉)
School of Environmental Science and Technology, Hanoi University of Science and Technology, Hanoi, Vietnam
e-mail: uan.dokhac@hust.edu.vn

M.-H. Tran
CH2017A-Environmental Engineering, School of Environmental Science and Technology, Hanoi University of Science and Technology, Hanoi, Vietnam

 
M. P. Shah (ed.), *Sustainable Industrial Wastewater Treatment and Pollution Control*,
https://doi.org/10.1007/978-981-99-2560-5_2

of wastewater from 0.5–3.7 to 0.5–2 $m^3$ of wastewater per $m^3$ of the produced milk (Slavov 2017). Averagely, wastewater could be estimated about 1 $m^3$ per ton of produced milk (Tavolaro and Oliveira 2006).

## *1.2 Characteritics of Milk Wastewater*

The main pollutant components of wastewater from milk products could contain mainly BOD (about 90%). It should be noted that, 1–3 L water was used to produce 1 L of milk. After the production process, wastewater could be generated about 2–2.5 L (Slavov 2017). Milk wastewater is usually white, neutral, or slightly alkaline. It should be noted that in low dissolved oxygen condition, the lactose was fermented easily into lactic acid, resulted in low pH for milk wastewater. This could decrease pH, resulted in causing the casein precipitation (Roy et al. 2020). Milk wastewater contains dissolved organic matter, suspended solids, sugars, and fats (Naito et al. 2020). Main parameters in untreated milk wastewater include biochemical oxygen demand (BOD), with an average of 0.8–2.5 kg/ton of wastewater; chemical oxygen demand (COD) usually about 1.5 times the BOD level; total suspended solids about 100–1000 (mg/L); phosphorus (10–100 mg/L); and nitrogen (about 6% of the BOD). The waste loads of specific milk components are: 1 kg milk fat contains 3 kg COD, 1 kg lactose contains 1.13 kg COD, 1 kg protein contains 1.36 kg COD (Lee et al. 2013). Milk wastewater is composed of highly biodegradable organic matter. Low molecular weight organic compounds such as sugars promote fungal growth in wastewater. Fats create a scum on the surface of the water, causing a lack of oxygen in the water. The milk wastewater could also cause an unpleasant odor. In addition, milk wastewater may contain some colorants that will change the color of the receiving water. Milk wastewater also has great turbidity (Ayats-Vidal et al. 2020; Hamann and Krömker 1997). Wastewater also contains detergents from the cleaning process of factories, machinery, and equipment (Tikariha and Sahu 2014).

## *1.3 Treatment of Milk Wastewater*

Milk wastewater can be treated by various methods (Ahmad et al. 2019; Licata et al. 2021). For example, milk wastewater was treated by the biotechnological technology (i.e. anerobic and aerobic processes) (Kaur 2021). Besides, the Sequential Batch Reactor (SBR) technology was also applied to treat milk wastewater (Yahi et al. 2014). SBR technology is developed on the basis of activated sludge technology. It is operated in batch test which is easily controlled over time. More important, it has a simple structure but it could achieve high treatment efficiency. In addition, SBR technology can effectively remove nitrogen and phosphorus. In recent years, MBR has become an advanced wastewater treatment technology (Do and Chu 2022; Uan et al.

2013). This technology includes a combination of membrane filter and a conventional biological treatment system (Chen and Uan 2013; Do et al. 2013). MBR technology could treat organic pollutants effectively (Banu et al. 2011). Wastewater treated by MBR technology could meet a requirement of national regulation for wastewater treatment (Rajesh Banu et al. 2009). Wastewater treated by MBR technology could be reused for plantation, car washing, fire prevention, or concrete mixing (Banu et al. 2009; Do et al. 2009, 2018). Milk wastewater contains high nutrient substancces. Therefore, choosing and applying appropriate technology for effective treatment is essential.

### *1.4 Objective of the Work*

A combined technology (SBR-MBR) could improve the efficiency of milk wastewater treatment. Therefore, it is necessary to study and analyze the important conditions to evaluate the efficiency of milk wastewater treatment of SBR technology combined with membrane filtration. It was compared with conventional SBR technology. In particular, it is necessary to determine the appropriate conditions to operate the milk wastewater treatment system with this technology for different milk wastewater sources. The factors affecting the system performance were investigated to find out suitable operating conditions for the system.

## 2 Technologies Applied to Milk Wastewater Treatment

### *2.1 Biological Anaerobic Technology*

The anaerobic process is applied to treat milk wastewater to produce biogas (Bella and Rao 2021). Biogas can be used as a source of heat or energy. This process produces less sludge. Thus, reducing the problems associated with sludge treatment. Anaerobic systems save energy during operation because aerators are not used in aerobic systems (Goli et al. 2019). Upflow anaerobic sludge blanket (UASB) system is widely used to treat milk wastewater (Xu et al. 2018). The principle of this process is to use anaerobic microorganisms to decompose organic compounds and collect biogas. UASB could be operated at high COD (2050 mg/L) and high organic load of 0.031 kg COD/$m^3$ per day. COD could be removed at high efficiency of 90% (Umetsu et al. 2004). UASB system can treat milk wastewater containing high fat content (868 mg/L) to achieve the COD efficiency of more than 50%. The average biogas production could be 179 $m^3$/day, in which methane could be accounted for 75%. Biogas can be recovered as a source of biogas (Arunadevi and Saravanaraja 2020). During operation, the anaerobic treatment system consumes less energy. However, this technology has still limitations such as long start-up time, using a large area, long treatment time. Sludge formation in the process is difficult to control. In addition, it is easily unstable due to loading shock. It is also affected by toxic substances (Passeggi et al. 2012). The UASB system normally contains mesophilic anaerobes. UASB could be incorporated with an anaerobic filter to treat milk wastewater. The yield of methane gas was obtained up to 0.354 $m^3$ CH4/kg COD which was corresponding to a hydraulic retention time of 1.7 days (Ji et al. 2020).

Anaerobic Sequencing Batch Reactor (ASBR) technology is used to treat high-strength milk wastewater. The ASBR system can be operated at low hydraulic retention time of 6h. At 58 °C, it could treat COD and BOD at high efficiencies of 62 and 75%, respectively. However, when the temperature was varied from 5 to 52 °C, and the hydraulic retention time varies from 6 to 24 h, it could remove organic matter at the efficiency ranging from 75% to 90% for BOD and 62% to 90% for COD (Jiraprasertwong et al. 2018). The ASBR system contains thermophilic bacteria that could improve the removal efficiency of organic matter in milk wastewater (Matsumoto et al. 2012).

## 2.2 *Biological Aerobic Technology*

Milk wastewater from the UASB tank was normally introduced into the aerotank tank. In the aeration condition, the dissolved and colloidal organic substances in the milk wastewater were oxidated by aerobic microorganisms. An aerated system was put in the aeration tank to provide oxygen for growing microorganisms. In this condition, aerobic microorganisms consumed the oxygen to utilize the organic compounds. As a result, the contaminants were partly converted to $CO_2$, $H_2O$, and sludge (Li and Zhang 2002).

Sequencing Batch Reactor (SBR) system could treat high COD concentrations in milk wastewater (400–2500 mg/L). The COD removal in SBR could reach over 90% (Wu et al. 2008). At the aeration time of 19 h, the SBR system could treat milk wastewater effectively (over 98%) for high COD variation (ranging from 400 to 7500 mg/L). However, the treatment efficiency was decreased with increasing the organic loading rate (Li and Zhang 2002). It was also reduced when the hydraulic retention time was decreased (Khalaf et al. 2021). It was reported that injection of oxygen into reactor could enhance the treatment of dairy wastewater (Martín-Rilo et al. 2018).

## 2.3 *Other Technology*

Electrochemical coagulation technology can be used to treat milk wastewater. This process can remove organic waste, turbidity, and color. In particular, the electrocoagulation process can effectively treat suspended colloidal particles in milk wastewater (Bazrafshan et al. 2013; Chezeau et al. 2020). Electrochemical coagulation technology has the ability to treat COD and fat up to 98–99%, and the electrolysis time is about 7 minutes (Sharma 2014). In recent years, biochar was used in treating wastewater effectively. Particularly, it was applied to absorb the ammonium in domestic wastewater (Vu and Do 2021; Vu et al. 2021a, b). Biochar could be used for milk wastewater based on the adsorption process (Bazrafshan et al. 2016).

The membrane filtration system was used to recover lactose and milk protein. This could also help reduce COD in milk wastewater (Reig and Vecino 2021). In addition, membrane technology could be operated at an average filtration capacity of about 11 $L/m^2h$ to treat milk wastewater effectively (Kumar et al. 2013). Application of SBR technology combined with membrane filtration could be an alternative solution to improve the efficiency of milk wastewater treatment.

# 3 Development of SBR-MBR Technology for Milk Wastewater Treatment

## *3.1 Operation of the SBR-MBR System*

A lab-scale SBR-MBR was developed to treat synthetic milk wastewater (Fig. 1). Working volume of SBR tank used in the study is 5 L (with the dimensions of diameter and height of 150 × 450 mm). Membrane was submerged in the SBR to form the SBR-MBR system. Membrane characteristics were shown in Table 1.

The air was supplied into the tank by a blower. The air flowrate was maintained at 2–5 L/min to supply oxygen to the system. Besides, the air flow could help to reduce membrane fouling. The working cycles of the SBR and SBR-MBR systems are shown in Table 2.

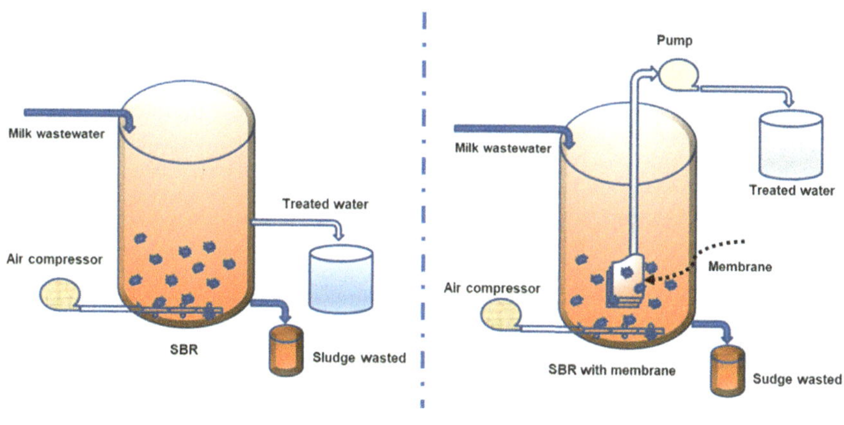


**Fig. 1** Schematic diagram of SBR and SBR-MBR for treating milk wastewater

**Table 1** Membrane used in the SBR-MBR system

| Photo of membrane module | Membrane type | Pore size, μm | Surface area, $m^2$ |
|---|---|---|---|
| 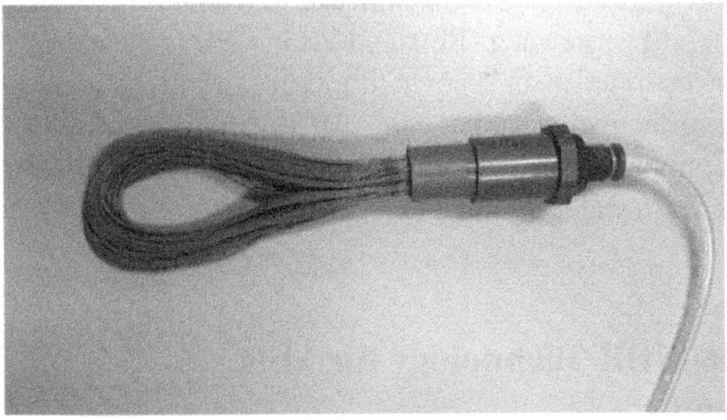 | MF, hollow fiber | 0.22 | 0.065 |

**Table 2** Working cycles of the SBR and SBR-MBR systems

| System | Operational cycles | | | | |
|---|---|---|---|---|---|
| | Wastewater filling | Aeration | Settling | Decanter | Sludge wasted |
| SBR | 0.5 h | 6–10 h | 1 h | 0.5 h | 0.5 h |
| SBR-MBR | 0.5 h | 6 h | Membrane filtration (1 h) | | 0.5 h |

The membrane module was submerged in the SBR tank. Filtration was carried out by a suction pump. A mini pump (using DC 6–12 V voltage, capacity of 5–12 W, flowrate of 1–2 L/min) was used in this work. The aeration system was located below the membrane module to provide oxygen for the treatment process. On the other hand, the air in the tank would create a flow, running along the surface of the membrane. This could limit the clogging of the membrane surface. In addition, the suction pump was operated in a cycle of 10 minutes of work, 2 minutes of pause.

## 3.2 *Air Supply Flow for Treatment System*

The air supply plays an important role in the biological processes in SBR (Struk-Sokołowska et al. 2018). In addition, the air supply also plays an important role in reducing membrane clogging in the SBR-MBR system (Bae et al. 2003). Therefore, it is necessary to calculate the amount of air required to the system. The amount of air was calculated based on the following conditions of the system (i.e. influent flowrate of milk wastewater of 4 L/day; the influent parameters including pH of 7.5; SS of 506 mg/L; BOD of 479 mg/L; COD of 1135 mg/L; TN of 95 mg/L; and TP of 18 mg/L). Several following conditions were assumed for calculation, such as (i) the activated sludge content in the inlet of the tank ($X_o = 0$); (ii) sludge retention time ($\theta_c = 30$ days); (iii) BOD = 0.42 COD; and (iv) wastewater temperature ($t = 25$ °C). The amount of sludge produced per day was calculated according to the growth rate of the sludge (Lateef et al. 2013).

$$Y_b = \frac{Y}{1 + k_d\theta c} = \frac{0.4}{1 + 0.03 \times 30} = 0.21 \text{ kg VS/day},$$

in which, $Y = 0.4$ g VSS/g BOD; $k_d = 0.03$ gCOD/gVSS per day.

In addition, the amount of activated sludge generated by reduction of BOD with the VSS increasing in a day could be determined by the following equation (Khalaf et al. 2021).

$$P_X = \frac{Y(S_0 - S)Q}{1 + k_d} = Y_b(S_0 - S)Q = 0.21 \times (479 - 47.9 \times 4 = 0.362 \text{ g VSS/day})$$

The required amount of air could be determined by calculating the theoretical oxygen required to remove BOD and nitrification (oxidize ammonium $NH_4^+$ to $NO_3^-$) (Tchobanoglous et al. 2003).

$$\begin{aligned} OC_0 &= Q(S_0 - S)/0.42 - 1.42P_X + 4.57\, Q(N_0/N) \\ &= 0.004 \times (479 - 47.9)/0.42 - 1.42 \times 0.362 + 4.57 \times 0.004\,(95 - 50) \\ &= 4.415 \text{ kgO}_2/\text{day} \end{aligned}$$

The actual amount of oxygen required for the SBR tank could be calculated according to the following equation (Li and Zhang 2002).

$$OC_t = OC_0\left(\frac{C_{s20}}{\beta.C_{sh} - C_d}\right)\frac{1}{1.024^{(t-20)}\alpha} = 4.415\left(\frac{9.08}{1 \times 8.24 - 2}\right)\frac{1}{1.024^5 \times 0.9} = 6.36 \text{ kg/day}$$

Therefore, the amount of air required could be calculated as below.

$$Q_{air} = \frac{OC_t}{OU} \times f = \frac{6.36}{2.1} \times 1.5 = 4.55 \text{ m}^3/\text{day},$$

in which, $U = O_u \times h = 7 \times 0.3 = 2.1$ g $O_2/m^3$, $O_u$ is the capacity to dissolve oxygen into wastewater, $O_u = 7$ g $O_2/m^3$.m; $h$ is the depth of the water level in the SBR tank, m.

Based on the above results, the air supply flow could be calculated as, $v_k = Q_{air}/(24 \times 60) = 4.55/1440 = 0.00315$ m$^3$/min $= 3.15$L/min. During operation, the air supply flowrate was adjusted to 3 L/min.

### 3.3 Cultivation of the Activated Sludge

Synthetic milk wastewater was prepared by mixing the Moc Chau pasteurized fresh milk (expired type) with tap water in a 1000 mL plastic bottles. The prepared milk wastewater has a very high concentration of pollutants compared to the actual milk wastewater. Therefore, the synthetic milk wastewater was diluted with a certain tap water to ensure the BOD and COD in the range of 499 and 1276 mg/L before filling in the experimental system. The seeding activated sludge was taken from the aerobic tank in a wastewater treatment system of Samsung Vietnam Company (Thai Nguyen, Vietnam). The sludge was put into a cylindrical glass (with a height of 460 mm, diameter of 65 mm). 800 mL of synthetic milk wastewater was added to reactor for sludge cultivation. Air was supplied continuously at a flowrate of 2 L/min. During cultivation of sludge, the temperature was adjusted in the appropriate range (25 °C). It is important to monitor the nutrients during sludge cultivation (Balasubramanian et al. 2018). After about 2 weeks, the seeding sludge grew significantly. It was then transferred to the SBR and SBR-MBR systems.

### 3.4 Working Process of Experimental System

A sludge retention time of 30 days was maintained in the experimental system. One batch test was carried out everyday in which, Phase 1 was from day 1 to day 22. During this time, the milk wastewater was treated by SBR technology. Phase 2 was from day 23 to day 30. The milk wastewater was treated by using SBR-MBR technology. In the SBR reactor, milk wastewater was treated according to the following operating cycles, i.e. (i) Filling phase: Milk wastewater was pumped into the SBR tank in 0.5 h. In this period, the influent wastewater was mixed with the activated sludge available in the reactor; (ii) Aeration phase: air flowrate was provided into the reactor. The aeration was maintained for 6–10 h for a biochemical reaction between milk wastewater and activated sludge. In the aeration phase, nitrification and organic matter decomposition processes were taken place. Those processes could remove BOD, COD, and nitrogen compounds (i.e. ammonium). In this process, the nitrification happened rapidly. The oxidation of ammonium ($NH_4^+$) was carried out by the species of bacteria (i.e. Nitrosomonas) which converted ammonium to nitrite ($NO_2^-$). Other bacteria such as nitrobacter would oxidize nitrite to nitrate ($NO_3^-$); (iii) Sedimentation phase: In this phase, no aeration was performed. The sludge was settled in 1 h completely. In this period, the denitrification process happened partly. It could create two layers of sludge in the tank, i.e. a supernatant was in the top layer, whereas the settled sludge was at the bottom; (iv) Decanter phase: After the sludge was settled, the supernatant was withdrawn by decanter for 0.5 h; (v) Sludge withdrawal phase: When the sludge retention time was over, a part of settled sludge was removed from the bottom of the tank in 0.5 h.

The membrane module used in the SBR-MBR system consists of many hollow fibers. The membrane pore sizes are very small (0.22 μm). Therefore, microorganisms could not be passed through the membrane (Bae et al. 2003). As a result, the sludge would be retained in the reactor completely. The treated water was pumped to a storage tank. During operation, the air blower provided oxygen for microorganisms. Besides, it helps to remove the sludge deposited on the membrane surface to minimize the membrane clogging. The effect of air supply flow on the treatment efficiency of the system was investigated by supplying different air flowrates, i.e. 2.0; 3.0; 4.0 L/min to treat the same wastewater volume of 4 L/d, at the same hydraulic retention time of 8 h. The effect of aeration time on the treatment efficiency of the system was also examined by maintaining at different aeration time of 6, 8, 10 h to treat the same amount of wastewater (4 L/d) at the same air flowrate of 3 L/min. During operations, several parameters (i.e. $BOD_5$, COD, TN, TP) were monitored and analyzed following the standard methods (APHA 2017).

## 3.5 *Effect of Air Flowrate on System Performances*

Figure 2 shows a variation of $BOD_5$, COD, TN, TP removal efficiencies at the air flowrate of 2 L/min, and at the aeration time of 8 h. In stage 1, the system performance was not stable for the first few days. It then became more stable. The effluent concentrations of $BOD_5$, COD, TN, TP tended to decrease rapidly after day 17. From day 18 onward, the effluent concentrations were almost unchanged. At the end of stage 1, the lowest effluent $BOD_5$, COD, TN, TP concentrations of the system were 97 mg/L, 298 mg/L, 46.1 mg/L, and 9.7 mg/L, respectively. At the air flowrate of 3 L/min, the effluent $BOD_5$, COD, TN, TP concentrations reduced to 48 mg/L, 183 mg/L, 37.3 mg/L, and 7.7 mg/L, respectively. In case of increasing air flowrate to 4 L/min, the effluent $BOD_5$, COD, TN, TP concentrations continuously decreased to 41 mg/L, 102 mg/L, 12.4 mg/L, and 4.4 mg/L, respectively.

In stage 2, the effluent concentrations had sharply decreased, and the system performance increased significantly. At the end of phase 2, the lowest effluent $BOD_5$, COD, TN, TP concentrations of the system were 46 mg/L, 119 mg/L, 15.9 mg/L, and 5.0 mg/L, respectively. In stage 2, the system performance was much better than that of stage 1. As seen from Figure 2, the system performance in the stage 2 (operated in 8 days) was more stable than that of stage 1 (operated in 22 days). The observations show that, application of membrane filtration enhanced the treatment efficiency of the system (Tan et al. 2021). The system performance increased significantly and faster compared to that of the system without the membrane filtration.

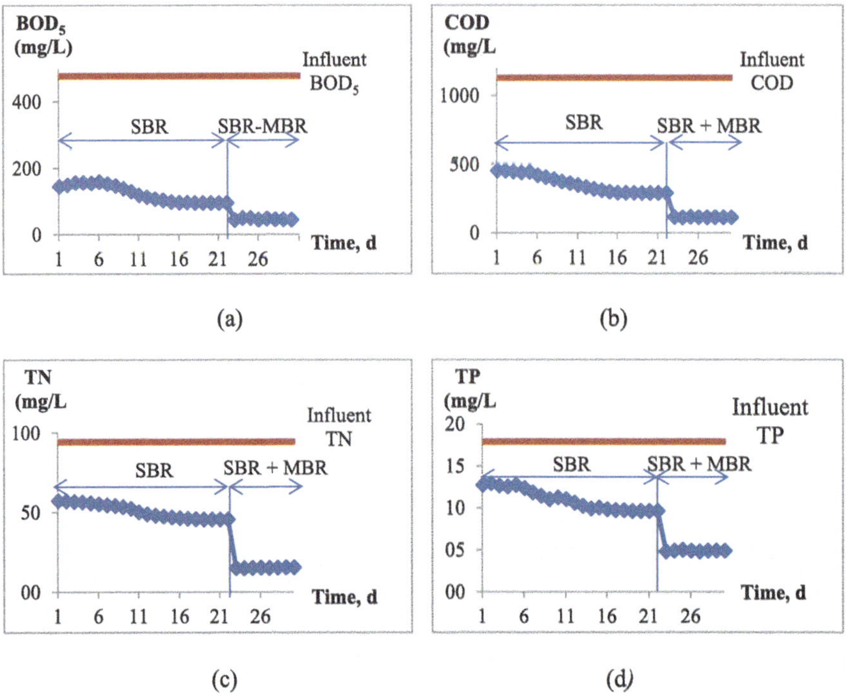


**Fig. 2** Effect of air flowrate on system performances **a**, for BOD removal, **b**, for COD removal, **c**, for TN removal, and **d**, for TP removal

### *3.6 Effect of Aeration Time on System Performance*

The $BOD_5$, COD, TN, TP removal efficiencies of the system operated at aeration time of 6 h and at air flowrate of 3 L/min were presented in Fig. 3.

As seen from the figure, in stage 1, from day 1 to 10, the system operation was not stable. It then became more stable after day 10. The concentrations of $BOD_5$, COD, TN, TP in the effluent decreased rapidly after day 17. At the end of stage 1, the $BOD_5$, COD, TN, TP in the effluent were 111 mg/L, 121 mg/L, 46.1 mg/L, and 8.2 mg/L, respectively. At the aeration time of 8 h, the concentrations $BOD_5$, COD, TN, TP in the effluent decreased to 18 mg/L, 40 mg/L, 13.5 mg/L, and 4.5 mg/L, respectively. In case of increasing aeration time to 10 h, the effluent $BOD_5$, COD concentrations decreased significantly to 24 mg/L, 41 mg/L, respectively. However, the effluent TN, TP concentrations were almost the same as the value of TN and TP in case of aeration time of 8 h (i.e. 12.4 mg/L and 4.2 mg/L, respectively).

In stage 2, the concentrations $BOD_5$, COD, TN, TP in the effluent reduced significantly compared to the values in stage 1. At the end of stage 2, the lowest concentrations $BOD_5$, COD, TN, TP in the effluent were 24 mg/L, 40 mg/L, 14.8 mg/L, and 4.8 mg/L, respectively. As seen from the figure, the system performance in stage

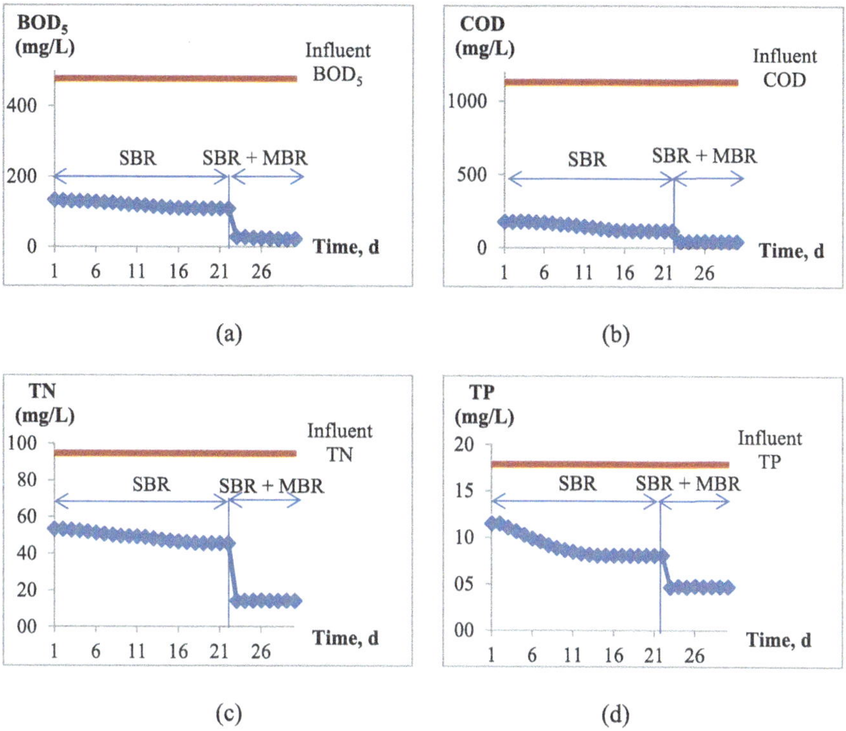


**Fig. 3** Effect of aeration time on system performance **a**, for BOD removal, **b**, for COD removal, **c**, for TN removal, and **d**, for TP removal

2 was much better than that in stage 1. Though the system operated in stage 2 was only 8 days, it was more stable than that of stage 1 (operated in 22 days). Therefore, using membrane filtration could increase the system performance. It enhanced significantly when compared to the system without the membrane filtration (Dereli et al. 2012; Tan et al. 2021).

## 4 Operational Problems and Troubleshooting

### *4.1 Sludge Foam on the Surface*

It was seen that, at low concentration of activated sludge in the reactor, the color of wastewater looked yellow-brown. White sludge foam appeared on the surface (Fig. 4).

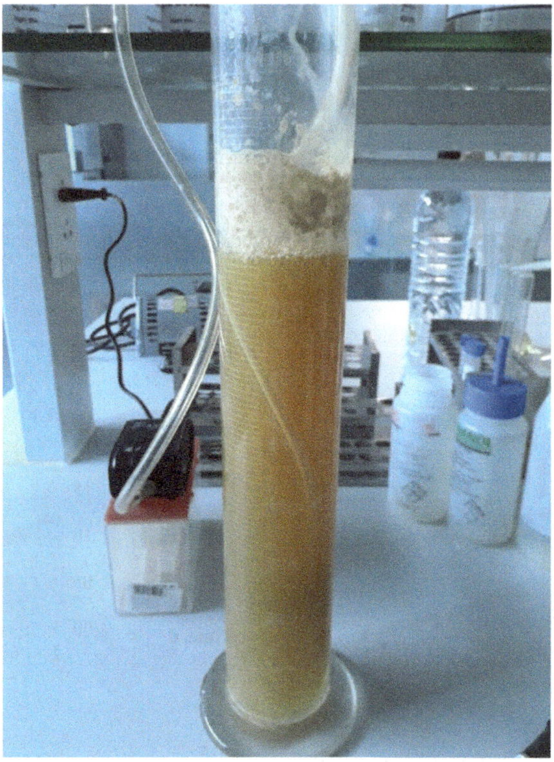

**Fig. 4** Cultivation of activated sludge with synthetic milk wastewater

The reason could be due to the low sludge content, the system was operated at overloading. Reduction of the organic load by diluting the wastewater before entering the tank could be used to overcome the floating foam. In addition, adding activated sludge to the tank to ensure the right F/M ratio could be an alternative solution. In case of high concentration of activated sludge, white foam floats also appeared on the surface of the reactor (Wu et al. 2008). The reason could be that the system was operated at under-loading. Self-decomposition of the activated sludge increased at low substrate loading. Therefore, supplying more substrate to the reactor could be a good solution. In addition, as seen from Fig. 4, the sludge foam was yellow. This could be due to a change in the microorganisms in reactor, i.e. growth of filamentous bacteria (e.g. Nocardia and M. Paricella) (Yao et al. 2019). Filamentous bacteria are hydrophobic (Henriet et al. 2017). Filamentous bacteria caused the sludge in the reactor to be difficult to settle. Most of the sludge was floated and making a foam layer in the surface of the reactor.

## 4.2 *Sludge Floats on the Surface*

During the sludge cultivation process, sludge was formed into large flocs and floated to the surface of reactor. This may be due to denitrification process. This phenomenon could be overcome by increasing the sludge circulation rate (Cisterna-Osorio et al. 2021). At the same time, reduction of sludge age could limit denitrification process. Another reason could be due to the high air flowrate which could make foam and sweep the sludge out of the reactor. Therefore, it is necessary to reduce the aeration flowrate into the reactor.

## 4.3 *Hard-To-Settle Sludge*

Characteristics of sludge cultivated by synthetic milk wastewater could be small size, fine floc, and slow settling ability. As a result, water after settling had become light yellow. This phenomenon happened in the early stage. The reason could be due to the low-activated sludge content in the reactor, while receiving high organic loading rate (Wilén et al. 2018). In this case, the system was operated at overloading condition. The influent wastewater should be diluted before adding to the reactor. Another way, i.e. adding more activated sludge, could be used to maintain an appropriate F/M ratio (i.e. F/M ~ 0.2–0.4 $d^{-1}$).

## 4.4 *System Interrupted or Intermittent Operation*

During the experiment, the system was interrupted sometimes. In this case, it is necessary to recover the activated sludge by transferring it to a sludge cultivation reactor. This will maintain a necessary condition for the activated sludge growth. In this period, it is necessary to supply enough substrates, nutrients, and aeration. Activated sludge content in the reactor should be checked before restarting the system. If the activated sludge content decreased, it is necessary to recover and increase the sludge content to the desired level (Corsino et al. 2022). At the same time, it is necessary to increase the aeration rate and increase the recycling sludge rate.

# 5 Conclusions

Several conclusions could be obtained from operating a lab-scale SBR technology combined with hollow fiber microfiltration membrane (SBR-MBR) (pore size of 0.22 μm, surface area of 0.065 $m^2$) to treat synthetic milk wastewater (4 L per day). The air flowrate affected the system performances. The treatment efficiencies of $BOD_5$,

COD, TN, TP were 97%, 95%, 86%, and 75%, respectively at 3 L/min of air supply for 8 h. The SBR-MBR could help to stabilize the microbial density compared to SBR technology for treating milk wastewater which contains relatively high concentration of nutrients and biodegradable organic matter. Membrane filtration could help to improve the removal efficiency of organic matter, resulted in reducing the treatment time. In particular, the membrane has a small pore size so that it could improve sludge separation efficiency. It could maintain high concentration of sludge in the reactor, therefore, it could enhance the removal efficiency. Moreover, it could be capable to remove harmful microorganisms.

The SBR-MBR could be applied to treat dairy wastewater with high efficiency. SBR-MBR technology is suitable for milk production plants with small area. However, SBR-MBR also has a disadvantage compared to the conventional SBR technology. More energy could be consumed due to using a suction pump for the membrane filtration. On the other hand, during operation, membrane could be clogged. Hence, the treatment capacity of the system could be reduced. Therefore, overcoming the above disadvantages could promote the SBR-MBR to be a promising technology to treat milk wastewater. Further study should be conducted to determine the appropriate parameters for large-scale system which could be applied for the treatment of milk wastewater in Vietnam.

**Acknowledgements** The authors greatly appreciate the support from Hanoi University of Science and Technology, Vietnam (No. 4461/QĐ-ĐHBK-SĐH for master project of 17AKT-KTMT04).

## References

Ahmad T et al (2019) Treatment and utilization of dairy industrial waste: a review. Trends Food Sci Technol 88:361–372. https://doi.org/10.1016/j.tifs.2019.04.003

APHA (2017) Standard methods for the examination of water and wastewater, 23rd edn. American Public Health Association, Washington DC, USA, Apha Washington

Arunadevi P, Saravanaraja M (2020) Two phase upflow anaerobic sludge blanket (UASB) reactor on the reduction of chemical oxygen demand in dairy effluent. Asian J Adv Res:15–23

Ayats-Vidal R, Valdesoiro-Navarrete L, García-González M, Asensio-De la Cruz O, Larramona-Carrera H, Bosque-García M (2020) Predictors of a positive oral food challenge to cow's milk in children sensitized to cow's milk. Allergol et Immunopathol 48:568–575. https://doi.org/10.1016/j.aller.2020.03.007

Bae T-H, Han S, Tak TM (2003) Membrane sequencing batch reactor system for the treatment of dairy industry wastewater. Process Biochem 39:221–231. https://doi.org/10.1016/S0032-9592(03)00063-3

Balasubramanian R, Sircar A, Sivakumar P, Anbarasu K (2018) Production of biodiesel from dairy wastewater sludge: a laboratory and pilot scale study. Egypt J Pet 27:939–943. https://doi.org/10.1016/j.ejpe.2018.02.002

Banu JR, Kaliappan S, Kumar A, Yeom IT, Uan DK (2011) Effect of low temperature thermochemical pretreatment on sludge reduction potential of membrane bioreactor treating primary treated dairy wastewater. Water Qual Res J 46:312–320. https://doi.org/10.2166/wqrjc.2011.026

Banu JR, Uan DK, Chung IJ, Kaliappan S, Yeom IT (2009) A study on the performance of a pilot scale A2/0-MBR system in treating domestic wastewater. J Environ Biol 30:959–963

Bazrafshan E, Kord Mostafapour F, Alizadeh M, Farzadkia M (2016) Dairy wastewater treatment by chemical coagulation and adsorption on modified dried activated sludge: a pilot-plant study. Desalin Water Treat 57:8183–8193. https://doi.org/10.1080/19443994.2015.1018331

Bazrafshan E, Moein H, Kord Mostafapour F, Nakhaie S (2013) Application of electrocoagulation process for dairy wastewater treatment. J Chem 2013:640139. https://doi.org/10.1155/2013/640139

Bella K, Rao PV (2021) Anaerobic digestion of dairy wastewater: effect of different parameters and co-digestion options—a review. Biomass Conv Biorefinery. https://doi.org/10.1007/s13399-020-01247-2

Boguniewicz-Zablocka J, Klosok-Bazan I, Naddeo V (2019) Water quality and resource management in the dairy industry. Environ Sci Pollut Res 26:1208–1216. https://doi.org/10.1007/s11356-017-0608-8

Chen JC, Uan DK (2013) Low dissolved oxygen membrane bioreactor processes (LDO-MBRs): a review. Int J Environ Eng 5:129–149. https://doi.org/10.1504/IJEE.2013.052917

Chezeau B, Boudriche L, Vial C, Boudjemaa A (2020) Treatment of dairy wastewater by electrocoagulation process: advantages of combined iron/aluminum electrodes. Sep Sci Technol 55:2510–2527. https://doi.org/10.1080/01496395.2019.1638935

Chu L, Yokogawa H, Kawaguchi T (2004) Dairy production in Vietnam: opportunities and challenges. J Fac Agric Kyushu Univ 49. https://doi.org/10.5109/4578

Cisterna-Osorio P, Calabran-Caceres C, Tiznado-Bustamante G, Bastias-Toro N (2021) Influent with particulate substrate, clean, innocuous and sustainable solution for bulking control and mitigation in activated sludge process. Water 13. https://doi.org/10.3390/w13070984

Corsino SF et al (2022) Integrated production of biopolymers with industrial wastewater treatment: effects of OLR on process yields, biopolymers characteristics and mixed microbial community enrichment. J Water Proc Eng 47:102772. https://doi.org/10.1016/j.jwpe.2022.102772

Dereli RK, Ersahin ME, Ozgun H, Ozturk I, Jeison D, van der Zee F, van Lier JB (2012) Potentials of anaerobic membrane bioreactors to overcome treatment limitations induced by industrial wastewaters. Biores Technol 122:160–170. https://doi.org/10.1016/j.biortech.2012.05.139

Do K-U, Chu X-Q (2022) Chapter 5—performances of membrane bioreactor technology for treating domestic wastewater operated at different sludge retention time. In: Shah M, Rodriguez Couto S, Biswas J (eds) Development in wastewater treatment research and processes. Elsevier, pp 107–122. https://doi.org/10.1016/B978-0-323-85583-9.00010-7

Do K-U, Banu JR, Chung I-J, Yeom I-T (2009) Effect of thermochemical sludge pretreatment on sludge reduction and on performances of anoxic-aerobic membrane bioreactor treating low strength domestic wastewater. J Chem Technol Biotechnol 84:1350–1355. https://doi.org/10.1002/jctb.2189

Do K-U, Kim J-H, Quang C (2018) Sludge characteristics and performance of a membrane bioreactor for treating oily wastewater from a car wash service station. Desal Water Treat 120:166–172. https://doi.org/10.5004/dwt.2018.22716

Do K-U, Rajesh Banu J, Kaliappan S, Yeom I-T (2013) Influence of the thermochemical sludge pretreatment on the nitrification of A/O reactor with the removal of phosphorus by simultaneous precipitation. Biotechnol Bioprocess Eng 18:313–320. https://doi.org/10.1007/s12257-012-0492-5

Goli A, Shamiri A, Khosroyar S, Talaiekhozani A, Sanaye R, Azizi K (2019) A review on different aerobic and anaerobic treatment methods in dairy industry wastewater. J Environ Treat Tech 6:113–141

Ha, Dang H (2021) Wastewater management and reuse wastewater for agriculture in Diem Thuy-Phu Binh-Thai Nguyen-Viet Nam. In: Proceedings of 15th international conference on Humanities and Social Sciences 2019 (IC-HUSO 2019) 11th–12th November 2019, Faculty of Humanities and Social Sciences, Khon Kaen University, Thailand

Hamann J, Krömker V (1997) Potential of specific milk composition variables for cow health management. Livest Prod Sci 48:201–208. https://doi.org/10.1016/S0301-6226(97)00027-4

Henriet O, Meunier C, Henry P, Mahillon J (2017) Filamentous bulking caused by *Thiothrix* species is efficiently controlled in full-scale wastewater treatment plants by implementing a sludge densification strategy. Sci Rep 7:1430. https://doi.org/10.1038/s41598-017-01481-1

Hoang V, Nguyen A, Hubbard C, Nguyen K-D (2021) Exploring the governance and fairness in the milk value chain: a case study in Vietnam. Agriculture 11:884

Ji S, Ma W, Wei Q, Zhang W, Jiang F, Chen J (2020) Integrated ABR and UASB system for dairy wastewater treatment: engineering design and practice. Sci Total Environ 749:142267. https://doi.org/10.1016/j.scitotenv.2020.142267

Jiraprasertwong A, Vichaitanapat K, Leethochawalit M, Chavadej S (2018) Three-stage anaerobic sequencing batch reactor (ASBR) for maximum methane production: effects of COD loading rate and reactor volumetric ratio. Energies 11:1543

Kaur N (2021) Different treatment techniques of dairy wastewater. Groundw Sustain Dev 14:100640. https://doi.org/10.1016/j.gsd.2021.100640

Khalaf AH, Ibrahim WA, Fayed M, Eloffy MG (2021) Comparison between the performance of activated sludge and sequence batch reactor systems for dairy wastewater treatment under different operating conditions. Alex Eng J 60:1433–1445. https://doi.org/10.1016/j.aej.2020.10.062

Kumar P, Sharma N, Ranjan R, Kumar S, Bhat ZF, Jeong DK (2013) Perspective of membrane technology in dairy industry: a review. Asian Australas J Animal Sci 26:1347–1358. https://doi.org/10.5713/ajas.2013.13082

Lateef A, Nawaz Chaudhry M, Ilyas S (2013) Biological treatment of dairy wastewater using activated sludge. ScienceAsia 39:179. https://doi.org/10.2306/scienceasia1513-1874.2013.39.179

Lee JM, Yoon JS, Jeon SA, Lee SY (2013) Sensitization patterns of cow's milk and major components in young children with atopic dermatitis. Asia Pac Allergy 3:179–185. https://doi.org/10.5415/apallergy.2013.3.3.179

Li X, Zhang R (2002) Aerobic treatment of dairy wastewater with sequencing batch reactor systems. Bioprocess Biosyst Eng 25:103–109. https://doi.org/10.1007/s00449-002-0286-9

Licata M, Ruggeri R, Iacuzzi N, Virga G, Farruggia D, Rossini F, Tuttolomondo T (2021) Treatment of combined dairy and domestic wastewater with constructed wetland system in Sicily (Italy). Pollutant removal efficiency and effect of vegetation. Water 13. https://doi.org/10.3390/w13081086

Martín-Rilo S, Coimbra R, Escapa C, Otero M (2018) Treatment of dairy wastewater by oxygen injection: occurrence and removal efficiency of a benzotriazole based anticorrosive. Water 10:155. https://doi.org/10.3390/w10020155

Matsumoto EM, Osako MS, Pinho SC, Tommaso G, Gomes TM, Ribeiro R (2012) Treatment of wastewater from dairy plants using Anaerobic Sequencing Batch Reactor (ASBR) following by Aerobic Sequencing Batch Reactor (SBR) aiming the removal of organic matter and nitrification. Water Prac Technol 7:wpt2012048. https://doi.org/10.2166/wpt.2012.048

Naito M et al (2020) Changes in casein component-specific antibody levels during oral immunotherapy for milk allergy. Pediatr Allergy Immunol 31:421–424. https://doi.org/10.1111/pai.13218

Passeggi M, López I, Borzacconi L (2012) Modified UASB reactor for dairy industry wastewater: performance indicators and comparison with the traditional approach. J Clean Prod 26:90–94. https://doi.org/10.1016/j.jclepro.2011.12.022

Rajesh Banu J, Uan DK, Yeom IT (2009) Nutrient removal in an A2O-MBR reactor with sludge reduction. Biores Technol 100:3820–3824. https://doi.org/10.1016/j.biortech.2008.12.054

Reig M, Vecino X (2021) Use of membrane technologies in dairy industry: an overview 10. https://doi.org/10.3390/foods10112768

Roy D, Ye A, Moughan PJ, Singh H (2020) Composition, structure, and digestive dynamics of milk from different species—a review. Front Nutr 7. https://doi.org/10.3389/fnut.2020.577759

Sharma D (2014) Treatment of dairy waste water by electro coagulation using aluminum electrodes and settling, filtration studies. Int J ChemTech Res 6:591–599

Slavov AK (2017) General characteristics and treatment possibilities of dairy wastewater—a review. Food Technol Biotechnol 55:14–28. https://doi.org/10.17113/ftb.55.01.17.4520

Struk-Sokołowska J, Rodziewicz J, Mielcarek A (2018) Effect of dairy wastewater on changes in COD fractions in technical-scale SBR type reactors. Water Sci Technol 2017:156–169. https://doi.org/10.2166/wst.2018.099

Tan LC, Peschard R, Deng Z, Ferreira ALM, Lens PNL, Pacheco-Ruiz S (2021) Anaerobic digestion of dairy wastewater by side-stream membrane reactors: comparison of feeding regime and its impact on sludge filterability. Environ Technol Innov 22:101482. https://doi.org/10.1016/j.eti.2021.101482

Tavolaro P, Oliveira CAF (2006) Evaluation of a GMP training of milkers in dairy goat farms in São Paulo, Brazil. Int J Environ Health Res 16:81–88. https://doi.org/10.1080/09603120500392608

Tchobanoglous G, Burton FL, Stensel HD, Metcalf, Eddy I, Burton F (2003) Wastewater engineering: treatment and reuse. McGraw-Hill Education

Tikariha A, Sahu O (2014) Study of characteristics and treatments of dairy industry waste water. J Appl Environ Microbiol 2:16–22

Uan DK, Yeom IT, Arulazhagan P, Rajesh Banu J (2013) Effects of sludge pretreatment on sludge reduction in a lab-scale anaerobic/anoxic/oxic system treating domestic wastewater. Int J Environ Sci Technol 10:495–502. https://doi.org/10.1007/s13762-012-0120-0

Umetsu K, Ohyama T, Kishimoto T, Takahashi J, Matsumoto N, Hamamoto O (2004) The performance of UASB reactor treatment of squeezed dairy manure. Nogyo Shisetsu (J Soc Agric Struct Japan) 35:1–8. https://doi.org/10.11449/sasj1971.35.1

Vu N-T, Do K-U (2021) Insights into adsorption of ammonium by biochar derived from low temperature pyrolysis of coffee husk. Biomass Conv Biorefinery. https://doi.org/10.1007/s13399-021-01337-9

Vu N-T, Ngo T-H, Nguyen T-T, Do K-U (2021a) Performances of coffee husk biochar addition in a lab-scale SBR system for treating low carbon/nitrogen ratio wastewater. Biomass Conv Biorefinery. https://doi.org/10.1007/s13399-021-01788-0

Vu N-T, Nguyen T-H-T, Do K-U (2021b) Chapter 29—removal of ammonium from aqueous solution by using dried longan peel as a low-cost adsorbent. In: Shah M, Rodriguez-Couto S, Mehta K (eds) The future of effluent treatment plants. Elsevier, pp 569–588. https://doi.org/10.1016/B978-0-12-822956-9.00029-5

Wilén B-M, Liébana R, Persson F, Modin O, Hermansson M (2018) The mechanisms of granulation of activated sludge in wastewater treatment, its optimization, and impact on effluent quality. Appl Microbiol Biotechnol 102:5005–5020. https://doi.org/10.1007/s00253-018-8990-9

Wu S, Zhu J, Miller C (2008) Dairy milking wastewater treatment using a lab-scale sequencing batch reactor (SBR). Trans ASABE 51:1057–1065. https://doi.org/10.13031/2013.24525

Xu S, Zhang L, Huang S, Zeeman G, Rijnaarts H, Liu Y (2018) Improving the energy efficiency of a pilot-scale UASB-digester for low temperature domestic wastewater treatment. Biochem Eng J 135:71–78. https://doi.org/10.1016/j.bej.2018.04.003

Yahi H, Madi N, Midoune K (2014) Contribution to biological treatment of dairy effluent by sequencing batch reactor (SBR). Desal Water Treat 52:2315–2321. https://doi.org/10.1080/19443994.2013.878254

Yao J, Liu J, Zhang Y, Xu S, Hong Y, Chen Y (2019) Adding an anaerobic step can rapidly inhibit sludge bulking in SBR reactor. Sci Rep 9:10843. https://doi.org/10.1038/s41598-019-47304-3

# Role of Microalgae in Wastewater Treatment and Their Role in Nutrient Recovery

**Dipannita Parial and Satarupa Dey**

## 1 Introduction

A large amount of wastewater is continuously been generated by urban, agricultural and industrial sectors. Most of these wastewater have high concentration of nitrogen, carbon and other organic elements which leads to the eutrophication of aquatic bodies in which they are released (Jin et al. 2014). Usually in any wastewater systems only 16% of the water is reused and only 35.8%, 35.8% and 35.7% of organics, NH4-N and TP can be recovered which clearly indicate that there are plenty of room for improvement in the treatment systems when the reuse and recovery of nutrients are considered. Moreover, the treatment processes should follow stringent rules and regulations issued by central and local governments to limit the discharge of pollutants. Also, they must be cost-effective and eco-friendly with an enhanced water reclamation and resource recycling (Jin et al. 2014).

The most important challenges of any treatment plant are nutrients removal (mainly nitrogen and phosphorus) along with meeting the criteria of discharge and water reuse (Jin et al. 2014). Most of the conventional water treatment methods such as activated sludge treatment, anaerobic-anoxic–oxic (A2/O), anaerobic-oxic (A/O), sequencing batch reactor (SBR), oxidation ditch, etc., have overall high COD removal ability. However, the main drawbacks of these processes are inefficient in nitrogen and phosphorus removal. Moreover, they also face challenges to meet the stringent nutrients discharge standards with high efficiency and low costs (Whitton et al. 2015) and are associated with high energy consumption, instability in treatment effect, long

D. Parial
Department of Botany, Sammilani Mahavidyalaya, Baghajatin, EM Bypass, Kolkata 700094, India

S. Dey (✉)
Department of Botany, Shyampur Siddheswari Mahavidyalaya, Howrah, West Bengal 711312, India
e-mail: dey1919@gmail.com



M. P. Shah (ed.), *Sustainable Industrial Wastewater Treatment and Pollution Control*, https://doi.org/10.1007/978-981-99-2560-5_3

process, carbon emission, excess sludge discharge and recyclable resource wasting. All these limitations are the main barrier to sustainable development and resource recycling.

Microalgae-based wastewater treatment (MBWT) process can be considered as one of the most promising, cost-effective and eco-friendly technologies for the advanced treatment and nutrient recovery of wastewater and has attracted considerable attention in the recent years. The feasibility of the usage of microalgae-based wastewater treatment has been recommended by different groups of researchers for having high potential of nutrient recovery (Whitton et al. 2015; Shah 2020). Moreover, the symbiotic relationship of microalgae and bacteria in the process was found to be a beneficial one as microalgae utilized $CO_2$ and generated $O_2$, which was concomitantly used by the heterotrophic bacteria to assimilate and degrade the carbon, nitrogen and phosphorus in organic solids. Also, the biomass produced by microalgae in wastewater treatment could be used as feedstock in biorefinery or other applications (Mudimu et al. 2014).

This chapter largely deals with an overview of the microalgae used for wastewater treatment. The different types of physical and chemical properties of the wastewater suitable for algal growth has been discussed in detail along with their ability of nutrient assimilation/removal and biomass productivity. The limitations of the microalgal treatment have also been highlighted along with the possible solution.

## 2 Composition of Wastewater

The composition of wastewater largely consists of solids including settleable particles and colloids which do not settle easily. The solid components of wastewater are usually less than 0.1% by weight which presents the major challenge in its treatment and disposal. The wastewater stream also consists of a large number of microbes and microalgae which may be pathogenic in nature and is also capable of consuming the organic constituents present which can lead to a rapid change in the quality of wastewater. The inputs of the wastewater vary greatly leading to a constant change in the composition of wastewater. Chemically, the wastewater is composed of organic and an inorganic components which are very complex in nature and it is difficult to completely define them. In domestic wastewater, the solid particles mostly consist of organic substances which are largely proteins, carbohydrates and fats together with the products of their decomposition. All these compounds are subject to decomposition by bacteria and are combustible. However, the inorganic solids are much inert and do not decay easily. The inorganic solids comprise of minerals, sand, gravel and silt that are responsible for the hardness and mineral content of the water. Suspended solids mostly comprise of 70% organic solids and 30% inorganic solids which include sand, grit, clay, fecal solids, paper, pieces of wood and particles of food and garbage and they can only be removed from the wastewater by physical or mechanical means, such as sedimentation or filtration. Apart from them, there are settleable solids which are of sufficient size and weight and can settle in a given period of time and colloidal suspended solids which do not settle readily. These colloidal suspended solids are not readily removed by physical or mechanical treatment facilities.

## 3 Conventional Sewage Treatment

Most treatment methods are mainly divided into three main parts such as physical method, chemical method and biological method. During physical method no gross chemical or biological changes are carried out, and it is a physical phenomenon. They are largely carried out by the process of sedimentation, which is related to the settling of solids by gravitational force and is routinely done at the beginning and end of wastewater treatment operations. This separation process is followed by chemical treatment which includes the process of chlorination. Chlorine is a strong oxidizing chemical mainly used to kill bacteria and slow down the rate of decomposition of the wastewater. Ozone treatment is another treatment method which is also widely used. For coagulation of solid wastes, addition of polyvalent metals are done to form an insoluble end product which is removed from wastewater. The typical coagulant includes lime (that can also be used in neutralization), certain iron containing compounds (such as ferric chloride or ferric sulphate), and alum (aluminium sulphate). In the biological method microorganisms are used which undergo biochemical decomposition of wastewaters to stable end products and carbon dioxide. The biological treatment methods are mainly targeted towards the organic solids. Finally, treatment to control odours, to retard biological activity or to destroy pathogenic organisms may also be needed.

There exist various types of wastewater treatment process having various sizes of treatment capacity. The main conventional treatments include activated sludge treatment, anaerobic-anoxic–oxic (A2/O), anaerobic-oxic (A/O), sequencing batch reactor (SBR), oxidation ditch, etc., that have overall high COD removal ability (Li et al. 2011; Shah 2021b). Most of these processes are not efficient in nitrogen and phosphorus removal. This is largely attributed to low mixed liquor volatile suspended solid (MLVSS), short sludge retention time (SRT) and different characteristics of wastewater sources; moreover, the design defects and poor management of the treatment facilities are the main cause of lowering the efficiency of wastewater treatment plants (Jin et al. 2014). According to Jin et al. (2014), the new trends in wastewater treatment industry around the world includes strengthening of pollutants elimination process, promoting energy saving and low-carbon operation and resource recycling, wastewater reclamation, nutrients and energy recovery. Sheik et al. (2014) proposed a concept called "wastewater biorefinery column" using traditional activated sludge wastewater treatment process to recover lipids, PHA, organic acids and other valuable resources however, this lacks economic and environmental sustainability for future development. Integration of microalgal system in wastewater treatment plants can be used for successful resource recycling and wastewater reclamation (Menger-Krug et al. 2012).

# 4 Microalgae in Wastewater Treatment: A Green Technology

Microalgae or microphytes are microscopic algae, too small to see with naked eyes. They grow very fast and have a simple morphology. They are unicellular, mostly photosynthetic and are well-known for their adaptable nature to grow on different (even adverse) growth conditions. They exist not only in aquatic but also in terrestrial ecosystem. Hence, they show huge species diversity, which can live in a wide range of environment. Microalgae may occur individually or in colonies and their sizes can range from a few micrometres to a few hundred micrometres. Moreover, they can be autotrophic, heterotrophic, mixotrophic or photoheterotrophic depending on the carbon source used in their metabolism as well as light conditions.

The word microalgae include both prokaryotic and eukaryotic photosynthetic organisms. Prokaryotic microalgae are represented singly by cyanophyta (also known as cyanobacteria), whereas eukaryotic microalgae include chlorophyta (green algae), rhodophyta (red algae), phaeophyta (brown algae), bacillariophyta (diatoms) and chrysophyta (golden-brown algae). In spite of their dissimilar cellular organization, they both perform oxygenic photosynthesis in similar way and are cultured using the same fundamentals and technologies.

Due to some unavoidable disadvantages of the conventional physicochemical water treatment methods, there is an urgent need of treating wastewater with green technological approach. Microalgae play a critical role in eliminating toxic pollutants from sewage water in a process known as bioremediation. Bioremediation by eukaryotic microalgae and cyanobacteria (phycoremediation) for the exclusion and/or biotransformation of heavy metals, xenobiotics and other toxic materials from wastewater is emerging rapidly due to its environmental sustainability and economic feasibility. They show significant role in tertiary treatment of urban sewage water not only due to its metabolic capabilities of removing toxic contaminants but also in producing secondary metabolites as exudates that prevent further pathogen growth. The basic functioning of microalgae in wastewater treatment is shown in Fig. 1.

The idea of wastewater management using microalgae was first proposed by Oswald in the year 1950 (Oswald and Gotass 1957; Shah 2021a; Oswald 1963). Energy production through harvesting and utilization of algal biomass was further proposed (Benemann et al. 1977). Shirai and his group (1998) investigated the treatment of soy sauce effluent by four types of microalgae and later on used the harvested algal biomass for ethanol production. Afterwards many scientists treated several urban and industrial wastewater using different microalgae which will be discussed in this chapter.

Domestic as well as industrial wastewater contain high amount of carbon, nitrogen, phosphorus and other compounds, which serve as the growth nutrients for microalgae and support the algal growth. Algae simply convert these nutrients into biomass. Algae remove pollutants from wastewater by the processes known as biosorption and bioaccumulation. Biosorption is a reversible, physicochemical process that involves the binding of metal ions or other organic compounds to the dead

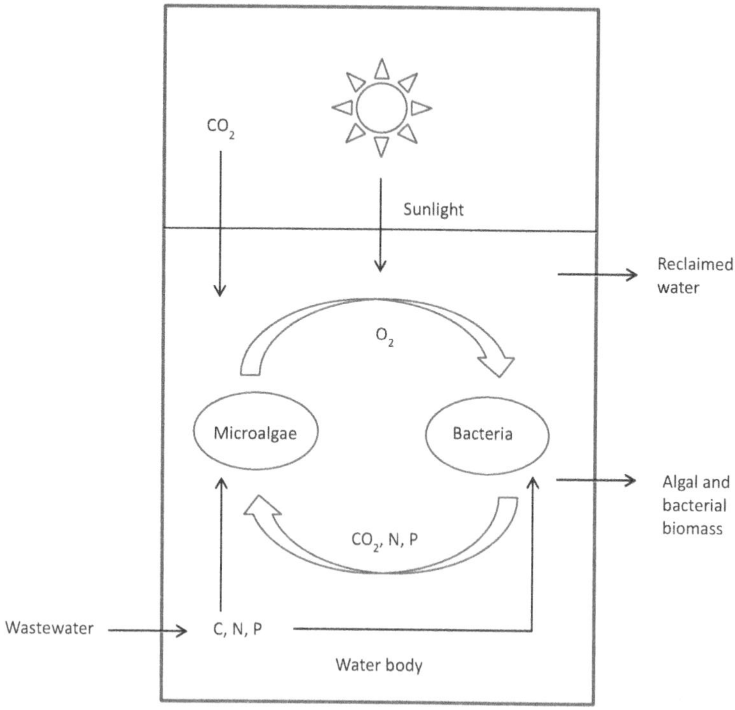


**Fig. 1** Basic operation principle of wastewater treatment using microalgae

or inactive cell wall of microalgae through adsorption, electrostatic interaction, ion exchange, chelation and micro-precipitation. Algal cell wall contains several polysaccharides and functional groups such as carboxyl, hydroxyl, sulphydryl, amino, etc., which help them to bind heavy metal and make them serve as suitable biological ion-exchange resin. Even though this quality of microalgae is exploited for removing toxic metals from wastewater, it can be used to remove organic compounds as well. Moreover, microalgal mats can also be used as a bio adsorbent to eliminate low concentration of hydrophobic organic pollutants from wastewater. Bioaccumulation, in contrast, comprises of two stages and it involves entry of contaminants inside the cell. Firstly, the metal ions bind to the cell surface passively, which is identical with the biosorption process and then they are actively transported at the cost of cellular energy to the interior of the cell, thereby becoming incorporated into the metabolic cycle of that organism. Mineralization is also a significant process in carbon sequestration. Considerable amounts of $CO_2$ can be sequestered by consumption and precipitation of carbonates and bicarbonates due to the alkaline medium developed by photosynthetic activity.

## 5 Removal of Nutrients

Satisfactory nutrient exclusion from wastewater is a major challenge and most of the traditional approaches are expensive and non-environmentally friendly (as they result in $CO_2$ emission and require use of chemicals). Generally municipal and agricultural wastewater contains maximum amount of nutrients (nitrogen, phosphorus and other minerals). If the untreated or partially treated wastewater is thrown away into natural water bodies, it will cause eutrophication, a serious threat to the aquatic life forms. Eutrophication manifests by lowering dissolved oxygen, production of harmful microalgal blooms, thereby leading to disturbances in the balance of the aquatic ecosystem that may result in the death of fish and can be detrimental to humankind as well. Henceforth, nutrient removal from sewage water is getting increased attention and is being strictly monitored worldwide.

The use of urban and industrial effluent as a nutrient feedstock to produce algal biomass has environmental and economic benefits. Wastewater nutrients such as nitrogen, phosphorus, ammonia, sulphur, iron are consumed by microalgae and help them to grow. Thus, nutrients can be removed satisfactorily with the biomass subsequently. Table 1 shows different microalgal species that have shown promising results for future in situ treatment of wastewater.

**Phosphorus**: Inorganic phosphorus may be present in wastewater in several ionic states depending on the pH of the medium ($H_3PO_4$ < 2.15; $H_2PO_4^-$ 2.15–7.20; $HPO_4^{2-}$ 7.20–12.33; and $PO_4^{3-}$ > 12.33). Inorganic P is considered as the most bioavailable form of phosphorus and microalgae are reported to preferably uptake $H_2PO^{4-}$ and $HPO4^{2-}$. Phosphate is assimilated by means of active transport through a symporter with $H^+$ or $Na^+$ ions providing the driving force. It is also reported that soluble organic P compounds are important source of bioavailable phosphorus (Li and Brett 2013; Borowitzka et al. 2016). Microalgae non-specifically hydrolyse these bound phosphate groups by the expression of extracellular membrane-bound as well as free phosphatases. Microalgae are also observed to assimilate phosphorus even at low concentrations, and appear as an interesting alternative to the post-treatment systems currently used, which include denitrifying filters that require a source of organic carbon and emit $CO_2$.

In microalgae, phosphorus is an essential element involved in several metabolic pathways and serves as an essential component for many structures such as DNA, RNA, cell membrane and ATP. Phosphorus is incorporated into organic compounds following phosphorylation of ADP and subsequent synthesis of ATP. This is an endergonic reaction, gaining energy from oxidation of respiratory substrates or photosynthetic electron transport chain. Finally, the synthesized ATP allows the transfer of phosphate groups to organic compounds by substrate-level phosphorylation (e.g. conversion of glucose to glucose-6-phosphate as seen in glycolysis). Moreover, in P rich condition, microalgae can accumulate excess P than they require for their survival. This additional P is stored as acid-insoluble polyphosphate granules in a mechanism called “luxury uptake” for future use.

**Table 1** Nutrient removal efficiencies of various microalgal strains from different wastewater types as reported in independent studies

| Microalgae | Wastewater type | Time | Nitrogen (%) | Phosphorus (%) | Carbon (%) | References |
|---|---|---|---|---|---|---|
| *Scenedesmus obliquus* | Secondary treatment effluent | 2 days | 96 | 55 | NA | Ruiz-Marin et al. (2010) |
| *Scenedesmus* sp. and natural bacteria population | Secondary treatment effluent | 91 days | 36 | 62 | 35 | He and Xue (2010) |
| *Chlamydomonas reinhardtii* | Secondary treatment effluent | 4 days | 99 | 98 | 30.2 COD | Su et al. (2012) |
| *Chlorella* sp., *Scenedesmus*, *Pediastrum*, *Nitzschia*, *Navicula*, *Crucigenia*, *Synedra* and bacteria | Waste water Lagoon effluent | 20 days | 75 | 23 | NA | Christenson and Sims (2012) |
| *Scenedesmus obliquus* and bacteria | Secondary treatment effluent | 130 days | 95 | 94 | 39 | Zamalloa et al. (2013) |
| Algal–bacterial consortium | PSW | 40 days | 70 | 85 | 90 | Posadas et al. (2013) |
| *Consortium of Woronichinia sp., Actuodesmus sp., Aulacoseira sp., Desmodesmus quadricaudatus, Nitzschia sp., Limnothrix redekei and Gomphonema parvulum* | PSW | 40 days | 92 | 96 | 89 | Posadas et al. (2014) |
| *Tetraselmis suecica* | Aquaculture waste water | | | | | Tetraselmis suecica |
| *Micratinium reisseri* | Waste water from mines | NA | 97 | 83 | 62 | Ji et al (2014) |

(continued)

**Table 1** (continued)

| Microalgae | Wastewater type | Time | Nitrogen (%) | Phosphorus (%) | Carbon (%) | References |
|---|---|---|---|---|---|---|
| *Chlorella vulgaris* and natural wastewater microbial community | Pre-PSW | 150 days | 70 | 52 | 53 | Choi (2015) |
| *Chlorella sorokiniana* | Pharmaceutical waste water | NA | 70 | 89 | | Escapa et al. (2015) |
| *Chlorella pyrenoidosa* and wastewater microbial community | Secondary treatment effluent | 18 days | 95 | 84 | 78 | Dahmani et al. (2016) |
| Consortia of *Chlorella and Phormidium sp* | Gray water | NA | 94 | 90 | 69 | Choudhary et al. (2017) |
| *Chlorella vulgaris* | Treated municipal waste water | 37 days | 61 | 71 | 21 COD | Tao et al. (2017) |
| *Tetraselmis sp.* | Aqueous phase waste water from biomass to energy generation process | NA | 98.5 | 98 | NA | Das et al. (2019) |
| *Picochlorum sp.* | Aqueous phase waste water from biomass to energy generation process | NA | 95.4 | 97.2 | 94.3 | Das et al. (2019) |

**Nitrogen**: Nitrogen is crucial for all living organisms including microalgae. It is the key element for nucleic acids, amino acids, chlorophyll, polyamines and alkaloids. Microalgae can assimilate nitrogen from a wide variety of organic (e.g. amino acids, urea, purines and nucleosides) and inorganic sources (e.g. $NH_4^+$, $NO_3$ and $NO_2$). The assimilation of organic nitrogen by microalgae can occur in both autotrophic as well as heterotrophic conditions. Among inorganic sources, microalgae preferentially uptake $NH_4^+$ as its assimilation and incorporation is energetically more favourable. Ammonium enters microalgal cell by a group of membrane transporter proteins belonging to the ammonium transporter family, an evolutionarily common protein expressed in bacteria, yeast, algae and higher plants. Inorganic nitrogen assimilation in microalgae is interrelated with their carbon metabolism that requires a carbon skeleton in the form of keto-acid to incorporate the absorbed nitrogen into organic compounds. After entry, ammonium can directly be incorporated into amino acids necessary for growth and other metabolic functions. On the contrary, $NO_3$ and $NO_2$ must be reduced to $NH_4^+$ by the enzymes nitrate reductase and nitrite reductase, respectively, for intracellular uptake. Furthermore, the transport of $NO_3$ into the cell is an energy-requiring process that consumes ATP directly.

**Carbon**: Like any other living organism, microalgae need carbon the most in order to live, grow and reproduce. It is the basic building block of life and virtually all biological molecules such as sugars, nucleotides, proteins, fats, etc., contain carbon. Microalgae can utilize inorganic carbon (mainly $CO_2$) photoautotrophically, as their primary carbon source. In water, gaseous $CO_2$ dissociates into carbonate ($CO_3^{2-}$) and bicarbonate ($HCO_3^-$) ions depending on the pH, temperature, cation concentration and salinity. $CO_2$ can easily diffuse across the plasma membrane of microalgal cells due to its non-polar nature, however $HCO_3^-$ involves active transport mechanisms. After entry, $HCO_3^-$ is quickly converted to $CO_2$ by carbonic anhydrase within the chloroplast to enable inorganic carbon fixation. Finally, the absorbed inorganic carbon enters the Calvin cycle leading to the synthesis of organic carbon.

The amount of carbon in wastewater is a key factor for microalgal growth. Therefore, shortage of inorganic carbon indirectly decreases nitrogen and phosphorus remediation by limiting microalgal growth. Many reports have shown that exogenous supply in the form of $CO_2$ or bicarbonate salt improves microalgal growth and in turn promotes nutrient removal (Yao et al. 2015; Qi et al. 2017). However, concentrations above an optimum range of $CO_2$ have been reported to reduce the efficacy of nutrient elimination by inhibiting microalgal respiration (Sforza et al. 2012) although the tolerance to $CO_2$ has been found to be strain-dependent (Wang et al. 2008; Zhao and Su 2014). Additionally, this approach is energy-expensive and may reduce the ability of microalgae to use and consequently remove the carbonaceous material from wastewater.

A number of facultative heterotrophic microalgae can also remove organic carbon from wastewater by utilizing organic carbonaceous material as carbon source either in a mixotrophic mode with $CO_2$ and light or in a strict heterotrophic method in the absence of light. Most of the organic carbonaceous material is composed of fibres and proteins with lesser amount of sugars. However, the presence of extremely heterogeneous biological carbonaceous material such as low-molecular weight butyric acid to more complex polycyclic aromatic hydrocarbons and synthetic polymers restricts its accessibility as a viable carbon source. However, heterotrophic microorganisms, for example bacteria and fungi, are reported to convert complex carbonaceous material to a suitable substrate in order to make it a viable carbon source for microalgae (Lowrey et al. 2015).

Addition of exogenous biodegradable organic carbon sources such as glucose, glycerol, acetate, ethanol, fructose, sucrose, lactose, etc., are also reported to increase the efficiency of nitrogen and phosphorus removal from wastewater although this strategy is cost-intensive (Yee 2015). To minimize the cost, waste organic carbon substrates such as food wastes (e.g. dairy waste and cane molasses), polysaccharide hydrolysate (produced from starch or straw) and high strength domestic or livestock wastewater (centrate) have been used as well.

## 6 Removal of Xenobiotic Compounds

Industries and other human activities discharge several toxic chemicals into the environment. Like bacteria and fungi, microalgae also play a critical role in dispersion, chemical transformation and bioaccumulation of many of those toxic xenobiotic compounds (as shown in Table 2) by adsorption, absorption and biodegradation.

**Polycyclic aromatic hydrocarbons (PAHs)** are chemical compounds comprising of carbon and hydrogen only and contain a number of aromatic rings. They are produced by the incomplete combustion of organic matter. Industrial processes and the extraction and use of fossil fuels are the dominant sources of PAHs in the environment. For example, coal-fired power plants are known to release a large amount of PAHs. Due to their high degree of toxicity (carcinogenic and/or teratogenic) and persistent nature, they are in urgent need to be controlled very intensely. A number of microalgal species have been reported to assimilate and degrade PAHs under certain conditions. These organisms are able to biotransform low-molecular weight PAHs to its non-toxic form. For example, naphthalene can be hydrolized to form two-ring PAHs, whereas phenanthrene, tricyclic PAH can be metabolized to its hydroxylated intermediate. Moreover, Lei et al. (2007) suggested that the presence of one PAH stimulated the removal of the other PAH. Cerniglia et al. (1979) indicated the existence of oxidation mechanisms for aromatic hydrocarbons in prokaryotic alga *Agmenellum quadruplicatum* that led to the production of 1-naphthol as the main degradation product of naphthalene.

**Monoaromatic hydrocarbons (MAHs)** such as BTEX (benzene, toluene, ethylbenzene and xylene) naturally occur in crude oil and are produced worldwide by the petrochemical industry. They are less reactive, hydrophobic and highly persistent compounds, well-known for their mutagenic and carcinogenic effect. These contaminants act as carbon sources for microalgae under ideal environmental condition.

**Chlorophenols** are another important group of extremely toxic and persistent xenobiotics having carcinogenic properties. These compounds are used in

**Table 2** Bioremediation of metals by different microalgal strains from wastewater

| Metal | Microalgal species | Media | Initial metal ion concentration | Biomass concentration | Temperature (°C) | pH | Time | Removal efficiency or accumulation | References |
|---|---|---|---|---|---|---|---|---|---|
| Cr III | *Spirogyra spp.* | Synthetic wastewater | 20–150 mg $L^{-1}$ | 1.0–3.0 g $L^{-1}$ | 25 | 5 | 15–180 min | 81.02% | Bishnoi et al. (2007) |
| Cr (VI) | *Scenedesmus incrassatulus* | Synthetic wastewater | 2.09 mg $L^{-1}$ | NA | 25 ± 2 | 8 | 13–16 days | 52.70% | Pena-Castro et al. (2004) |
| Cr (VI) | *Chlorella vulgaris* | Synthetic wastewater | 50 mg $L^{-1}$ | NA | 25 | 1.5 | 180 h | 43.3 mg $g^{-1}$ biomass | Xie et al. (2014) |
| Cr VI | *Spirulina platensis* | Artificial medium and wastewater | 2.925 mg $L^{-1}$ | 0.15 g $L^{-1}$ | 30 | NA | 28 days | 60.92% | Magro et al. (2012) |
| Cr (VI) | *Chlorella vulgaris* | Chromium solution | 50 mg $L^{-1}$ | 1 g $L^{-1}$ biomass | 25 | 0.5–5 | 4 h | 50.7–80.3% | Sibi (2016) |
| Cr (VI) | *Scenedesmus sp* | Chromium solution | 10 mg $L^{-1}$ | NA | 30 | 2.65 | 0.5–5 h | 92.89% | Pradhan et al. (2019) |
| Cu | *Chlorella vulgaris and Spirulina maxima* | Secondary effluent | 56.8 ppb | 2–6 mL/150 mL | 26–28 | 7.96 | 10 days | 81.70% | Chan et al. (2014) |
| Cu | *Scenedesmus incrassatulus* | Synthetic wastewater | 6.57 mg $L^{-1}$ | NA | 25 ± 2 | 8 | 13–16 days | 31.70% | Pena-Castro et al. (2004) |
| Zn | *Chlorella vulgaris and Spirulina maxima* | Secondary effluent | 12.8 ppb | 2–6 mL/150 mL | 26–28 | 7.96 | 10 days | 94.10% | Chan et al. (2014) |

(continued)

**Table 2** (continued)

| Metal | Microalgal species | Media | Initial metal ion concentration | Biomass concentration | Temperature (°C) | pH | Time | Removal efficiency or accumulation | References |
|---|---|---|---|---|---|---|---|---|---|
| Zn | *Chlorella vulgaris, Chlorella sorokiniana, Scenedesmus quadricauda and 8 more* | Synthetic wastewater | 30 mg $L^{-1}$ | 5 × 108 cells/mL | NA | 6.8–7.2 | 5 min | 80% (C. sorokiniana & C. vulgaris); 99% (S. quadricauda) | Chong et al. (2000) |
| Zn | *Chlorella sp.* | Wastewater | 1–50 mg $L^{-1}$ | 538 × 105 cells $mL^{-1}$ | 28 | 8.1–8.6 | 288 h | 60–70% | Kumar and Goyal (2010) |
| Cd | *Scenedesmus incrassatulus* | Synthetic wastewater | 7.36 mg $L^{-1}$ | NA | 25 ± 2 | 8 | 13–16 days | 24.10% | Pena-Castro et al. (2004) |
| Cd | *Chlamydomonas reinhardtii* | Biosorption medium | 100 mg $L^{-1}$ | 800 mg $L^{-1}$ | 5–35 | 6 | 60 min | 42.6 ± 0.54 mg $g^{-1}$ dry biomass | Tüzün et al. (2005) |
| Cd | *Scenedesmus sp* | Synthetic medium | 0.5 mg $L^{-1}$ | NA | 25 | 6.2–6.5 | 24 h | 73% | Travieso et al. (1999) |
| | *Chlorella vulgaris* | Synthetic medium | 5 mg $L^{-1}$ | NA | 25 | 6.2–6.5 | 24 h | 66% | Travieso et al. (1999) |
| | *Spirulina platensis* | aqueous solution | 100 g $L^{-1}$ | 5 g $L^{-1}$ | 25 | 6 | 1 h | 92.76% | Sayadi et al. (2019) |
| | *Chlorella vulgaris* | | 100 g $L^{-1}$ | 5 g $L^{-1}$ | 25 | 6 | 1 h | 87.52% | Sayadi et al. (2019) |
| Pb | *Scenedesmus quadricauda* | Synthetic wastewater | 50 mg $L^{-1}$ | 200 mg $L^{-1}$ | NA | 5 | 60 min | 82% | Mirghaffari et al. (2015) |

(continued)

**Table 2** (continued)

| Metal | Microalgal species | Media | Initial metal ion concentration | Biomass concentration | Temperature (°C) | pH | Time | Removal efficiency or accumulation | References |
|---|---|---|---|---|---|---|---|---|---|
| Pb | *Chlamydomonas reinhardtii* | Biosorption medium | 100 mg $L^{-1}$ | 800 mg $L^{-1}$ | 5–35 | 5 | 60 min | 96.3 ± 0.86 mg $g^{-1}$ dry biomass | Tüzün et al. (2005) |
| Pb | *Chlorella kessleri* | synthetic wastewater | 10 ppm | 1.5 g $L^{-1}$ | ~ 28 | 6 | 2 h | >95% | Sultana et al. (2020) |
| Pb | *Chlorella vulgaris* | Battery ibdustry wastewater | NA | 2 g $L^{-1}$ | 25 | 6 | 1 h | 89.26% | Malakootian et al. (2019) |
| Hg | *Phaeodactylum tricornutum* | Seawater enriched | 120 µg $L^{-1}$ | NA | NA | 7.5 | 6 days | 2,229 mg $g^{-1}$ biomass | Deng et al. (2013) |
| Hg | *Chlamydomonas reinhardtii* | Biosorption medium | 100 mg $L^{-1}$ | 800 mg $L^{-1}$ | 5–35 | 6 | 60 min | 72.2 ± 0.67 mg $g^{-1}$ dry biomass | Tüzün et al. (2005) |
| Hg | *Chlorella vulgaris* | Wastewater | 0.3–0.8 mg $L^{-1}$ | 1 mg $L^{-1}$ | 18–21 | "Proper pH for the microalga" | 24 h | 79–86% | Fard and Mehrnia (2017) |
| Ni | *Chlorella vulgaris, Chlorella sorokiniana, Scenedesmus quadricauda and 8 more* | Synthetic wastewater | 30 mg $L^{-1}$ | 5 × 108 cells/mL | NA | | 5 min | 98% (S. quadricauda); 46.7% (C. vulgaris); 33.3% (C. sorokiniana) | Chong et al. (2000) |
| Ni | *Chlorella vulgaris* | Metal solution | 10–40 µg/mL | NA | NA | 6.8–7.2 | 24 h | 33–41% | Wong et al. (2000) |

the production of dyes, fungicides, herbicides, pesticides and wood preservatives. *Anabaena* sp. and *Aulosira fertilissima* showed a marked ability to accumulate dichlorodiphenyltrichloroethane (DDT), fenitrothion and chlorpyrifos (insecticides). They are reported to bioconcentrate and degrade DDT into its two main metabolites: dichlorodiphenyldichloroethylene (DDD) and dichlorodiphenyldichloroethane (DDE) (Lal et al. 1987).

## 7 Removal of Heavy Metals

Out of all the contaminants, heavy metals have drawn the most attention due to its tremendous toxic nature. Heavy metals are discharged in the environments by anthropogenic activities such as mining, production of batteries, paints, metal alloys, from leather tanning, tincture wood preservatives, electroplating industries and also from the production and use of fossil fuels. Metal ions affect aquatic life and are incorporated into the food chain that ultimately results in biomagnification. Biomagnification results in several health issues such as cancer, kidney problems, liver failure, birth defects, respiratory disorders and heart diseases in human.

A number of microalgal strains are reported to tolerate heavy metals and their potential to absorb metals is shown in Table 3. They are capable of accumulating substantial amounts of heavy metal ions (15 mg $g^{-1}$ biomass) from aqueous solutions that help them challenge alternative treatment methods. Different polysaccharides and functional groups present in the fibrous and amorphous microalgal cell wall interact and sequester heavy metals. While, most of the techniques utilize adsorption mechanism by dried and dead biomass, the possibility to assimilate heavy metals into living or dead cells has also been studied. Moreover, microalgae synthesize chelating agents (capable of binding heavy metals) that control metal ion concentration at intracellular level, thereby neutralizing associated toxicity.

Like nutrients, microalgae absorb metal ions firstly by rapid adsorption followed by a much slower chemisorption process. Investigation on cellular distribution revealed the presence of large amount of metal ions in cell wall and an insoluble fraction within cell. Various microalgal species such as *Chlorella vulgaris, Scenedesmus* sp., *Chlorococcum* sp., *Lyngbya spiralis, Tolypothirx tenuis, Stigonema* sp., *Phormidium molle, Aphanothece halophytica* and *Chroococcus paris* are reported to eliminate Hg (II), Cd (II) and Pb (II) from aqueous solution (Tüzün et al. 2005). Another microalga *Dictyosphaerium chlorelloides* is found to be Cr (III) tolerant and has shown significant amount of metal ion accumulation in its cell wall, cytoplasm, vacuoles and chloroplasts (Pereira et al. 2013). Although Cr (III) is an important nutrient for fat and sugar metabolism in organism, it can be oxidized to its more lethal form by $MnO_2$ or bacteria. Cr (VI) has mutagenic and carcinogenic properties and causes several diseases, such as bronchogenic carcinoma, asthma, dermatitis and pneumonitis. The toxicity of Cr (VI) is 500–1000 times higher in a living cell compared to that of Cr (III) (Han et al. 2007). Biotransformation of Cr (VI) into less toxic Cr (III) is reported by Shen et al. (2013) in a detoxification assay using

**Table 3** Elimination of different xenobiotic compounds from wastewater by microalgae

| Xenobiotics | Microalgal Species | Degradation* or Tolerance** | Time (days/hours) | References |
|---|---|---|---|---|
| γ-Hexachlorocyclohexane | *Anabaena azotica* | 48.80% | 5 days | Zhang et al. (2012) |
| Benzo(a)pyrene | *Selenastrum capricornutum* | 99% | 15 h | García de Llasera et al. (2016) |
| | *Scenedesmus acutus* | 95% | 72 h | |
| Fluoranthene | *Chlorella vulgaris, Scenedesmus platydiscus, Scenedesmus quadricauda,* and *Selenastrum capricornutum* | 78% | 7 days | Lei et al. (2007) |
| Pyrenre | | 48% | | |
| Gasoline | *Tetraselmis chuii* | significant tolerance | 96 h | Paixão et al. (2007) |
| p-Chlorophenol | *Chlorella vulgaris and Coenochloris pyrenoidosa* | 100% | 5 days | Lima et al. (2004) |
| Bisphenol A | *Chlorella fusca* | 90% | 7 days | Hirooka et al. (2005) |
| 2,4-Dichlorophenol | *Chlorella fusca var. vacuolata* | 23% | 4 days | Tsuji et al. (2003) |
| 2,4-dichlorophenol | *Tetraselmis marina* | > 1 mM of 2,4-DCP removed in a 2 L photobioreactor | 6 days | Petroutsos et al. (2008) |
| Phenol | *Chlorella sp., Scenedesmus obliquus, Stichococcus sp. and Phormidium sp.* | >85% | 28 days | Safonova et al. (2004) |
| PCB (Aroclor 1242) | *Chattonella subsalsa, Chattonella marina var. marina* and *Chattonella marina var. ovata* | No cytotoxic effects at 0.7 mg $L^{-1}$ Aroclor 1242 | 10 days | Niestroy et al. (2014) |
| 2,4,6-Trinitrotoluene (TNT) | *Anabaena sp.* | 95% | 33 days | Pavlostathis and Jackson (2002) |

(continued)

**Table 3** (continued)

| Xenobiotics | Microalgal Species | Degradation* or Tolerance** | Time (days/hours) | References |
|---|---|---|---|---|
| Pyrimethanil (fungicide) | *Scenedesmus obliquus* and *Scenedesmus quadricauda* | 10% | 4 days | Dosnon-Olette et al. (2010) |
| Dimethomorph (fungicide) | | 24% | | |
| Isoproturon (herbicide) | | 58% | | |
| Atrazine (herbicide) | *Chlorella vulgaris* and *Synechococcus elongatus* | 83–90% | | González-Barreiro et al. (2006) |
| Terbutryn (herbicide) | | 85–93% | | |
| Fluroxypyr (herbicide) | *Chlamydomonas reinhardtii* | 57% | 5 days | Zhang et al. (2011) |

*Chlorella vulgaris*. This study showed the involvement of secondary alcohols in the reduction of Cr (VI) to Cr (III), with -$NH_2$ and –COOH as the main functional groups associated with the biosorption and biofixation processes. Another microalga *Ostreococcus tauri* is reported to convert inorganic arsenic (As) into a less toxic organic form that can easily be incorporated into biogeochemical cycles and can stimulate the volatilization of As through biomethylation (Zhang et al. 2013). Heavy metal adsorption by active microalgae varies due to alterations in the electronegativity of the heavy metals ions as shown by Sultana et al. (2020).

## 8 Microalgae Culture and Harvesting Techniques

A number of microalgae cultivation practices are followed during wastewater treatment. The most common is the suspended growth system, either in open or closed systems (photobioreactor). Corrugated raceway pond and high rate algal pond are examples of open cultivation systems. Closed microalgae cultivation (in photobioreactors) reduces undesirable contamination, loss of water by evaporation, and loss of injected $CO_2$. Among several photobioreactors, biocoil, horizontal tubular and vertical photobioreactors are usually used for wastewater treatment. To avoid the challenges of biomass harvesting in suspended growth system, microalgae can also be attached to a solid support medium where the attached cells contact wastewater. The nutrients are absorbed by the attached microalgal biomass and consequently more biomass are produced. However, the development cost of the support medium is high making this attached growth technique unsuitable for commercial use. Furthermore microalgae can be cultured in either batch or semi-continuous manner to treat wastewater. Most of the lab-scale microalgal bioremediation experiments are conducted in batch mode. Additionally, wastewater can be treated either using a single microalgal strain (monoculture) or using a consortium (association of two or more strains). It has been observed that a mixed culture of microalgal strains is more beneficial since wastewater is contaminated with different types of contaminants and the limitation of one microalgal strain could be overcome by other.

Removal of biomass from the treated wastewater is a vital step in microalgal bioremediation of wastewater. In most cases, a two-phase harvesting procedure is followed in which initially the harvesting is done by sedimentation, flocculation and filtration to get the biomass slurry which is then concentrated using a centrifuge.

## 9 Advantages of Using Microalgae

Microalgae-based technologies have a number of unique benefits. Their lucrative biodegradation techniques and the capability to survive under extreme environmental condition encourage researchers to explore newer strains suitable for wastewater management. It is an environment-friendly, viable alternative to the conventional treatments that are widely used currently. It is particularly attractive due to a number of reasons discussed below.

Firstly, microalgae serve as the primary producers of aquatic ecosystems. They maintain earth's sustainability by generating $O_2$ through photosynthesis and consuming the greenhouse gas $CO_2$ at the same time. Microalgae eliminate impurities directly from wastewater by absorbing organic or inorganic nutrients like carbon, nitrogen and phosphate which subsequently contribute to their growth. Furthermore, they enhance sedimentation rate by acting as flocculants during wastewater treatment.

Wastewater treated by an algal–bacterial consortium reduces the complexity and energy of the treatment process. Microalgae also support aerobic microorganisms by supplying oxygen which replaces the mechanical aeration procedure. They create the perfect environment for bacterial development and breakdown of organic substances by oxygenation. However, bacteria cannot remove mineral nutrients (such as phosphorus) efficiently, that ultimately lead to eutrophication in freshwater ecosystems. Therefore, this contaminated water must be treated further before it is discharged into natural water bodies, which makes the whole process cost-intensive. Anaerobic processes are also less effective and expensive as it requires external oxygen for the mineralization of pollutants. Microalgae serve as the best candidate here due to its $O_2$ evolving and nutrient assimilation properties. Microalgae also support co-metabolic degradation by secreting extracellular bio-surfactant compounds into the environment, which eventually enhances the bio-availability of the pollutants.

This technology has also a lower greenhouse gas emission rate. For example, most of the nitrogen is assimilated by the microalgae instead of being converted to oxides of nitrogen. Negligible emission of $N_2O$ by microalgae in combination with associated microorganisms in wastewater treatment has been reported by Guieysse et al. (2013) and Fagerstone et al. (2011).

As microalgae are aquatic, they do not require any terrestrial habitat for cultivation. Furthermore, they can be used for any type of wastewater, as they grow in both fresh and saline water. Microalgae are easy to handle, fast-growing organism producing large amount of biomass in a very short time. They require only light, $CO_2$ and minerals to grow and in turn synthesize large amount of protein, carbohydrate and lipid which can be processed to several industrially important value-added products.

Microalgal biomass are used in different commercial fields such as human nutrition, in cosmetics products, pigments, bio-fertilizers, medicinal, pharmaceutical, biofuel industry, as animal and aquatic feedstock, for the synthesis of antioxidant, antimicrobial, antiviral, antibacterial and anticancer drugs, etc. They also synthesize a number of value-added products such as docosahexaenoic acid (DHA), carotenoids enzyme polymer, lipid, natural dye, polyunsaturated fatty acid, peptide, acetylic acids, β-carotene, agars, agarose, alginates, carrageenans, polyunsaturated fatty acids, vitamin B, ketocarotenoid, astaxanthin, lutein and sterols which are used in several industrial products. Therefore, this green technology gives a dual purpose for algae cultivation, water purification and biomass production for future use.

## 10 Use of Harvested Microalgae

Wastewater-grown microalgal biomass can be used as bio-fertilizers as they enhance the nitrogen and phosphorus content of soil, in addition to many minerals (Ca, K, Fe, Mn, etc.) and various plant stimulating compounds required by the terrestrial plants. Although pathogens and other micropollutants associated with microalgae are the barriers for such application. Harvested microalgae with low lipid content are anaerobically digested to yield biogas such as methane. Numerous studies have also shown biodiesel production by growing lipid rich microalgae in wastewater. The potential biogas yield from microalgal biomass was reported to be as high as 200–600 mL/g organic content (Mudimu et al. 2014). Wastewater treated microalgae are used to produce alcohol, lactic acid, bioplastics. Furthermore, glycerol is produced as a by-product of biodiesel production. Microalgae and cyanobacteria produce exopolysaccharides which can be used as a gelling agent, thickener, stabilizer and as biolubricants and anti-inflammatory agents. Due to the high nutritional value, harvested microalgae can be used as animal feed. In the diluted swine manure wastewater, the possibility of producing omega-3 fatty acid-rich microalgae biomass as a source of feed ingredient has been explored. Microalgal biomass from aquaculture effluent can also be used as food for fish.

## 11 Challenges

Microalgal growth in wastewater is influenced by a number of factors such as pH, temperature, turbidity, light, $CO_2$, $O_2$ and nutrient concentration. It has also been observed that microalgal growth rate, nutrient recycling and wastewater treatment efficacy are affected in the absence of an external $CO_2$ supply. The presence of heavy metals, toxic organic compounds, bacterial pathogens and predators (zooplankton) inhibits on microalgae growth. In heterotrophic or mixotrophic cultivation, light restriction can disturb microalgal growth and ultimately the bioremediation efficacy. In certain cases, additional supply of organic carbon or the requirement of aeration and pumping systems to the culture make the whole technique costly. Another key challenge is the non-sterile environment associated with the process. Additionally, microalgal growth is hampered in large scale, deep, open cultivation systems mainly due to light saturation at the top layer. Although it can be overcome by higher flow velocity, it will again make the process energy-expensive.

Even though microalgae have very high efficacy in eliminating impurities from wastewater, rapid exposure to very high concentration of toxic compounds can affect the microalgal culture and consequently the phycoremediation process. Presence of high concentration of ammonia can also be toxic to the microalgae. To conquer these challenges, more investigation in microalgal metabolic engineering is required. In addition, appropriate strain or growth conditions should be chosen that enable autoflocculation or bioflocculation of the biomass to avoid expensive harvesting techniques.

# References

Benemann JR, Weissman JC, Koopman BL, Oswald WJ (1977) Energy production with microalgae. Nature 19:19–23

Bishnoi NR, Kumar R, Kumar S, Rani S (2007) Biosorption of Cr (III) from aqueous solution using algal biomass *Spirogyra* spp. J Hazard Mater 145:142–147

Borowitzka MA, Beardall J, Raven J (2016) The Physiology of microalgae. Springer International Publishing, Springer

Cerniglia CE, Gibson DT, Baalcn CV (1979) Algal oxidation of aromatic hydrocarbons: formation of 1-naphthol from naphthalene by *Agmenellum quadruplicatum*, strain PR-6. Biochem Biophys Res Commun 88:50–58

Chan A, Salsali H, Mc Bean E (2014) Heavy metal removal (copper and zinc) in secondary effluent from wastewater treatment plants by microalgae. ACS Sustain Chem Eng 2:130–137

Choi H (2015) Intensified production of microalgae and removal of nutrient using a microalgae membrane bioreactor (MMBR). Appl Biochem Biotechnol 175:2195–2205

Chong AM, Wong YS, Tam NF (2000) Performance of different microalgal species in removing nickel and zinc from industrial wastewater. Chemosphere 41(1–2):251–257

Choudhary P, Prajapati SK, Kumar P, Malik A, Pant KK (2017) Development and performance evaluation of an algal biofilm reactor for treatment of multiple wastewaters and characterization of biomass for diverse applications. Bioresour Technol 224:276–284

Christenson LB, Sims RC (2012) Rotating algal biofilm reactor and spool harvester for wastewater treatment with biofuels by-products. Biotechnol Bioeng 109:1674–1684

Dahmani S, Zerrouki D, Ramanna L, Rawat I, Bux F (2016) Cultivation of *Chlorella pyrenoidosa* in outdoor open raceway pond using domestic wastewater as medium in arid desert region. Bioresour Technol 219:749–752

Das P, Thaher M, Khan S, Abdul Quadir M, Al-Jabri H (2019) The effect of culture salinity on the harvesting of microalgae biomass using pilot-scale Tangential-Flow-Filter membrane. Bioresour Technol 293:122057

Deng GF, Zhang TW, Yang LM, Wang QQ (2013) Studies of biouptake and transformation of mercury by a typical unicellular diatom *Phaeodactylum tricornutum*. Chinese Sci Bull 58:256–265

Dosnon-Olette R, Trotel-Aziz P, Couderchet M, Eullaffroy P (2010) Fungicides and herbicide removal in *Scenedesmus* cell suspensions. Chemosphere 79:117–123

Escapa C, Coimbra RN, Paniagua S, García AI, Otero M (2015) Nutrients and pharmaceuticals removal from wastewater by culture and harvesting of *Chlorella sorokiniana*. Bioresour Technol 185:276–284

Fagerstone KD, Quinn JC, Bradley TH, De Long SK, Marchese AJ (2011) Quantitative measurement of direct nitrous oxide emissions from microalgae cultivation. Environ Sci Technol 45(21):9449–9456

Fard GH, Mehrnia MR (2017) Investigation of mercury removal by Micro-Algae dynamic membrane bioreactor from simulated dental waste water. J Environ Chem Eng 5:366–372

García de Llasera MP, Olmos-Espejel J, Díaz-Flores G, Montaño-Montiel A (2016) Biodegradation of benzo(a)pyrene by two freshwater microalgae *Selenastrum capricornutum* and *Scenedesmus acutus*: a comparative study useful for bioremediation. Environ Sci Pollut Res Int 23(4):3365–3375

Gonzalez-Barreiro O, Rioboo C, Herrero C, Cid A (2006) Removal of triazine herbicides from freshwater systems using photosynthetic microorganisms. Environ Pollut 144:266–271

Guieysse B, Plouviez M, Coilhac M, Cazali L (2013) Nitrous Oxide ($N_2O$) production in axenic *Chlorella vulgaris* microalgae cultures: evidence, putative pathways, and potential environmental impacts. Biogeosciences 10:6737–6746

Han X, Wong YS, Wong MH, Tam NFY (2007) Biosorption and bioreduction of Cr(VI) by a microalgal isolate, *Chlorella miniata*. J Hazard Mater 146:65–72

He S, Xue G (2010) Algal-based immobilization process to treat the effluent from a secondary wastewater treatment plant (WWTP). J Hazard Mater 178:895–899
Hirooka T, Nagase H, Uchida K, Hiroshige Y, Ehara Y, Nishikawa JI, Nishihara T, Miyamoto K, Hirata Z (2005) Biodegradation of bisphenol A and disappearance of its estrogenic activity by the green alga *Chlorella fusca* var. *vacuolata*. Environ Toxicol Chem 24:1896–1901
Ji M-K, Kabra AN, Salama ES, Roh HS, Kim JR, Lee DS, Jeon B-H (2014) Effect of mine wastewater on nutrient removal and lipid production by a green microalga *Micratinium reisseri* from concentrated municipal wastewater. Bioresour Technol 157:84–90
Jin L, Zhang G, Tian H (2014) Current state of sewage treatment in China. Water Res 66:85–98. https://doi.org/10.1016/j.watres.2014.08.014
Kumar R, Goyal D (2010) Waste water treatment and metal ($Pb^{2+}$, $Zn^{2+}$) removal by microalgal based stabilization pond system. Indian J Microbiol 50:34–40
Lal S, Lal R, Saxena DM (1987) Bioconcentration and metabolism of DDT, fenitrothion and chlorpyrifos by the blue-green algae *Anabaena* sp. and *Aulosira fertilissima*. Environ Pollut 46:187–196
Lei AP, Hu ZL, Wong YS, Tam NFY (2007) Removal of fluoranthene and pyrene by different microalgal species. Bioresour Technol 98:273–280
Li Y, Chen YF, Chen P, Min M, Zhou W, Martinez B, Zhu J, Ruan R (2011) Characterization of a microalga Chlorella sp. well adapted to highly concentrated municipal wastewater for nutrient removal and biodiesel production. *Bioresour Technol* 102(8):5138–5144. https://doi.org/10.1016/j.biortech.2011.01.091
Li B, Brett MT (2013) The influence of dissolved phosphorus molecular form on recalcitrance and bioavailability. Environ Pollut 182:37–44
Lima SAC, Raposo MFJ, Castro PML, Morais RM (2004) Biodegradation of p-chlorophenol by a microalgae consortium. Water Res 38:97–102
Lowrey J, Brooks MS, McGinn PJ (2015) Heterotrophic and mixotrophic cultivation of microalgae for biodiesel production in agricultural wastewaters and associated challenges - a critical review. J Appl Phycol 27:1485–1498
Magro CD, Deon MC, De Rossi A, Reinehr CO, Hemkemeier M, Colla LM (2012) Chromium (VI) biosorption and removal of chemical oxygen demand by Spirulina platensis from wastewater-supplemented culture medium. J Environ Sci Health A Tox Hazard Subst Environ Eng 47(12):1818–1824
Malakootian M, Yousefi Z, Limoni Z (2019) Removal of lead from battery industry wastewater by *Chlorella vulgaris* as green micro-algae. Desalin Water Treat 141:248–255
Menger-Krug E, Niederste-Hollenberg J, Hillenbrand T, Hiessl H (2012) Integration of microalgae systems at municipal wastewater treatment plants: implications for energy and emission balances. Environ Sci Technol 46(21):11505–11514. https://doi.org/10.1021/es301967y
Mirghaffari N, Moeini E, Farhadian O (2015) Biosorption of Cd and Pb ions from aqueous solutions by biomass of the green microalga, *Scenedesmus quadricauda*. J Appl Phycol 27:311–320
Mudimu O, Rybalka N, Bauersachs T, Born J, Friedl T, Schulz R (2014) Biotechnological screening of microalgal and cyanobacterial strains for biogas production and antibacterial and antifungal effects. Metabolites 4:373–393
Oswald W, Gotaas H (1957) Photosynthesis in sewage treatment. Trans Am Soc Civ Eng 122:73–105
Niestroy J, Martínez AB, Band-Schmidt CJ (2014) Analysis of concentration-dependent effects of copper and PCB on different *Chattonella* spp. microalgae (raphidophyceae) cultivated in artificial seawater medium. EXCLI J 13:197–211
Oswald WJ (1963) High rate pond in waste disposal. Dev Ind Biotechnol 4:112–119
Paixao JF, Nascimento IA, Pereira SA, Leite MBL, Carvalho GC, Silveira Jr JSC, Reboucas M, Matias RA, Rodrigues ILP (2007) Estimating the gasoline components and formulations toxicity to microalgae (*Tetraselmis chuii*) and oyster (*Crassostrea rhizophorae*) embryos: an approach to minimize environmental pollution risk. Environ Res 103:365–374
Pavlostathis SG, Jackson GH (2002) Biotransformation of 2, 4,6-trinitrotoluene in a continuous-flow Anabaena sp. system. Water Res 36(7):1699–1706

Pena-Castro JM, Martínez-Jeronimo F, Esparza-Garcia F, Canizares-Villanueva RO (2004) Heavy metals removal by the micro-alga *Scenedesmus incrassatulus* in continuous cultures. Bioresour Technol 94:219–222

Pereira M, Bartolome MC, Sanchez-Fortun S (2013) Bioadsorption and bioaccumulation of chromium trivalent in Cr(III)-tolerant microalgae: a mechanisms for chromium resistance. Chemosphere 93:1057–1063

Petroutsos D, Katapodis P, Samiotaki M, Panayotou G, Kekos D (2008) Detoxification of 2,4-dichlorophenol by the marine microalga *Tetraselmis marina*. Phytochemistry 69(3):707–714

Posadas E, García-Encina PA, Soltau A, Domínguez A, Díaz I, Muñoz R (2013) Carbon and nutrient removal from centrates and domestic wastewater using algal bacterial biofilm bioreactors. Bioresour Technol 139:50–58

Posadas E, García-Encina PA, Domínguez A, Díaz I, Becares E, Blanco S, Muñoz R (2014) Enclosed tubular and open algal–bacterial biofilm photobioreactors for carbon and nutrient removal from domestic wastewater. Ecol Eng 67:156–164

Pradhan D, Sukla LB, Mishra BB, Devi N (2019) Biosorption for removal of hexavalent chromium using microalgae *Scenedesmus* sp. J Clean Prod 209:617–629

Qi F, Xu Y, Yu Y, Liang X, Zhang L, Zhao H, Wang H (2017) Enhancing growth of *Chlamydomonas reinhardtii* and nutrient removal in diluted primary piggery wastewater by elevated $CO_2$ supply. Water Sci Technol 75:2281–2290

Ruiz-Marin A, Mendoza-Espinosa LG, Stephenson T (2010) Growth and nutrient removal in free and immobilized green algae in batch and semi-continuous cultures treating real wastewater. Bioresour Technol 101:58–64

Safonova E, Kvitko KV, Iankevitch MI, Surgko LF, Afti IA, Reisser W (2004) Biotreatment of industrial wastewater by selected algal-bacterial consortia. Eng Life Sci 4:347–353

Sayadi MH, Rashki O, Shahri E (2019) Application of modified *Spirulina platensis* and *Chlorella vulgaris* powder on the adsorption of heavy metals from aqueous solutions. J Environ Chem Eng 7(3). https://doi.org/10.1016/j.jece.2019.103169

Sforza E, Cipriani R, Morosinotto T, Bertucco A, Giacometti GM (2012) Excess $CO_2$ supply inhibits mixotrophic growth of *Chlorella protothecoides* and *Nannochloropsis salina*. Bioresour Technol 104:523–529

Shah MP (2020) Microbial bioremediation & biodegradation. Springer

Shah MP (2021a) Removal of emerging contaminants through microbial processes. Springer

Shah MP (2021b) Removal of refractory pollutants from wastewater treatment plants. CRC Press

Sheik AR, Muller EE, Wilmes P (2014) A hundred years of activated sludge: time for a rethink. Front Microbiol 5:47. https://doi.org/10.3389/fmicb.2014.00047

Shen QH, Zhi TT, Cheng LH, Xu XH, Chen HL (2013) Hexavalent chromium detoxification by nonliving *Chlorella vulgaris* cultivated under tuned conditions. Chem Eng J 228:993–1002

Shirai F, Kunii K, Sato C, Teramoto Y, Mizuki E, Murao S, Nakayama S (1998) Cultivation of microalgae in the solution from the desalting processes of soy sauce waste treatment and utilization of the algal biomass for ethanol fermentation. World J Microbiol Biotechnol 14:839–842

Sibi G (2016) Biosorption of chromium from electroplating and galvanizing industrial effluents under extreme conditions using *Chlorella vulgaris*. Green Energy Environ 1:172–177

Su Y, Mennerich A, Urban B (2012) Synergistic cooperation between wastewater-born algae and activated sludge for wastewater treatment: influence of algae and sludge inoculation ratios. Bioresour Technol 105:67–73

Sultana N, Hossain SMZ, Mohammed ME, Irfan MF, Haq B, Faruque MO, Razzak SA, Hossain MM (2020) Experimental study and parameters optimization of microalgae based heavy metals removal process using a hybrid response surface methodology-crow search algorithm. Sci Rep 10:15068

Tao Q, Gao F, Qian CY, Guo XZ, Zheng Z, Yang ZH (2017) Enhanced biomass/ biofuel production and nutrient removal in an algal biofilm airlift photobioreactor. Algal Res 21:9–15

Travieso L, Cañizares RO, Borja R, Benítez F, Domínguez AR, Dupeyrón R, Valiente YV (1999) Heavy metal removal by microalgae. Bull Environ Contam Toxicol 62:144–151
Tsuji N, Hirooka T, Nagase H, Hirata K, Miyamoto K (2003) Photosynthesis-dependent removal of 2, 4-dichlorophenol by *Chlorella fusca* var. *vacuolata*. Biotech Lett 25:241–244
Tüzün I, Bayramoğlu G, Yalçın E, Başaran G, Çelik G, Arıca MY (2005). Equilibrium and kinetic studies on biosorption of Hg (II), Cd(II) and Pb(II) ions onto microalgae *Chlamydomonas reinhardtii*. J Environ Manage 77:85–92
Wang B, Li Y, Wu N, Lan CQ (2008) $CO_2$ bio-mitigation using microalgae. Appl Microbiol Biotechnol 79:707–718
Whitton R, Ometto F, Pidou M, Jarvis P, Villa R, Jefferson B (2015) Microalgae for municipal wastewater nutrient remediation: mechanisms, reactors and outlook for tertiary treatment. Environ Technol Rev 4(1):133–148
Wong JPK, Wong YS, Tam NFY (2000) Nickel biosorption by two *chlorella* species, *C. Vulgaris* (a commercial species) and *C. Miniata* (a local isolate). Bioresour Technol 73:133–137
Xie Y, Li H, Wang X, Ng IS, Lu Y, Jing K (2014) Kinetic simulating of Cr(VI) removal by the waste *Chlorella vulgaris* biomass. J Taiwan Inst Chem Eng 45:1773–1782
Yao L, Shi J, Miao X (2015) Mixed wastewater coupled with $CO_2$ for microalgae culturing and nutrient removal. PLoS ONE 10(9):e0139117
Yee W (2015) Feasibility of various carbon sources and plant materials in enhancing the growth and biomass productivity of the freshwater microalgae *Monoraphidium griffithii* NS16. Bioresour Technol 196:1–8
Zamalloa C, Boon N, Verstraete W (2013) Decentralized two-stage sewage treatment by chemical-biological flocculation combined with microalgae biofilm for nutrient immobilization in a roof installed parallel plate reactor. Bioresour Technol 130:152–160
Zhang H, Hu C, Jia X, Xu Y, Wu C, Chen L, Wang F (2012) Characteristics of $\gamma$-hexachlorocyclohexane biodegradation by a nitrogen-fixing cyanobacterium, *Anabaena azotica*. J Appl Phycol 24:221–225
Zhang S, Qiu CB, Zhou Y, Jin ZP, Yang H (2011) Bioaccumulation and degradation of pesticide fluroxypyr are associated with toxic tolerance in green alga *Chlamydomonas reinhardtii*. Ecotoxicology 20:337–347
Zhang SY, Sun GX, Yin XX, Rensing C, Zhu YG (2013) Biomethylation and volatilization of arsenic by the marine microalgae *Ostreococcus tauri*. Chemosphere 93:47–53
Zhao B, Su Y (2014) Process effect of microalgal-carbon dioxide fixation and biomass production: a review. Renew Sust Energ Rev 31:121–132

# Role of Microalgae in Integrated Wastewater Remediation and Valorization of Value-Added Compounds

**Rayanee Chaudhuri, Nageshwari Krishnamoorthy, and Balasubramanian Paramasivan**

## 1 Introduction

Population growth and industrialization, due to advancements in technologies, have increased water demand over the years. Freshwater is an unevenly distributed, limited natural resource. Presently, different regions around the globe (especially countries in the South-Asian region) are already under severe to moderate water stress, which is predicted to worsen by 2030. To meet the current water demand, often practiced excessive groundwater withdrawal may not be a sustainable option as it lowers the water table, thereby reducing the available surface water. More importantly, research showed only 10–20% of this total freshwater withdrawal actually gets consumed, while the rest 80–90% goes back to the environment as wastewater. Without proper treatment, direct discharge of this wastewater is not a safe practice, as it contains different toxic compounds and several nutrients. Other than this, essential nutrients like phosphorous are non-renewable and 80% of the total phosphate (P) (majority of which comes from the extraction of P rock) ends up either in the wastewater or in the agricultural runoff, ultimately polluting the nearby waterbodies causing eutrophication and disturbing their ecosystem (Solovchenko et al. 2019). To mitigate all these pressing issues, adopting a circular economy approach (i.e., reclamation and recycling of these limiting resources) is necessary to achieve sustainability.

The conventional wastewater reclamation techniques, which mostly focus only on nutrient removal but not recovery, show a significant energy requirement. For example, Acién et al. (2016) stated that one of the largest European wastewater treatment companies, Aqualia, annually removes around 25,000 tons of nitrogen and 5,000 tons of phosphorous, requiring up to 0.5 kWh $m^{-3}$ energy. Due to this significant energy consumption, the reclamation of 1000 L wastewater costs around

R. Chaudhuri · N. Krishnamoorthy · B. Paramasivan (✉)
Smart Agricultural Lab, Department of Biotechnology and Medical Engineering, National Institute of Technology Rourkela, Rourkela, Odisha 769008, India
e-mail: biobala@nitrkl.ac.in



M. P. Shah (ed.), *Sustainable Industrial Wastewater Treatment and Pollution Control*,
https://doi.org/10.1007/978-981-99-2560-5_4

0.2 €, which can go up to 5–8 €, if other advanced nutrient removal processes (e.g., advanced oxidation) are added (Fernández et al. 2018). It is mention-worthy that annually 0.5 Mt of microalgal biomass can be produced, using this significant amount of nutrients (Acién et al. 2016). Moreover, significant environmental impacts (due to greenhouse gas or GHG emission and wastage of recyclable limited resources) have pushed the scientific community to look for alternative low-carbon emitting treatment methods that integrate wastewater reclamation with resource recovery, promoting the concept of circular economy. In this regard, microalgal bioremediation can be an attractive alternative.

Microalgae, an important source of 3rd generation biofuel, uptake nutrients in the presence of sunlight and convert them into several valuable bioactive components (such as polyunsaturated fatty acids, polysaccharides, and pigments like chlorophyll and carotenoid), which has nutritional or therapeutic value. Due to their higher surface to volume ratio, microalgae can assimilate a significant amount of nutrients, making them an attractive source of biofertilizer. Due to the above-mentioned reasons, commercial cultivation of microalgal strains like *Arthrospira* sp. and *Chlorella* sp. has gained much attention. However, for commercial-level production of microalgal biomass, huge production cost often comes as an obstacle. Supplying sufficient nutrients is important to achieve a high algal productivity rate. Fernández et al. (2018) stated that, around 10, 1, and 200 tons of nitrogen, phosphate, and $CO_2$, respectively, are required for producing 100 tons of microalgal biomass. Nutrients that are supplied in the form of synthetic fertilizer, not only increase the production cost but also create significant environmental impacts during their production. For example, the production of synthetic N-fertilizers emits a significant amount of greenhouse gases. Hence, to reduce the production cost, use of nutrient-rich effluents (containing high nutrient and organic load) from different sources can help in integrating the wastewater reclamation and nutrient recovery, achieving environmental and economical sustainability. Additionally, being an autotrophic organism, they have high $CO_2$ capturing capacity combined with the other flue gases from the atmosphere, which helps in carbon sequestering and reduction of greenhouse gas (GHG) emissions. These benefits make the microalgae-based wastewater remediation technique an attractive, eco-friendly, and sustainable technology, which helps in attaining the circular economy goal. Therefore, the present book chapter discusses about how microalgal bioremediation can solve the drawbacks of conventional wastewater treatment technologies. In addition, it elucidates the possible strategies to enhance their efficiency, their limitations, and various ways of valorization of the cultivated microalgal biomass in detail.

## 2 Microalgal Cultivation Systems

The cultivation system is one of the influential factors that affect the total microalgal yield and bioremediation capacity. Algal cultivation can be divided into different categories based on their preference for carbon sources and dependency on light (Fig. 1). Photoautotrophic cultivation uses inorganic carbon sources ($CO_2$) in the presence of light for biomass growth. For large-scale cultivation in an open cultivation system, photoautotrophic strains are the most preferred ones as they can utilize both organic and inorganic carbon sources and also switch between the two depending on the light availability. In contrast, the heterotrophic cultivation system solely relies on organic carbon sources and mixotrophic can utilize both organic and inorganic carbon. While cultivating a widespread microalgal strain, *Scenedesmus* sp., Kamalanathan et al. (2018) showed that heterotrophic cultivation increases the growth rate and biomass concentration compared to photoautotrophic cultivation. A shoot up in the cell growth and productivity was observed under the exposure of light in the heterotrophic culture coupled with higher protein and carbohydrate content in the biomass. Compared with heterotrophic and autotrophic cultivation, Sajadian et al. (2018) achieved higher biomass production (257% and 158%, respectively) with mixotrophic cultivation, along with a maximum cell dry weight of 3.91 g $L^{-1}$. However, heterotrophic cultivation showed the highest lipid accumulation (i.e., 48.68% of the cell weight). Another such example is the cultivation of a novel microalgal strain *Leptolyngbya subtilis* JUCHE by Das et al. (2021), under heterotrophic conditions, where 4.66 times higher lipid concentration was obtained compared to autotrophic mode. Cultivation of *Chodatella* sp. in swine effluent under a mixotrophic condition showed 14 times increased biomass production and 5.6 times higher lipid productivity than the autotrophic cultivation mode (Li et al. 2014). From the above examples, it can be concluded that the oleaginous microalgal strains, in mixotrophic or heterotrophic conditions, can help in achieving a higher biomass and lipid accumulation rate compared to photoautotrophic cultivation.

Other than carbon source, microalgal cultivation can also be categorized into open and close cultivation systems depending on the circulation of growth media. Close

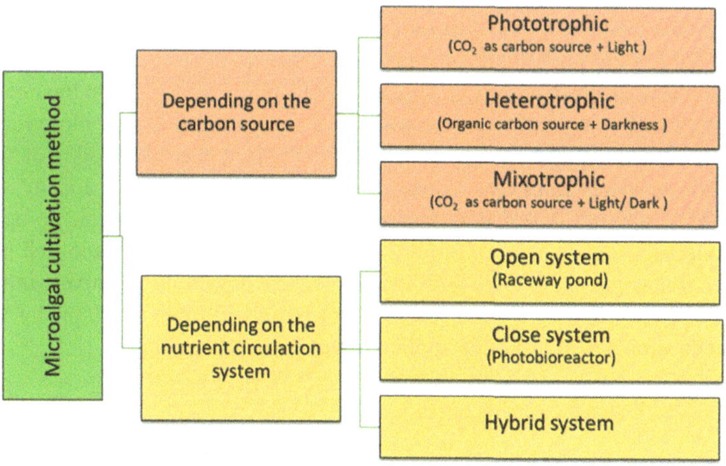


**Fig. 1** Various microalgal cultivation systems

cultivation systems like photobioreactors are commonly used for lab-scale cultivation for growing pure cultures. They offer high light availability, precise control, high volumetric biomass productivity, less risk of contamination but increase the operational cost by increasing the energy requirements (i.e., for recirculation and sparging $CO_2$). For example, the cultivation of *Scenedesmus almeriensis* in a 30,000 L photobioreactor by Acién et al. (2012) showed a daily energy consumption of 96 kWh and 240 kWh (i.e., 21 and 53.54% of the total energy consumption) for aeration and recirculation, respectively. Moreover, difficulties in scaling up make this kind of cultivation process less favorable.

On the other side, open systems like raceway ponds show maximum sunlight utilization efficiency with lower energy requirement and maintenance cost. Due to this, such open cultivation technique is preferable for microalgal biofuel production as its capital cost is at least ten times lower than the photobioreactor (Renuka et al. 2021). However, due to the environmental exposure, it is associated with a high contamination risk, making the quality control of the biomass problematic. In both open and close types of cultivation, the mode of operation is batch, and therefore, maintaining a constant exponential growth phase is difficult.

Other than the open and close system, there are two-phase hybrid systems that combine both open and close cultivations for attaining higher yield. The hybrid cultivation system involves a close cultivation in the photobioreactor during the growth phase under the controlled environment (to reduce the chances of contamination) followed by open cultivation under a nutrient-stressed condition to induce the lipid accumulation process (Brennan and Owende 2010). In a comparative study, Narala et al. (2016) showed that hybrid systems achieved a higher growth rate and biomass productivity than individual systems, hence can be highly beneficial for the cultivation of oleaginous microalgae. In addition, life cycle studies indicated lower environmental impact of hybrid cultivation in comparison with open and closed (Brennan and Owende 2010). Besides, different types of modified photobioreactors like membrane photobioreactor (MPBR) and anaerobic membrane bioreactor (AnMBR) have recently gained much attention from scientists, due to their capacity to remove a wide range of contaminants. After comparing different bioreactors for wastewater treatment, Marbelia et al. (2014) reported the advantages of MPBR regarding its nutrient removal and microalgal biomass production capacity. Aslam et al. (2022) reported AnMBR's efficiency for the removal of organic compounds and simultaneous production of biofertilizer and energy from methane-rich biogas. However, issues like membrane fouling and increase in salinity negatively influence the overall performance of such reactors and require further development on the preconcentration area, before the application of these technologies.

## 3 Strategies for Microalgal Cultivation in Wastewater

Depending on the source, physicochemical characteristics of wastewater vary. For example, municipal wastewater contains comparatively lower nutrient content but a significant amount of heavy metals. Whereas, effluent from the livestock or food processing industries contains a significant amount of nutrients, along with ammonia (as a nitrogen (N) source). These characteristics of wastewater (i.e., carbon (C)/N/P ratio, pollutant load) and cultivation conditions (light intensity and photoperiod, C source) influence the microalgal productivity, as well as the nutrient removal efficiency, and hence, optimizing these factors are important. Some of the wastewater physicochemical characteristics that significantly affect microalgal cultivation are briefly discussed below.

### *3.1 Dilution and Other Wastewater Pretreatments*

Effluent characteristics like total suspended solids (TSS) and the microbial count influence the microalgal growth in a wastewater remediation system. The presence of suspended particles at a high concentration hinders the light penetration in wastewater, restricting microalgal growth. Several pretreatment methods (like filtration, dilution, UV radiation, autoclaving, and chemical treatments) can be used to solve such issues, ensuring significant microalgal growth. Chu et al. (2015) showed that the pretreatment of effluent from starch processing anaerobic digester with 53 μm polyester filter bags can reduce the TSS below 70 mg $L^{-1}$, which can in turn ensure better microalgal growth. Other than filtration, different chemical flocculants (i.e., $Al_2(SO_4)_3$, $Fe_2\ (SO_4)_3$) can also be used to reduce the TSS in wastewater. High pollutant concentration (commonly observed in the effluents from livestock or food processing industries) can also negatively affect microalgal growth. Hence, in such cases, dilution can help reduce the pollutant load. For example, Ding et al. (2015) obtained the highest biomass growth rate along with 83, 92, and 90% removal of ammonia, phosphate, and COD, respectively, with 5 times diluted dairy effluent within a cultivation period of 6 days. Sriram and Seenivasan (2012) confirmed the growing ability of microalgal strains like *Euglena viridis*, *Chlorella sorokiniana* in 4 and 8 times diluted swine wastewater. However, in other microalgal strains like *Scenedesmus obliquus* and different Chlorella strains like *Chlorella sorokiniana* and *Chlorella vulgaris*, microalgal growth was observed only in 8 times diluted solution.

Regarding the influence of microbes on microalgal growth, Marjakangas et al. (2015) showed *Chlorella vulgaris* CY5 cultivated in 5 times diluted swine wastewater reaches to the highest biomass concentration at a sterile (effluent autoclaved at 121 °C for 20 min) cultivation environment. It is reported that, compared to the unsterile culture media, higher lipid productivity (i.e., 117 g $L^{-1}$) was observed in the sterile media in a 12-day time period. However, in case of large-scale wastewater treatment, sterilization techniques like autoclaving can increase the total treatment cost, making the process economically challenging and energy intensive. UV radiation, chlorination, or other disinfecting agents can be used as alternative sterilization techniques.

### 3.2 Wastewater Mixtures

High pollutant load and limited nutrient availability lower the microalgal wastewater remediation process. Hence, diluting a nutrient-rich effluent with another comparatively less polluted one can be beneficial for achieving better microalgal growth and higher bioremediation efficiency. For example, in dairy wastewater, significantly less ammonium content has been a limiting factor for algal growth. Mixing the slaughterhouse's ammonium-rich effluent with the dairy wastewater can help overcome this issue. Cultivation of microalgae in such mixed wastewater showed high nutrient removal efficiency along with a protein and lipid-enriched microalgal biomass, which can be used as an alternative source of food and energy. Zheng et al. (2017) showed an enhanced microalgal biomass production of 2.85 g $L^{-1}$ and nutrient removal efficiency of >90% by mixing effluents from swine farms and breweries at a ratio of 1:5. Among other cultivated strains, *Selenastrum minutum* was found to be the highest biomass producing microalgae with up to 37% lipid content and 90% nutrient removal efficiency (Gentili 2014).

### 3.3 C/N and N/P Ratio

Other crucial factors influencing microalgal growth are the C/N and N/P ratio of the culture medium. While using wastewater as nutrient media, this ratio varies depending on the effluent source. Microalgae use inorganic carbon sources like $CO_2$ or other flue gases that are often present in wastewater. On the other side, when cultivating bacteria or fungi with the microalgae, concentration of the organic carbon plays an important role. Gao et al. (2018) stated that the protein content of microalgal biomass is inversely proportional to the C/N ratio and an increase in protein content can be observed when sufficient N is supplied. Under nutrient-stressed condition or higher C/N or C/P conditions, carbohydrate and natural lipid accumulation increases in the microalgal cell. When Li et al. (2020) cultivated a unicellular microalgal strain *Porphyridium purpureum* at different C/N ratios, the results showed that a higher ratio under sufficient nitrogen availability can accelerate the carbohydrate and fatty acids like ω-6 PUFA accumulation process, while a limited nitrogen availability induces the degradation of phycobiliprotein.

For co-cultivation of *Scenedesmus obliquus* and indigenous bacterial strain *Bacillus megaterium*, Li et al. (2021) showed that a higher C/N ratio was favorable for achieving higher nutrient and COD removal efficiency. The authors achieved 85.98, 81.03, and 65.48%, of COD, phosphorus, and ammonium removal, respectively, with a C/N/P ratio of 106/16/1. Xu et al. (2017) showed that at an optimized C/N ratio of 5:1, algal monoculture was able to attain COD, nitrogen, and phosphorus removal efficiencies of 78%, 77%, and 73%, respectively. However, combined algal-fungal and algal–bacterial cultivation systems were able to achieve higher (around $\geq$80%) COD and total nitrogen removal efficiency.

For achieving the optimal C/N or C/P ratio, the researchers have explored mixing of two different types of wastewater. Zheng et al. (2017) mixed low C/N ratio livestock wastewater (i.e., effluent from swine farm) with high C/N wastewater from the brewery industry at a 1:5 ratio to achieve an optimal C/N ratio of 7.9. The microalgae cultivated in such mixed effluent offered a biomass yield of 2.85 g $L^{-1}$ with a 100% ammonium removal efficiency and 96, 90, and 93% removal efficiencies of total nitrogen, phosphorus, and COD, respectively.

### 3.4 $CO_2$ as Carbon Source

Supplying $CO_2$ externally can increase photosynthesis and microalgal productivity through complete utilization of the available light. It enhances the treatment efficiency, especially under a C limited condition. For example, for the case of domestic wastewater, which has a lower C/N ratio than intracellular ratio, supplying a carbon source externally (pure $CO_2$ or in the form of flue gas) helps in enhancing the microalgal productivity (Park et al., 2013). When $CO_2$ gets dissolved in the nutrient media, it releases $H^+$ ion, which helps in maintaining the pH value of the nutrient media as shown in Eq. (1).

$$CO_{2(gas)} \leftrightarrow CO_{2(aq)} \leftrightarrow H_2CO_3 \leftrightarrow HCO_3^- + H^+ \leftrightarrow CO_3^{2-} + H^+ \quad (1)$$

Park and Craggs (2010) reported that the external supply of $CO_2$ in a high rate algal pond (HRAP) can enhance the organic pollutant removal up to 95% by maintaining the pH without affecting the dissolved oxygen (DO) level. Externally, $CO_2$ can be supplied through air, flue gas, or in the purified form. Compared to the pure $CO_2$, the application of sparged $CO_2$ in the form of flue gas in a raceway pond showed a better wastewater treatment performance, along with microalgal biomass containing 64.8% carbon, 12.6% nitrogen, and 2.4% phosphorus (Posadas et al. 2015).

Besides, carbon availability also depends on the delivery process, which includes direct bubbling, sparging through porous and non-porous membranes, or by introducing microbubbles. Zheng et al. (2018) stated that direct bubbling or sparging of $CO_2$ does not allow the complete utilization of the supplied carbon source. However, for the case of microbubbles of purified $CO_2$, the carbon availability increases while traveling through the nutrient medium due to size reduction and rapid dissolution. However, pumping the microbubbles requires a significantly high amount of energy and increases the production cost of the microalgal biomass. Hence, it can be concluded that technology for supplying $CO_2$ externally should be site specific and dependent on the purity level or the type of $CO_2$ source.

### 3.5 Light Intensity and Photoperiod

Intensity and duration of the light play a vital role in microalgal growth rate, biomass productivity, nutrient uptake, and the lipid accumulation efficiency. For microalgal cultivation, solar or artificial lights are absorbed and turned into microalgal biomass via photosynthesis. Several studies prove that higher light intensity combined with an external supply of $CO_2$ has the capability to enhance the biomass productivity, along with a higher nutrient removal efficiency.

Microalgal cultivation with longer than 12 h light–dark phase cycle can increase the phosphate removal efficiency. Powell et al. (2009) have mentioned that light intensity positively influences the phosphate accumulation process. Higher intensity not only increases the biomass content but also offers higher lipid content. Similarly, an extended photoperiod (i.e., more than 12 h) can significantly enhance microalgal growth. Sirisuk et al. (2018) showed that a photoperiod of 18 h a day is the optimum for cultivating *Isochrysis galbana*, whereas for *Nannochloropsis salina* and *Phaeodactylum tricornutum*, the optimal photoperiod was 24 h. Nzayisenga et al. (2020) showed that after 15 days of cultivation of two microalgal strains *Desmodesmus* sp. and *Scenedesmus obliquus* with a photo-intensity of 300 $\mu E\ m^{-2}\ s^{-1}$, the biomass concentration was found to be 1.4 and 1.2 $g\ L^{-1}$, respectively, which was significantly higher than the microalgae cultivated with lower photo-intensity. The authors also mentioned that increased light intensity influenced the fatty acid composition by increasing the oleic acid and decreasing the linolenic acid concentration.

Different wavelengths of light can also influence microalgal growth and lipid accumulation. Sirisuk et al. (2018) observed an enhanced microalgal growth with a combination of red and blue light. Scientists also reported that under green light stress, *Isochrysis galbana, Nannochloropsis salina*, and *Phaeodactylum tricornutum* showed a lipid content of 63.3, 49.4, and 62.0%, respectively.

## 4 Microalgal Nutrient Uptake Mechanisms

To understand the microalgal wastewater remediation process and achieve better nutrient removal efficiency, it is essential to have a thorough knowledge of the microalgal nutrient uptake mechanisms. Other than macronutrients like N and P, microalgae require micronutrients like iron, calcium, or magnesium for their growth. However, as N and P are the most essential and growth-limiting elements, their uptake processes are emphasized here as follows:

Nitrogen, one of the most crucial macronutrients required for microalgal cell growth and for developing various biological components like protein, chlorophylls, genetic material, and ATP/ADP is obtained from different nutrient sources in different forms. Microalgae take up the inorganic nitrogen through the plasma membrane, reduce the oxidized form into ammonium ($NH_4^+$), and finally, incorporate it into amino acids like L-glutamine (Fig. 2) (Cai et al. 2013). Ammonium, which

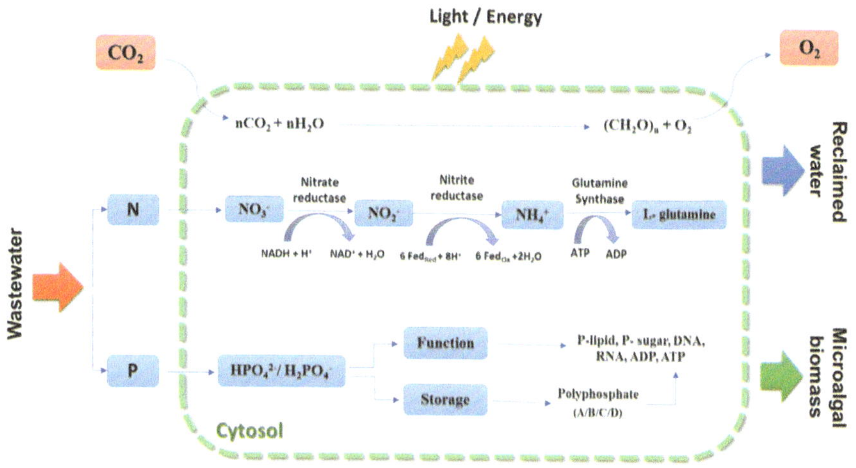


**Fig. 2** Microalgal nutrient uptake mechanisms

requires lesser uptake energy, is the most preferred nitrogen form and therefore gets consumed first. However, very high $NH_4^+$ concentration (i.e., >80–100 mg $L^{-1}$) can restrict microalgal growth (McDowell et al. 2020; Shah 2020). Nitrate ($NO_3^-$), being highly thermodynamically stable, is predominant in effluents and considered as the second most consumed form of nitrogen. Other than these, cyanobacteria can use atmospheric nitrogen and convert it into ammonia by means of fixation.

Phosphorus, another essential macronutrient for microalgal growth, is taken up through the plasma membrane, by active transport, in the form of di- and mono-hydrogen phosphates (i.e., $HPO_4^{2-}$ or $H_2PO_4^-$) and incorporated into the microalgal biomass (Cai et al. 2013; Solovchenko et al. 2019). Microalgae use this P to synthesize phospholipid (present in cell membrane) and phospho-sugar for DNA and RNA formation (Solovchenko et al. 2016, 2019). The ability of the microalgae to take up more phosphate than its requirement at an intermediate growth stage is known as "Luxury uptake." By nature, majority of the microalgal strains are acclimatized to phosphate-limited environment and hence, show the capability of luxury uptake whenever phosphorus becomes available. In phosphate-rich environment, luxury uptake of P occurs by rapidly increasing the phosphate content and microalgae store that phosphate into four different types of inorganic polyphosphates (A, B, C, and D) for future use (Fig. 2) (Solovchenko et al. 2019; Shah 2021b). Environmental parameters like light intensity, temperature, and phosphate concentration influence this luxury uptake and storage (Powell et al. 2009).

Lastly, the carbon present in every living organism, which is taken up in different organic and inorganic forms, depends on the nature of cultivation. For the autotrophs, inorganic C present in the form of $CO_2$ is taken up by the microalgae and fixed during photosynthesis, whereas for heterotrophs, organic compounds have to be supplied as a C source. Both the traits (auto and heterotrophic nature) can be observed in mixotrophic cultivation. Table 1 summarizes all the nutrient uptake mechanisms in

**Table 1** Equations used for explaining microalgal nutrient uptake mechanism

| Microalgal nutrient uptake equations | References |
|---|---|
| $8CO_2 + 7.37\,H_2O + 0.22\,NH_4 + 0.02\,H_2O_4P^- + 14.71$ photons $\rightarrow 2CO_{0.48}H_{1.83}N_{0.11}P_{0.01}$ (Chlamydomonas reinhardtii) $+ C_6H_{12}O_6 + 8.25O_2$ | Küçük et al. (2015) |
| $16\,NH_4^+ + 92\,CO_2 + 92\,H_2O + 14\,HCO_3^- + HPO_4^{2-} \rightarrow C_{106}H_{263}O_{110}N_{16}P + 106\,O_2$ | Ebeling et al. (2006) |
| $16\,NO_3^- + 124\,CO_2 + 140\,H_2O + HPO_4^{2-} \rightarrow C_{106}H_{263}O_{110}N_{16}P + 138\,O_2 + 18\,HCO_3^-$ | |

the form of equations. Other than nitrogen, phosphorus, and carbon, the presence of micronutrients like iron or other heavy metals also influences microalgal growth and can also show an inhibitory effect when present in a high concentration.

## 5 Integrated Nutrient Removal and Wastewater Remediation by Microalgae

Cultivating microalgae in wastewater is one of the newly explored low-cost bioremediation processes, in addition to lowering the additional nutrient (i.e., N and P) requirement for microalgal growth. Kothari et al. (2013) used *Chlamydomonas polypyrenoideum* for treating dairy-industry effluent, which showed a successful reduction of 90% nitrate and 70% phosphorus, coupled with 3.8 g of dry microalgal biomass containing 42% (w/w) lipid, after 10 days of cultivation. Such literature references for an integrated cultivation of algae and wastewater treatment along with the culture conditions are shown in Table 2.

Native microalgal strains are often found to be the most robust and efficient for the nutrient removal process. Pham and Bui (2020) treated concentrated nutrient-rich effluent from a fertilizer plant in Vietnam using a microalgal strain, *Scenedesmus* sp., isolated from the wastewater. The results showed that in a 10-day batch culture with 60 mg $L^{-1}$ inoculum, the authors were able to produce 70.2 mg (dry microalgal biomass) $L^{-1}$ and significantly reduce the nutrient load ($NH_4^+$: 92.8 ± 3.5%, $NO_3^-$: 83.6 ± 4.5%, TP: 95.8 ± 2.8%) including BOD and COD by 83.7 ± 2.3 and 92.6 ± 4.0%, respectively. This study also indicated that autoflocculation, an important property of microalgae, can enhance the bioremediation process. The highest flocculation activity achieved was 88.3%, with 60 mg $L^{-1}$ microalgal inoculum. In addition to the use of native strains, microalgal consortium has found to be a better option for practical application. Co-existing of different microalgal strains can enhance the nutrient removal process and also makes the system more robust. For example, when a consortium of two common robust microalgal species *Chlorella* sp. and *Scenedesmus* sp.

**Table 2** Process conditions of wastewater treatment and nutrient removal using microalgae

| Wastewater source | Algal species | Mode of cultivation/reactor volume | Pre-treatment/cultivation conditions | Time (days) | Biomass/value-added product yield | Pollutant/nutrient removal | Reference |
|---|---|---|---|---|---|---|---|
| Dairy industry | *Chlamydomonas polypyrenoideum* | Batch cultivation (500 ml) | Light intensity: 10 W $m^{-2}$<br>Photoperiod: 12:12 h light: dark<br>Temperature: 28 ± 2 °C | 10 | Biomass:3.8 g dry mass<br>Lipid: 42% (w/w) | 90% $NO_3^-$<br>74% $NO_2^-$<br>70% $PO_4^{3-}$<br>90% $NH_4^+$<br>61% $Cl^-$<br>58% $F^-$ | Kothari et al. (2013) |
| Textile effluent | *Chlorella vulgaris* | Batch cultivation (500 ml) | Light intensity: 69 lx<br>Photoperiod: 12:12 h light: dark<br>Temperature: 25 ± 1 °C | 15 | Biomass: 270,009 cells $ml^{-1}$ | 88.67% P<br>59.47% COD<br>75.68% color | El-kassas and Mohamed, (2014), Shah (2021a, b) |
| Wastewater | *Chlorella* sp. and *Scenedesmus* sp. consortium | Batch cultivation (250 ml) | Light intensity: 2080 lx<br>Photoperiod: 16:8 h light: dark<br>Temperature: 17.5 ± 1 °C<br>Aeration intensity: 2 L $min^{-1}$ | 21 | Biomass: 3.04 g $L^{-1}$ | 99.7–99.9% P<br>88.6–98.4% N | Koreiviene et al. (2014) |
| Digested piggery wastewater | *Arthrospira platensis* ZJWST-S1 | Raceway pond (250 L) | Light intensity: 115 µmol $m^{-2}$ $s^{-1}$<br>Photoperiod: 15:9 h light: dark<br>Temperature: 20–25 °C<br>Agitation intensity: 30 rpm $min^{-1}$ | 9 | Biomass: 45.2–64.7 g $m^{-2}$ $d^{-1}$ | 9.4–11.6% N<br>19.5–22.2% P<br>50% COD | Liu et al. (2015) |

(continued)

**Table 2** (continued)

| Wastewater source | Algal species | Mode of cultivation/reactor volume | Pre-treatment/cultivation conditions | Time (days) | Biomass/value-added product yield | Pollutant/nutrient removal | Reference |
|---|---|---|---|---|---|---|---|
| Municipal wastewater | *Coelastrum microporum* | Photobioreactor (760 ml) | Light intensity: 120 ± 10 μmol $m^{-2}$ $s^{-1}$<br>Photoperiod: 12:12 h light: dark<br>Temperature: 20 °C | 12 | – | 87.84% N<br>88.67% P<br>59.47% COD | Lee et al. (2015) |
| Slaughterhouse wastewater | *Phormidium tergestirum* | High-rate algal ponds (75 L) | Light intensity: 4500 ± 150 lx<br>Photoperiod: 12:12 h light: dark<br>Temperature: 25 ± 2 °C | 10 | Biomass: 7.1 g $m^{-2}$ $day^{-1}$<br>Lipid:13.9 ± 0.9%<br>Carbohydrate:12.6 ± 0.3%<br>Protein: 57.7 ± 6.1% | 56.7 ± 21.3% P<br>91.7 ± 13.1% COD | Hernández et al. (2016) |
| Slaughterhouse wastewater | *Anabaena* sp. | High-rate algal ponds (75 L) | Light intensity: 40,000 ± 7500 lx<br>Photoperiod: 13:11 h light: dark<br>Temperature: 20 ± 6 °C | 10 | Biomass: 10.7 g $m^{-2}$ $day^{-1}$<br>Lipid: 14.2 ± 1.0%<br>Carbohydrate: 21.7 ± 0.4%<br>Protein: 46.5 ± 4.9% | 90.7 ± 24.6% P<br>86.4 ± 7.5% COD | Hernández et al. (2016) |
| Municipal wastewater | Microalgal consortium | Batch cultivation | Light intensity: 130 μmol $m^{-2}$ $s^{-1}$<br>Temperature: 25 °C<br>Aeration: 0.8 L air $L^{-1}$ $min^{-1}$ with 2% $CO_2$ | 12 | Biomass:147 ± 1 mg $L^{-1}$ $day^{-1}$<br>Lipid: 56 ± 1 mg $L^{-1}$ $day^{-1}$ | 99% N<br>82% P | Aketo et al. (2020) |

(continued)

**Table 2** (continued)

| Wastewater source | Algal species | Mode of cultivation/reactor volume | Pre-treatment/cultivation conditions | Time (days) | Biomass/value-added product yield | Pollutant/nutrient removal | Reference |
|---|---|---|---|---|---|---|---|
| Fertilizer Plant effluent | *Scenedesmus* sp. | Batch cultivation (800 ml) | Light intensity: 95 μmol photons $m^{-2}$ $s^{-1}$<br>Photoperiod: 12:12 h light: dark<br>Temperature: 27 ± 1 °C<br>Agitation: 70 rpm<br>Agitation: 120 rpm with 1.5% $CO_2$<br>Microalgal inoculum: 60 mg $L^{-1}$ | 10 | Dry biomass: 70.2 ± 4.9 g $L^{-1}$ | 92.8 ± 3.5% $NH_4^+$<br>83.6 ± 4.5% $NO_3^-$<br>97.0 ± 1.7% $PO_4^{3-}$<br>95.8 ± 2.8% P<br>92.6 ± 4.0% COD | Pham and Bui (2020) |
| Primary-treated municipal wastewater | *Muriellopsis* sp. | Raceway pond (800 L) | Light intensity: 2000–2500 μmol $m^{-2}$ $s^{-1}$<br>Photoperiod: 12:12 h light: dark<br>Temperature: 26 °C | 4 | Biomass: 0.10 g $L^{-1}$ $day^{-1}$<br>Protein: 51%<br>Carotenoid: 0.6%<br>Lutein: 0.3% | 84% N<br>93% P | Cavieres et al. (2021) |

was co-cultivated using municipal wastewater, significant results were obtained. The consortium successfully achieved total N and P removal efficiencies of 88.6–96.4% and around 99%, respectively, within a cultivation period of 21 days. Along with the nutrient removal, the microalgal consortium was able to fix 0.65–1.37 g ($CO_2$) $L^{-1}$ per day with a maximum microalgal biomass concentration of 3.04 g $L^{-1}$ (Koreiviene et al. 2014). However, sometimes co-cultivation of different strains might compete for nutrients and show an inhibitory effect on the microalgal growth. For example, when Fergola et al. (2007) co-cultivated *Chlorella vulgaris* with another microalgal strain *Pseudokirchneriella subcapitata*, the former inhibited the growth of latter by excreting a secondary metabolite named chlorellin. Therefore, it can be concluded from the above discussions that while using consortium, it is crucial to ensure that the microalgal strains present in the consortium can coexist in synergy.

Microalgal nutrient uptake process is correlated with the microalgal biomass composition, where the molar ratio of N and P ranges from 8:1 to 45:1. This is one of the reasons why microalgae always show higher N removal capacity than the P (Whitton et al. 2015). For example, Fernández et al. (2018) stated that microalgal nutrient recovery under optimal conditions has an annual fixing capacity of 450 ton $CO_2$, 25 ton N, and 2.5 ton of P per hector, respectively, along with production of 200 ton of microalgal biomass. When Aketo et al. (2020) treated municipal effluent with several microalgal strains, *Parachlorella kessleri* showed a significant microalgal growth coupled with daily lipid productivity of $56 \pm 1$ mg $L^{-1}$ and N and P removal of 99 and 82%, respectively. Studies also show that secondary treated wastewater can be a good source of nutrients for culturing microalgae. Mandal and Mallick (2011) have cultivated freshwater microalgae, *Scenedesmus obliquus*, using poultry litter mixed with settling tank effluent of the secondary treated municipal wastewater (STMWW) and fish-pond wastewater. STMWW mixed with 15 g $L^{-1}$ poultry litter provided the highest biomass growth of $2.0 \pm 0.2$ g $L^{-1}$.

Apart from these lab-scale studies, results from the pilot-scale studies also need to be analyzed, as they portray a better image of the efficiency of this microalgal bioremediation technology at a real-time scenario. In one such study by Cavieres et al. (2021), *Muriellopsis* sp. was cultivated in an 800-L raceway pond. Results showed around 84, 93, and 99.9% reduction in N, P, and total coliform, respectively, along with the other compounds within 4 days of treatment. The authors also mentioned that the treated effluent met the local irrigation water standards.

## 6 Microalgal Remediation of Heavy Metal and Other Pollutants

Heavy metal and other persistent organic pollutant (POP) contamination through different anthropogenic activities and natural processes is a serious environmental issue around the globe. Due to their high persistence and non-biodegradable nature, they remain in the environment for an extended period, and once entered into the food chain, these pollutants accumulate via biomagnification, causing a severe threat to humankind. Microalgae, having advantages like fast growth with a minimal nutrient requirement, ability to survive and grow under extreme environmental conditions, and good absorption capacity due to the high surface area to volume ratio, show a high potential in the wastewater bioremediation process.

Heavy metal, a common contaminant present in domestic and industrial effluents, can be classified into a few categories like toxic heavy metal, precious heavy metals, and radioactive heavy metals. Hence, it is crucial not only to remove but also to recover those heavy metals. Unlike other conventional energy-intensive technologies, heavy metal remediation with live or dead microalgal biomass through biosorption and later recovering those metals is an environment-friendly and cost-effective technology. The presence of different groups like carboxyl, amino, hydroxyl, and sulfhydryl give a negative surface charge which helps in the rapid binding of all the positively charged metal ions on the microalgal biomass, followed by a slow intercellular uptake (Panda et al. 2021). Among the several freshwater microalgal species showing the phycoremediation capacity, the most common ones are *Chlorella* sp., *Scenedesmus* sp., *Spirulina* sp., *Chlamydomonas* sp., and *Tetraselmis* sp. Yang et al. (2015) used an oleaginous microalgal strain, *Chlorella minutissima* UTEX234, for bioremediation of wastewater combined with lipid production. The microalgal strain showed the ability to reduce 84, 84, 74, and 62% of the total copper, manganese, cadmium, and zinc, respectively. It is also noteworthy that instead of inhibition, the presence of 0.4 mM cadmium and copper induced the lipid accumulation process. Tatarová et al. (2021) used three microalgal strains *Chlorella vulgaris, Scenedesmus obliquus,* and *Chlamydomonas reinhardtii,* for the bioremediation of radioactively contaminated spent solution, where all the strains were able to reduce the contamination level at a significant level. The authors mentioned that *Chlorella vulgaris* successfully removed 90% of the Americium-241 and 70% of the Plutonium-239 within 3 h of treatment.

Other than heavy metals, POPs like organochlorines emerging from various chemical pesticides are also considered as hazardous pollutants due to the high stability, poor biodegradability, and long half-life. Microalgae (including cyanobacteria) have shown a great potential for bioremediation of such toxic compounds. Baglieri et al. (2016) used two algal strains *Chlorella vulgaris* and *Scenedesmus quadricauda* for the bioremediation of the wastewater collected from a greenhouse consisting of 5 bioactive pesticides (metalaxyl, pyrimethanil, fenhexamid, iprodione, and triclopyr). After 56 days of treatment, a significant reduction in N, P, and the bioactive toxic compounds was observed. However, authors mentioned that, for the case of pyrimethanil, instead of real degradation, removal was achieved by biosorption. Kurade et al. (2016) showed that among four tested microalgal strains, *Chlamydomonas pitschmannii, Chlamydomonas Mexicana, Chlorella vulgaris,* and *Scenedesmus obliquus, Chlorella vulgaris* had the highest diazinon removal capacity (i.e., 94%), where initial concentration was 20 ppm. The presence of diazinon also showed an increase in the production of different bioactive compounds like chlorophyll, carotenoid, and other antioxidants. However, a concentration beyond 40 ppm showed a significant impact on microalgal growth. Bai and Acharya (2016) reported that under seven days of cultivation, *Nannochlorpsis* sp. was able to completely remove triclosan (an antibiotic, commonly present in different cosmetics and pharmaceuticals), by accumulating triclosan into its intercellular space.

Overall, both live and dead microalgal cells can be used for the removal of heavy metals from the wastewater. However, live microalgal cells show better efficiency than the dead ones, as live microalgal cells can uptake those metals along with the nutrient in addition to the physical biosorption process. Cultivation of microalgae for bioremediation through accumulation, biosorption, and phycoremediation depends on three factors such as (1) the toxicity, concentration, and availability of the contaminant; (2) the environmental or abiotic factors; and (3) microalgal strains (or consortia), their removal/ degradation capacity, and tolerance level. To increase the efficiency of the bioremediation process, further research should focus on finding more suitable microalgal strains (i.e., robust strains with higher tolerance level and higher degradation/ removal capacity) and the fate of the produced microalgal biomass containing those heavy metals or other pollutants.

## 7 Valorization of Microalgal Biomass

Microalgae (including cyanobacteria) have been considered as an attractive source of the third generation of biofuel. Other than such low-value high-volume products, microalgae can be used to extract high-value low-volume products such as different proteins, lipids, vitamins, pigments, and other bioactive compounds. After extracting such valuable compounds, due to their nutrient adsorption capacity, the spent microalgal biomass can be used as a potential slow-release biofertilizer or biostimulant, enhancing the crop yield, or for biochar production as described in Fig. 3. Microalgal biomass can also be used for the production of methane (using anaerobic digestion method), bioplastic, or animal feed for fish or other livestock. All these valorization processes help in making microalgal wastewater treatment a cost-effective, environment-friendly, sustainable bioremediation technique.

### *7.1 Production of Biofuel*

In the last few decades, the rapid industrial development caused a sharp increase in energy demand. Fossil fuels, which are currently being used as a major source of energy, cause environmental pollution, and excessive extraction of such non-renewable energy source will exhaust the energy supply chain in the near future. This has forced the scientists to look for alternative energy sources, and in this context, microalgal bio-oil, the 3rd generation biofuel, has gained much attention due to its several advantageous properties. Microalgae, having a short lifespan and excellent ability to capture carbon along with the other flue gas, can be quickly grown in wastewater utilizing the unused nutrients present in the spent solutions. The high

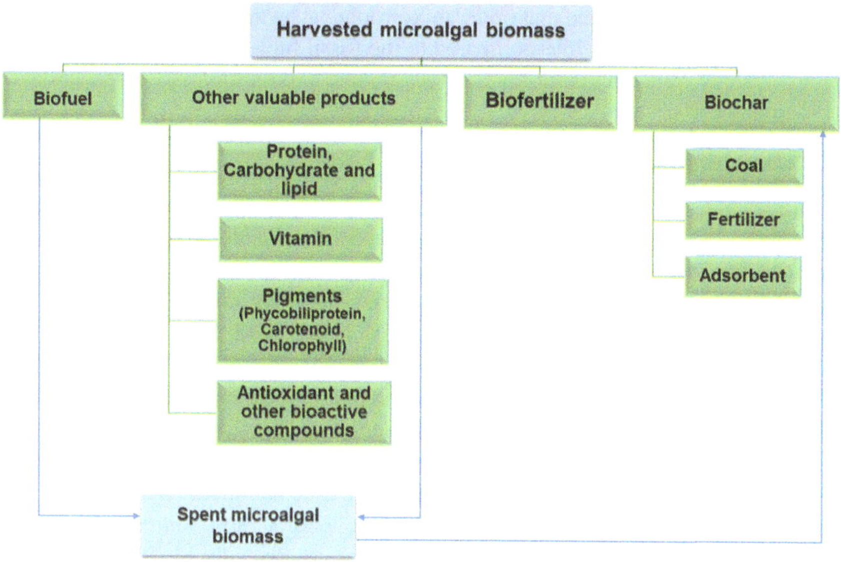


**Fig. 3** Different valorization processes of the harvested microalgal biomass

lipid accumulation capacity of the microalgae makes it an attractive feedstock for producing high-quality biofuel. Simultaneous production of biofuel (i.e., mainly, bioethanol and biodiesel) along with the microalgal wastewater remediation reduces the production cost and makes both the processes economically profitable. Moreover, the fact that the combustion of microalgal biofuel produces lesser amount of $CO_2$ compared to the conventionally used fossil fuel makes microalgal biofuel a more attractive and environment-friendly energy source.

For biodiesel production, different microalgal strains like *Chlorella* sp., *Scenedesmus* sp., and *Chlamydomonas* sp. are being frequently used. When two common microalgal strains *Chlorella* sp. and *Scenedesmus* sp. were co-cultivated by Silambarasan et al. (2021) at a 75% diluted swine wastewater, 1.78 g $L^{-1}$ biomass concentration was obtained along with the total N, P, and COD removal efficiencies of 94, 95, and 83%, respectively. Authors also reported a lipid content of 34.83% of the dry algal biomass, in which palmitic acid and oleic acid were found to be the two major fatty acids containing 51.75 and 35.45%, of the total lipid, respectively. Novel microalgal strain, *Chlamydomonas debaryana* IITRIND3, cultivated using dairy wastewater as a cheap nutrient source showed a daily lipid productivity of 87.5 $\pm$ 2.3 mg $L^{-1}$ with a significant amount of COD, total N, and P removal of 78.57, 87.56, and 82.17%, respectively (Arora et al. 2016). Regarding the environmental impact aspect, Adesanya et al. (2014) showed cultivation of *Chlorella vulgaris* in a hybrid bioreactor (containing an airlift photobioreactor integrated with a raceway pond) for commercial-level biofuel production. The system showed a 42% reduction in global warming potential and 38% fossil energy requirements compared to the fossil fuel. It was also stated that under mixotrophic growth conditions, these factors can be further reduced to 75% compared to the fossil fuel.

## 7.2 Production of Biofertilizer

Increasing worldwide population has increased food demand. To meet this demand, chemical fertilizers are often added to the soil to increase soil fertility and overall crop yield. However, it was observed that, after a certain point, addition of chemical fertilizer is not proportional to the increase in the yield. Excessive use of such synthetic fertilizers causes leaching, leading to several other environmental problems like eutrophication and soil acidification. Moreover, it also contributes to the direct emission of nitrous oxide. In this regard, nutrient-rich microalgal biomass possesses the potential to act as a slow-release biofertilizer, which can be used as a sustainable and eco-friendly alternative to the conventionally used chemical fertilizers.

Several N-fixing cyanobacterial strains like *Nostoc* sp., *Anabaena* sp., and *Scytonema* sp. have shown very high potential as eco-friendly biofertilizers (Chittora et al. 2020). Cyanobacteria enhance the yield by increasing nitrogen supply and enhancing the water holding capacity and aeration. Apart from this consortium of different microalgal strains also have shown good result when used as a biofertilizer. A microalgal consortium of *Chlorella* sp., *Parachlorella* sp., *Cyanobacteria Lyngbya* sp., and *Phormidium* sp. cultivated with the highly polluted parboiled rice mill effluent has shown high P, $NH_4^+$ removal efficiencies of 93 and 100%, respectively (Mukherjee et al. 2016). Along with the nutrient load reduction, the microalgae were successfully able to accumulate polyphosphate into their biomass and release P in the soil in the presence of P solubilizing organisms. Other than this, Silambarasan et al. (2021) used a consortium of two microalgal strains *Chlorella* sp. and *Scenedesmus* sp. for lipid production. After extraction of the lipid, the residual biomass was used as a biofertilizer along with the inorganic one for cultivation of *Solanum lycopersicum*, which showed better growth and higher crop yield.

Besides enhancing the soil fertility, due to the presence of several bioactive compounds, including phytohormones, the application of microalgal extract as foliar spray or for seed pretreatment has shown promising growth stimulating activity. Application of microalgal biomass has showed a positive effect on seed germination and plant growth. For example, when a native microalgal consortium was used for treating tomato seeds by Supraja et al. (2020), faster germination rate and better plant growth were observed. The authors also observed a 46% increase in plant growth when 60% microalgal extract was used as a foliar spray on the tomato plants. Coppens et al. (2016) observed that the addition of microalgal biomass enhances the carbohydrate and carotenoid levels in tomatoes.

However, using microalgal biomass alone as fertilizer would not be economically sustainable as other synthetic and organic slow-release fertilizer is already available in the market at a much cheaper price, for an equivalent amount of nutrient. For example, for 1 kg N, the production cost of the required microalgal biomass would be €289, whereas for the same amount of N, market available organic slow-release fertilizer and synthetic fertilizer will cost €11 and €7.9, respectively (Coppens et al. 2016). To ensure the economic sustainability, besides using cheap nutrient sources (for microalgal biomass production), microalgal biofertilizer can be mixed with comparatively cheaper organic or synthetic fertilizers in an optimized ratio.

### 7.3 *Production of Biochar*

Microalgae capture atmospheric $CO_2$ in the presence of sunlight for the photosynthesis process and help in carbon sequestration. Being a renewable and sustainable source, microalgal biomass can be used as a very suitable candidate for biochar production. Biochar is produced from the thermal decomposition of different organic feedstock, including microalgae, in the presence of limited or oxygen deficient conditions (Amin et al. 2016). Algal biochar can be produced by various thermal decomposition processes like pyrolysis (slow, fast, and intermediate), torrefaction, and gasification (Lee et al. 2020). Temperature plays a vital role in the quantity and quality of the produced biochar. Conventionally used slow pyrolysis, where thermal decomposition occurs at a 300–700 °C temperature for a prolonged period, produces a higher yield, whereas in fast pyrolysis, thermal decomposition is done at 500–1000 °C temperature, with a very short residence time (i.e., <2 s), and produces 75% bio-oil along with 12% biochar and 13% non-condensable gas (Bridgwater 2011). Because of the higher yield and cost-effectiveness, slow pyrolysis is often used as an effective biochar production method in small-scale farms (Lee et al. 2020). Fast pyrolysis, which produces a significant amount of bio-oil, is used to produce alternative energy (Bridgwater 2011).

The addition of biochar to the soil has shown improvement in the soil properties like increase in the soil carbon content and enhanced microbial biodiversity in the root zone. Regarding soil amendment, due to cation adsorption capacity, the application of algal biochar enhances the soil's nutrient retention capacity. Microalgal biochar, having a significant amount of N and other minerals, can be used as a biofertilizer. Biochar produced from microalgal biomass *Chlorella vulgaris* FSP-E, by thermal decomposition at 500 °C with an yield of 26.9%, consisted of a significant amount of carbon (61.32%) along with 9.76 and 0.02% of N and sulfur (S), respectively (Yu et al. 2018).

When Chang et al. (2014) produced biochar from chlorella-based residues by slow pyrolysis, the N, C, and hydrogen content of the biochar was found to be 10.48, 50.46, and 7.54%, respectively. After analyzing the surface properties of the biochar, authors suggested that such microalgal biochar can be used as a porous biofertilizer. In another experiment, Torri et al. (2011) showed that biochar produced by slow pyrolysis (at 350 °C temperature for 20 min) from the spent microalgal biomass is rich in N, which can work as an alternative to the N source for the soil. Other than pyrolysis, hydrothermal carbonization of *Arthrospira platensis,* at 190 °C for one hour, produced biochar containing 61.9 ± 0.6% carbon along with 8.7 ± 0.1% of N and 0.8 ± 0.1% of S (Yao et al. 2016).

Overall, the production of biochar from microalgal biomass (including the spent biomass after the extraction of lipid and other valuable compounds) and utilization as soil amendment to increase the yield is a strongly recommended carbon sequestering process. However, the process needs to be optimized to maximize the production yield. In this regard, other than conventional pyrolysis method, newly immerged thermochemical decomposition methods like microwave-assisted pyrolysis or hydrothermal carbonization should be investigated thoroughly for further development.

## 7.4 Production of Other Value-Added Compounds Such as Food Supplements, Pigments, and Therapeutic Agents

In addition to biofuel, biochar, and biofertilizer, microalgae can also produce different high-value bioactive compounds (Table 3). Among these, polyunsaturated fatty acid (PUFA) is considered to be one of the important nutrient sources as they have several positive effects on human health. Usually, the lipid content in microalgal biomass varies from 20 to 50% of dry biomass weight, which can be increased up to 85% when subjected to different stressed conditions, depending on the producing microalgal species. Four types of ω-3 PUFAs play a very important role in various human metabolism processes including α-linolenic acid (ALA, 18:3), eicosapentaenoic acid (EPA, 20:5), docosahexaenoic acid (DHA, 22:6), and stearidonic acid (SDA, 18:4). *Monoraphidium* sp. has been found to be a potential source for ALA, which works as a precursor for the production of ω-3 fatty acids (Lin et al. 2019). DHA and EPA, are two essential ω-3 fatty acids, which cannot be produced by the human body and should be taken through regular diet due to their nutritional properties. Among the few such fatty acid producing organisms, microalgal strains such as *Phaeodactylum* sp. and *Monodus* sp. were found to be rich in EPA, whereas strains like *Schizochytrium* sp. and *Crypthecodinium cohnii* are known for their high DHA content Ward and Singh (2005). Marine microalgal strains *Pavlova lutheri* and *Isochrysis* sp., contain 10.9% and 26.8% saponifiable SDA (w/w), respectively (Guil-Guerrero 2007). *Porphyridium purpureum* (red alga) can produce arachidonic acid (ARA), under stressed conditions (high salinity and limited nutrient and optimum temperature and pH) along with a considerable amount of lipid (40%). *Parietochloris incisa* is also considered a rich source of ARA (Bigogno et al. 2002).

Microalgal biomass produces carotenoids and chlorophyll, which are not only used as pigments like but also as bioactive compounds. Carotenoid acts as a source of nutrients in a regular diet and is a potential therapeutic agent. Based on their structure, they can be divided into two major groups: (1) carotenes (e.g., α- and ß-carotene, lycopene), and (2) xanthophylls (e.g., zeaxanthin, lutein, and β-cryptoxanthin). *H.*

**Table 3** Microalgae-derived value-added products and their applications

| Category/ subcategory | Biologically active compound | Microalgal strain | Usage | Reference |
|---|---|---|---|---|
| Carotenoid (Yellow photosynthetic pigment) | Astaxanthin | *Haematococcus pluvialis, Chlorella* sp., *Chlamydomonas nivalis* | Immune system stimulation, cardiovascular protective, anti-obesity | Brennan and Owende (2010), Hejazi and Wijffels (2004) |
| | β-carotene | *Dunaliella salina* | Vital nutrient for vision | Hejazi and Wijffels (2004) |
| | Lutein | *Dunaliella salina, Chlorella pyrenoidosa* | Anti-inflammatory, cardiovascular protective, prevention against age-related macular degeneration | García-Gonzáleza et al. (2005) |
| | Zeaxanthin | *H. pluvialis, Chlorella saccharophila* | Prevention against age-related macular degeneration | Singh et al. (2013) |
| PUFA | Eicosapentaenoic acid (EPA) | *Nannochloropsis* sp., *Phaeodactylum* sp., *Monodus* sp., *Nitzschia* spp. | Nutrient, food supplements, aquaculture | Ward and Singh (2005) |
| | Docosahexaenoic acid (DHA) | *Schizochytrium* sp. *and Crypthecodinium cohnii* | Anti-inflammatory, treatment of rheumatism and for treatment of skin diseases | Ward and Singh (2005) |
| | Arachidonic acid (ARA) | *Porphyridium purpureum, Parietochloris incisa* | Anti-inflammatory, Anti-allergen and anti-pathogenic agent, Maintain membrane fluidity and flexibility | Bigogno et al. (2002) |
| | Stearidonic acid (SDA) | *Pavlova lutheri, Isochrysis* sp. | Anti-inflammation, anticancer, antithrombotic agent | Guil-Guerrero (2007) |
| | α-Linolenic acid (ALA) | *Monoraphidium* sp. | Precursor for the synthesis of longer chain n-3 PUFA in human | Lin et al. (2019) |

*pluvialis* was found to be an excellent source of astaxanthin and can produce up to 1–8% (w/w) (Hejazi and Wijffels 2004). Another common pigment chlorophyll, which is an antioxidant and anticancer agent, can be divided into different categories like a, b, c, d, and e. Among them, chlorophyll-a is most abundantly available in plants and other freshwater algae, whereas the other b, c, d, and e all are restricted only to some algal strains. *Chlorella* sp. contains almost 7% chlorophyll of its biomass. Chlorophyceae (green algae) and Cyanophyceae (blue-green algae) are the two most significant sources of chlorophyll-a.

However, one significant bottleneck of producing these biorefinery products is the low efficacy of product recovery from the biomass due to their rigid cell wall, which also can increase the downstream process cost. Additionally, when cultivated in wastewater, difficulties in maintaining pure strains and higher chances of contamination makes the production of these compounds very tricky. More light should be shed on the extraction and purification methods, in order to get the maximum possible product, at a minimum cost. Moreover, there is very limited information available on the therapeutic use of the PUFAs. So, extensive study needs to be done to improve the knowledge on their applications in the pharmaceutical and agricultural industries.

## 8 Limitations of Microalgal-Based Wastewater Treatment

In a conventional urban wastewater treatment plant, removal of nutrients accounts for 50% of the total wastewater treatment cost (Acién et al. 2016). On the other hand, the cultivation of microalgae in the partially treated wastewater offers cheaper nutrient source coupled with simultaneous wastewater reclamation. Microalgae-based bioremediation promises several advantages including nutrient recovery, reduction of greenhouse gas emission, production of feedstock for bioethanol, biofertilizer, and other high-value bioactive compounds. However, there are some drawbacks of these technology which need to be addressed for practical implementation of this technology.

Land requirement is one of the major concerns of using microalgal bioremediation. Compared to the conventional treatment for 30,000–50,000 people equivalent wastewater, microalgal treatment (in open ponds) requires 30–50% higher land (Acién et al. 2016). On the contrary, photobioreactors offer very high productivity and require much lesser area. Yet, the additional energy requirement of such photobioreactors is 10 times higher than HRAP owing to recirculation and aeration. A life cycle assessment (LCA) study by Stephenson et al. (2010) showed a similar result when airlift tubular bioreactor's performance was compared with the raceway pond, where higher energy was spent for mixing by paddlewheels in case of the latter.

Organic pollutant-rich wastewater can inhibit the microalgal growth. Hence, microalgal nutrient removal treatment is found to be best suited only for the tertiary stage of wastewater treatment, which indicates its limited treatment capacity. Besides, the presence of some bacteria or protozoa in unsterilized wastewater, might restrict the microalgal growth. For this purpose sterilization of the media, which accounts for 20–30% of the total operation cost, can be eliminated by selecting robust microalgal strains, with high tolerance. Using such robust strains, along with high nutrient removal capacity, can be beneficial for treating raw wastewater due to their ability to adapt to the harsh and unfavorable conditions.

The biomass valorization process is an important step to make the whole process economically sustainable. Harvesting, which is the first step of this valorization process, requires a significant amount of energy, increasing the overall production cost by 20–30% (Zhu et al. 2020; Krishnamoorthy et al. 2021). Moreover, the harvesting process also influences the other downstream processes. The lack of efficient and economic downstream process is another major challenge. One of the recent developments in this regard is the co-cultivation of filamentous fungi with different microalgal strain, which can increase the harvesting efficiency, by immobilizing microalgal cell through mycelial interaction. One of such example is cultivation of *Aspergillus fumigatus* with *Chlorella protothecoides* by Muradov et al. (2015). This novel co-cultivation system showed $8.96 \pm 2.1$ g $L^{-1}$ biomass concentration, along with $2.042 \pm 0.44$ g $L^{-1}$ lipid. The authors also mentioned that co-cultivation showed higher yield and better remediation efficiency.

Therefore, in order to optimize a novel cultivation system, ensuring both economic and environmental sustainability, further research is needed. Along with the novel cultivation strategy, finding more robust microalgal strains with high lipid accumulation and nutrient removal capacity will help to enhance the efficiency of the bioremediation process. In addition, new extremophilic microalgal strains, which can survive under extreme environmental conditions that will increase the robustness of the microalgal-based remediation, have to be researched. More focus on the efficient and economically beneficial microalgal harvesting method will aid in making the process more cost-effective. Regarding the valorization of the microalgal biomass, identification of more valuable bioactive compounds will enhance the economic stability of the process. Lastly, more life cycle analysis studies are needed to evaluate the environmental impact of such microalgal-based treatment process on an industrial scale level.

## 9 Conclusion and Future Perspectives

Integrating microalgae in the biological wastewater treatment plays a vital role in establishing a circular economy approach, by simultaneously producing different economically valuable products.

After taking a closer look into such combined systems, a few conclusions drawn are stated below:

- Depending on the source, characteristics or composition of the wastewater can vary. For example, dairy wastewater is deficient in ammonia, which is one of the vital nitrogen sources, determining the microalgal growth. On the other side, livestock wastewater is rich in ammonia. Therefore, mixing these two different types of wastewaters can be a solution for better nutrient removal or production of significant amount of microalgal biomass.
- For wastewater remediation, microalgal strains should be screened on the basis of faster growth rate, robustness, and adoptability in higher C: N: P ratio of wastewater.
- Closed cultivation systems like membrane photobioreactors (MPBR) are found to be suitable for the microalgal cultivation and wastewater treatment process. However, while scaling up the process, issues like membrane fouling can pose as main challenges. Currently, MPBR combined with other immobilized elements and utilization of renewable energy sources (like solar energy) are under investigation.
- Open cultivating systems like HRAP show higher productivity rate. Hence, those are found to be apt for producing biofuel using wastewater as a nutrient source. However, more studies are needed to optimize operating environmental parameters to increase nutrient removal efficiency.
- Harvesting and other downstream processing of the produced microalgal biomass requires a significant amount of energy, which increases the total production cost. In this regard, microalgal cultivation with filamentous fungi makes the harvesting process easier and more energy-efficient. Additionally, using different renewable energy like solar energy, can reduce the cost of dewatering or drying of the harvested biomass.
- Identifying more high-value low-volume and low-value high-volume algal-derived products and valorizing the microalgal biomass will make the whole wastewater treatment process economically more sustainable and beneficial.
- Most of the microalgal bioremediation studies are done at a lab-scale. Lack of data from the commercial-level study and the generalized facts derived from the lab-scale data are obstacles in the field-level implementation of this technology.

**Acknowledgements** The authors thank the Department of Biotechnology and Medical Engineering of National Institute of Technology Rourkela for providing the necessary research facility. The authors greatly acknowledge the Ministry of Science and Technology, Government of India, for sponsoring the Ph.D. program of the first author.

## References

Acién FG, Fernández JM, Magán JJ, Molina E (2012) Production cost of a real microalgae production plant and strategies to reduce it. Biotechnol Adv 30(6):1344–1353. https://doi.org/10.1016/j.biotechadv.2012.02.005

Acién FG, Gómez-Serrano C, Morales-Amaral MM, Fernández-Sevilla JM, Molina-Grima E (2016) Wastewater treatment using microalgae: how realistic a contribution might it be to significant urban wastewater treatment? Appl Microbiol Biotechnol 100(21):9013–9022. https://doi.org/10.1007/s00253-016-7835-7

Adesanya VO, Cadena E, Scott SA, Smith AG (2014) cultivation system. Bioresour Technol 163:343–355. https://doi.org/10.1016/j.biortech.2014.04.051

Aketo T, Hoshikawa Y, Nojima D, Yabu Y, Maeda Y, Yoshino T, Takano H, Tanaka T (2020) Selection and characterization of microalgae with potential for nutrient removal from municipal wastewater and simultaneous lipid production. J Biosci Bioeng 129(5):565–572. https://doi.org/10.1016/j.jbiosc.2019.12.004

Amin FR, Huang Y, He Y, Zhang R, Liu G, Chen C (2016) Biochar applications and modern techniques for characterization. Clean Technol Environ Policy 18(5):1457–1473. https://doi.org/10.1007/s10098-016-1218-8

Arora N, Patel A, Sartaj K, Pruthi PA, Pruthi V (2016) Bioremediation of domestic and industrial wastewaters integrated with enhanced biodiesel production using novel oleaginous microalgae. Environ Sci Pollut Res. https://doi.org/10.1007/s11356-016-7320-y

Aslam A, Khan SJ, Shahzad A (2022) Science of the total environment Anaerobic membrane bioreactors (AnMBRs) for municipal wastewater treatment-potential benefits, constraints, and future perspectives: an updated review. Sci Total Environ 802:149612. https://doi.org/10.1016/j.scitotenv.2021.149612

Baglieri A, Sidella S, Barone V, Fragalà F, Silkina A, Nègre M, Gennari M (2016) Cultivating Chlorella vulgaris and Scenedesmus quadricauda microalgae to degrade inorganic compounds and pesticides in water. Environ Sci Pollut Res. https://doi.org/10.1007/s11356-016-6996-3

Bai X, Acharya K (2016) Removal of trimethoprim, sulfamethoxazole, and triclosan by the green alga Nannochloris sp. J Hazard Mater 315:70–75. https://doi.org/10.1016/j.jhazmat.2016.04.067

Bigogno C, Khozin-Goldberg I, Adlerstein D, Cohen Z (2002) Biosynthesis of arachidonic acid in the oleaginous microalga Parietochloris incisa (chlorophyceae): radiolabeling studies. Lipids 37(2):209–216. https://doi.org/10.1007/s11745-002-0882-6

Brennan L, Owende P (2010) Biofuels from microalgae—A review of technologies for production, processing, and extractions of biofuels and co-products. Renew Sustain Energy Rev 14:557–577. https://doi.org/10.1016/j.rser.2009.10.009

Bridgwater AV (2011) Review of fast pyrolysis of biomass and product upgrading. Biomass Bioenerg 38:68–94. https://doi.org/10.1016/j.biombioe.2011.01.048

Cai T, Park SY, Li Y (2013) Nutrient recovery from wastewater streams by microalgae: status and prospects. Renew Sustain Energy Rev 19:360–369. https://doi.org/10.1016/j.rser.2012.11.030

Cavieres L, Bazaes J, Marticorena P, Riveros K, Medina P, Sepúlveda C, Riquelme C (2021) Pilot-scale phycoremediation using Muriellopsis sp. For wastewater reclamation in the Atacama Desert: microalgae biomass production and pigment recovery. Water Sci Technol 83(2):331–343. https://doi.org/10.2166/wst.2020.576

Chang Y, Tsai W, Li M (2014) Chemical characterization of char derived from slow pyrolysis of microalgal residue. J Anal Appl Pyrol, 8–13. https://doi.org/10.1016/j.jaap.2014.12.004

Chittora D, Meena M, Barupal T, Swapnil P (2020) Cyanobacteria as a source of biofertilizers for sustainable agriculture. Biochem Biophys Rep, 22 (November 2019). https://doi.org/10.1016/j.bbrep.2020.100737

Chu HQ, Tan XB, Zhang YL, Yang LB, Zhao FC, Guo J (2015) Continuous cultivation of Chlorella pyrenoidosa using anaerobic digested starch processing wastewater in the outdoors. Biores Technol 185:40–48. https://doi.org/10.1016/j.biortech.2015.02.030

Coppens J, Grunert O, Van Den Hende S, Vanhoutte I, Boon N, Haesaert G, De Gelder L (2016) The use of microalgae as a high-value organic slow-release fertilizer results in tomatoes with increased carotenoid and sugar levels. J Appl Phycol 28(4):2367–2377. https://doi.org/10.1007/s10811-015-0775-2

Das S, Nath K, Chowdhury R (2021) Comparative studies on biomass productivity and lipid content of a novel blue-green algae during autotrophic and heterotrophic growth. Environ Sci Pollut Res 28:12107–12118. https://doi.org/10.1007/s11356-020-09577-4

Ding J, Zhao F, Cao Y, Xing L, Liu W, Mei S, Li S (2015) Cultivation of microalgae in dairy farm wastewater without sterilization. Int J Phytorem 17(3):222–227. https://doi.org/10.1080/15226514.2013.876970

Ebeling JM, Timmons MB, Bisogni JJ (2006) Engineering analysis of the stoichiometry of photoautotrophic, autotrophic, and heterotrophic removal of ammonia-nitrogen in aquaculture systems. Aquaculture 257(1–4):346–358. https://doi.org/10.1016/j.aquaculture.2006.03.019

El-kassas HY, Mohamed LA (2014) Bioremediation of the textile waste effluent by Chlorella vulgaris. Egypt J Aquat Res 40(3):301–308. https://doi.org/10.1016/j.ejar.2014.08.003

Fergola P, Cerasuolo M, Pollio A, Pinto G, DellaGreca M (2007) Allelopathy and competition between Chlorella vulgaris and Pseudokirchneriella subcapitata: experiments and mathematical model. Ecol Model 208(2–4):205–214. https://doi.org/10.1016/j.ecolmodel.2007.05.024

Fernández FGA, Gómez-Serrano C, Fernández-Sevilla JM (2018) Recovery of nutrients from wastewaters using microalgae. Front Sustain Food Syst 2(September):1–13. https://doi.org/10.3389/fsufs.2018.00059

Gao B, Liu J, Zhang C, Van De Waal DB (2018) Biological stoichiometry of oleaginous microalgal lipid synthesis: the role of N: P supply ratios and growth rate on microalgal elemental and biochemical composition. Algal Res 32(April):353–361. https://doi.org/10.1016/j.algal.2018.04.019

García-Gonzáleza M, Morenoa J, Manzano JC, Florencio FJ, Guerrero GM (2005) Production of Dunaliella salina biomass rich in 9-cis-β-carotene and lutein in a closed tubular photobioreactor. J Biotechnol 115(1):81–90

Gentili FG (2014) Bioresource Technology Microalgal biomass and lipid production in mixed municipal, dairy, pulp and paper wastewater together with added flue gases. Biores Technol 169:27–32. https://doi.org/10.1016/j.biortech.2014.06.061

Guil-Guerrero JL (2007) Stearidonic acid (18:4n–3): metabolism, nutritional importance, medical uses and natural sources. Eur J Lipid Sci Technol 109(12):1226–1236. https://doi.org/10.1002/ejlt.200700207

Hejazi MA, Wijffels HR (2004) Milking of microalgae.pdf. Trends Biotechnol 22(4):189–194.

Hernández D, Riaño B, Coca M, Solana M, Bertucco A, García-González MC (2016) Microalgae cultivation in high rate algal ponds using slaughterhouse wastewater for biofuel applications. Chem Eng J 285:449–458. https://doi.org/10.1016/j.cej.2015.09.072

Kamalanathan M, Chaisutyakorn P, Gleadow R, Beardall J (2018) A comparison of photoautotrophic, heterotrophic, and mixotrophic growth for biomass production by the green alga Scenedesmus sp. (Chlorophyceae). Phycologia 57(3):309–317. https://doi.org/10.2216/17-82.1

Koreiviene J, Valčiukas R, Karosiene J, Baltrenas P (2014) Testing of Chlorella/Scenedesmus microalgae consortia for remediation of wastewater, CO2 mitigation and algae biomass feasibility for lipid production. J Environ Eng Landsc Manag 22(2):105–114. https://doi.org/10.3846/16486897.2013.911182

Kothari R, Prasad R, Kumar V, Singh DP (2013) Production of biodiesel from microalgae Chlamydomonas polypyrenoideum grown on dairy industry wastewater. Bioresour Technol 144(October 2018):499–503. https://doi.org/10.1016/j.biortech.2013.06.116

Krishnamoorthy N, Unpaprom Y, Ramaraj R, Maniam GP, Govindan N, Arunachalam T, Paramasivan B (2021) Recent advances and future prospects of electrochemical processes for microalgae harvesting. J Environ Chem Eng 9(5):105875

Küçük K, Tevatia R, Sorgüven E, Demirel Y, Özilgen M (2015) Bioenergetics of growth and lipid production in Chlamydomonas reinhardtii. Energy 83:503–510. https://doi.org/10.1016/j.energy.2015.02.054

Kurade MB, Kim JR, Govindwar SP, Jeon BH (2016) Insights into microalgae mediated biodegradation of diazinon by Chlorella vulgaris: microalgal tolerance to xenobiotic pollutants and metabolism. Algal Res 20:126–134. https://doi.org/10.1016/j.algal.2016.10.003

Lee CS, Lee SA, Ko SR, Oh HM, Ahn CY (2015) Effects of photoperiod on nutrient removal, biomass production, and algal-bacterial population dynamics in lab-scale photobioreactors treating municipal wastewater. Water Res 68:680–691. https://doi.org/10.1016/j.watres.2014.10.029

Lee XJ, Ong HC, Gan YY, Chend W-H, Mahlia TMI (2020) State of art review on conventional and advanced pyrolysis of macroalgae and microalgae for biochar, bio-oil and bio-syngas production. Energy Convers Manage 210(March):112707. https://doi.org/10.1016/j.enconman.2020.112707

Li Y, Tsai W, Hsu Y, Xie M, Chen J (2014) Comparison of autotrophic and mixotrophic cultivation of green microalgal for biodiesel production. Energy Procedia 52:371–376. https://doi.org/10.1016/j.egypro.2014.07.088

Li D, Liu R, Cui X, He M, Zheng S, Du W, Gao M, Wang C (2021) Co-culture of bacteria and microalgae for treatment of high concentration biogas slurry. J Water Process Eng 41(March):102014. https://doi.org/10.1016/j.jwpe.2021.102014

Li S, Ji L, Chen C, Zhao S, Sun M, Gao Z, Wu H, Fan J (2020) Bioresource technology efficient accumulation of high-value bioactive substances by carbon to nitrogen ratio regulation in marine microalgae Porphyridium purpureum. Bioresource Technol 309(February), 123362

Lin Y, Ge J, Zhang Y, Ling H, Yan X, Ping W (2019) Monoraphidium sp. HDMA-20 is a new potential source of α-linolenic acid and eicosatetraenoic acid. Lipids Health Dis 18(1):1–10. https://doi.org/10.1186/s12944-019-0996-5

Liu R, Guo Q, Zheng W, Chen L, Luo J (2015) Cultivation of an Arthrospira platensis with digested piggery wastewater. Water Sci Technol 72(10):1774–1779. https://doi.org/10.2166/wst.2015.353

Mandal S, Mallick N (2011) Waste utilization and biodiesel production by the green microalga Scenedesmus obliquus. Appl Environ Microbiol 77(1):374–377. https://doi.org/10.1128/AEM.01205-10

Marbelia L, Bilad MR, Passaris I, Discart V, Vandamme D, Beuckels A, Muylaert K, Vankelecom IFJ (2014) Membrane photobioreactors for integrated microalgae cultivation and nutrient remediation of membrane bioreactors effluent. Bioresour Technol J 163:228–235. https://doi.org/10.1016/j.biortech.2014.04.012

Marjakangas JM, Chen CY, Lakaniemi AM, Puhakka JA, Whang LM, Chang JS (2015) Simultaneous nutrient removal and lipid production with Chlorella vulgaris on sterilized and non-sterilized anaerobically pretreated piggery wastewater. Biochem Eng J 103:177–184. https://doi.org/10.1016/j.bej.2015.07.011

McDowell D, Dick JT, Eagling L, Julius M, Sheldrake GN, Theodoridou K, Walsh PJ (2020) Recycling nutrients from anaerobic digestates for the cultivation of Phaeodactylum tricornutum: a feasibility study. Algal Res 48(March):101893. https://doi.org/10.1016/j.algal.2020.101893

Mukherjee C, Chowdhury R, Sutradhar T, Begam M, Magdalene S, Kumar S, Ray K (2016) Parboiled rice effluent: a wastewater niche for microalgae and cyanobacteria with growth coupled to comprehensive remediation and phosphorus biofertilization. ALGAL 19:225–236. https://doi.org/10.1016/j.algal.2016.09.009

Muradov N, Taha M, Miranda AF, Wrede D, Kadali K, Gujar A, Stevenson T, Ball AS, Mouradov A (2015) Fungal-assisted algal flocculation: application in wastewater treatment and biofuel production. Biotechnol Biofuels 8(24). https://doi.org/10.1186/s13068-015-0210-6

Narala RR, Garg S, Sharma KK, Thomas-hall SR (2016) Comparison of microalgae cultivation in photobioreactor open raceway pond, and a two-stage hybrid system. Front Energy Res 4(August):1–10. https://doi.org/10.3389/fenrg.2016.00029

Nzayisenga JC, Farge X, Groll SL, Sellstedt A (2020) Effects of light intensity on growth and lipid production in microalgae grown in wastewater. Biotechnol Biofuels 13(1):1–8. https://doi.org/10.1186/s13068-019-1646-x

Panda S, Mishra S, Akcil A, Kucuker MA (2021) Microalgal potential for nutrient-energy-wastewater nexus: innovations, current trends and future directions. Energy Environ 32(4):604–634. https://doi.org/10.1177/0958305X20955187

Park JBK, Craggs RJ (2010) Wastewater treatment and algal production in high rate algal ponds with carbon dioxide addition. Water Sci Technol 61(3):633–639. https://doi.org/10.2166/wst.2010.951

Park JBK, Craggs RJ, Shilton AN (2013) Enhancing biomass energy yield from pilot-scale high rate algal ponds with recycling. Water Res 47(13):4422–4432. https://doi.org/10.1016/j.watres.2013.04.001

Pham TL, Bui MH (2020) Removal of nutrients from fertilizer plant wastewater using Scenedesmus sp.: formation of bioflocculation and enhancement of removal efficiency. J Chem. https://doi.org/10.1155/2020/8094272

Posadas E, del Mar Morales M, Gomez C, Acién FG, Muñoz R (2015) Influence of pH and $CO_2$ source on the performance of microalgae-based secondary domestic wastewater treatment in outdoors pilot raceways. Chem Eng J 265:239–248. https://doi.org/10.1016/j.cej.2014.12.059

Powell N, Shilton A, Chisti Y, Pratt S (2009) Towards a luxury uptake process via microalgae—defining the polyphosphate dynamics. Water Res 43(17):4207–4213. https://doi.org/10.1016/j.watres.2009.06.011

Renuka N, Ratha SK, Kader F, Rawat I, Bux F (2021) Insights into the potential impact of algae-mediated wastewater beneficiation for the circular bioeconomy: a global perspective. J Environ Manage 297:113257. https://doi.org/10.1016/j.jenvman.2021.113257

Sajadian SF, Morowvat MH, Ghasemi Y (2018) Investigation of autotrophic, heterotrophic, and mixotrophic modes of cultivation on lipid and biomass production in Chlorella vulgaris. Nat J Physiol Pharm Pharmacol 8(4):594–599. https://doi.org/10.5455/njppp.2018.8.0935625122017

Shah MP (2020) Microbial Bioremediation & Biodegradation. Springer

Shah MP (2021a) Removal of refractory pollutants from wastewater treatment plants. CRC Press

Shah MP (2021b) Removal of emerging contaminants through microbial processes. Springer

Silambarasan S, Logeswari P, Sivaramakrishnan R, Incharoensakdi A, Cornejo P, Kamaraj B, Chi NTL (2021) Removal of nutrients from domestic wastewater by microalgae coupled to lipid augmentation for biodiesel production and influence of deoiled algal biomass as biofertilizer for Solanum lycopersicum cultivation. Chemosphere 268:129323. https://doi.org/10.1016/j.chemosphere.2020.129323

Singh D, Puri M, Wilkens S, Mathur AS, Tuli DK, Barrow CJ (2013) Characterization of a new zeaxanthin producing strain of Chlorella saccharophila isolated from New Zealand marine waters. Biores Technol 143:308–314. https://doi.org/10.1016/j.biortech.2013.06.006

Sirisuk P, Ra CH, Jeong GT, Kim SK (2018) Effects of wavelength mixing ratio and photoperiod on microalgal biomass and lipid production in a two-phase culture system using LED illumination. Bioresour Technol 253:175–181. https://doi.org/10.1016/j.biortech.2018.01.020

Solovchenko A, Verschoor AM, Jablonowski ND, Nedbal L (2016) Phosphorus from wastewater to crops: an alternative path involving microalgae. Biotechnol Adv 34(5):550–564. https://doi.org/10.1016/j.biotechadv.2016.01.002

Solovchenko AE, Ismagulova TT, Lukyanov AA, Vasilieva SG, Konyukhov IV, Pogosyan SI, Lobakova ES, Gorelova OA (2019) Luxury phosphorus uptake in microalgae. J Appl Phycol 31(5):2755–2770. https://doi.org/10.1007/s10811-019-01831-8

Sriram S, Seenivasan R (2012) Microalgae cultivation in wastewater for nutrient removal. J Algal Biomass Utilization 3(2):9–13

Stephenson AL, Kazamia E, Dennis JS, Howe CJ, Scott SA, Smith AG (2010) Life-cycle assessment of potential algal biodiesel production in the United Kingdom: a comparison of raceways and air-lift tubular bioreactors. Energy Fuels 4:4062–4077. https://doi.org/10.1021/ef1003123

Supraja KV, Behera B, Balasubramanian P (2020) Efficacy of microalgal extracts as biostimulants through seed treatment and foliar spray for tomato cultivation. Ind Crops Prod 151(March):112453. https://doi.org/10.1016/j.indcrop.2020.112453

Tatarová D, Galanda D, Kuruc J, Gaálová B (2021) Phytoremediation of 137Cs, 60Co, 241Am, and 239Pu from aquatic solutions using Chlamydomonas reinhardtii, Scenedesmus obliquus, and Chlorella vulgaris. Int J Phytorem 23(13):1376–1381. https://doi.org/10.1080/15226514.2021.1900061

Torri C, Samorì C, Adamiano A, Fabbri D, Faraloni C, Torzillo G (2011) Preliminary investigation on the production of fuels and bio-char from Chlamydomonas reinhardtii biomass residue after biohydrogen production. Biores Technol 102(18):8707–8713. https://doi.org/10.1016/j.biortech.2011.01.064

Ward OP, Singh A (2005) Omega-3/6 fatty acids: alternative sources of production. Process Biochem 40(12):3627–3652. https://doi.org/10.1016/j.procbio.2005.02.020

Whitton R, Ometto F, Pidou M, Jarvis P, Villa R, Jefferson B (2015) Microalgae for municipal wastewater nutrient remediation: mechanisms, reactors and outlook for tertiary treatment. Environ Technol Rev 4(1):133–148. https://doi.org/10.1080/21622515.2015.1105308

Xu J, Wang X, Sun S, Zhao Y, Hu C (2017) Effects of influent C/N ratios and treatment technologies on integral biogas upgrading and pollutants removal from synthetic domestic sewage, pp 1–13. https://doi.org/10.1038/s41598-017-11207-y

Yang J, Cao J, Xing G, Yuan H (2015) Bioresource Technology Lipid production combined with biosorption and bioaccumulation of cadmium, copper, manganese and zinc by oleaginous microalgae Chlorella minutissima UTEX2341. Biores Technol 175:537–544. https://doi.org/10.1016/j.biortech.2014.10.124

Yao C, Wu P, Pan Y, Lu H, Chi L, Meng Y, Cao X, Xue S, Yang X (2016) Bioresource technology evaluation of the integrated hydrothermal carbonization-algal cultivation process for enhanced nitrogen utilization in Arthrospira platensis production. Biores Technol 216:381–390. https://doi.org/10.1016/j.biortech.2016.05.110

Yu KL, Show PL, Ong HC, Ling TC, Chen WH, Salleh MAM (2018) Biochar production from microalgae cultivation through pyrolysis as a sustainable carbon sequestration and biorefinery approach. Clean Technol Environ Policy 20(9):2047–2055. https://doi.org/10.1007/s10098-018-1521-7

Zheng H, Liu M, Lu Q, Wu X, Ma Y, Cheng Y, Addy M, Liu Y, Ruan R (2017) Balancing carbon/nitrogen ratio to improve nutrients removal and algal biomass production in piggery and brewery wastewaters. Biores Technol. https://doi.org/10.1016/j.biortech.2017.10.057

Zheng Q, Xu X, Martin GJO, Kentish SE (2018) Critical review of strategies for $CO_2$ delivery to large-scale microalgae cultures. Chin J Chem Eng 26(11):2219–2228. https://doi.org/10.1016/j.cjche.2018.07.013

Zhu L, Hu T, Li S, Nugroho YK, Li B, Cao J, Show P-L, Hiltunen E (2020) Effects of operating parameters on algae Chlorella vulgaris biomass harvesting and lipid extraction using metal sulfates as flocculants. Biomass Bioenergy 132:105433

# Plants and Microorganisms as Useful Tool for Accumulation and Detoxification of Heavy Metals from Environment

**Sandra Pérez-Álvarez, Eduardo Fidel Héctor Ardisana, Marco Antonio Magallanes-Tapia, Manuel García Ulloa Gómez, Ana Elsi Ulloa Pérez, María Esther González Vega, and Víctor Hugo Villarreal Ramirez**

## 1 Introduction

Pollution of ecosystems such as terrestrial and aquatic by several sources has turn into an important problem around the world. Contaminants, like heavy metals, can go into the body throughout the food and induce some chronic diseases and death (Babu et al. 2021). Anthropogenic activities are an important reason for heavy metals' accumulation in soil and aquatic systems such as sewage sludge, application of some fertilizers like phosphate, some pesticides and burning of fossil fuel (Rafique and Tariq 2016). They are not decomposable, so they can continue in the ecosystems for a long period (Yan et al. 2020).

Heavy metals are chemicals that belong to a metallic group with a moderately high density, atomic numbers and atomic weights, and the most abundant in the ecosystems are lead (Pb), nickel (Ni), arsenic (As), cadmium (Cd), mercury (Hg) copper (Cu), chromium (Cr) and zinc (Zn) (Ashraf et al. 2019).

Consequently, at the moment, several remediation techniques exist based mainly on physio-chemical or mechanical approaches like the incineration of contaminated soil, washing soils, application of electric field, excavation and solidification (DalCorso et al. 2019; Shah 2021a), but they are expensive, non-efficient when the

---

S. Pérez-Álvarez (✉) · V. H. V. Ramirez
Facultad de Ciencias Agrícolas y Forestales, Universidad Autónoma de Chihuahua, Chihuahua, México
e-mail: spalvarez@uach.mx

E. F. H. Ardisana
Instituto de Posgrado, Universidad Técnica de Manabí, Manabí, Ecuador

M. A. Magallanes-Tapia · M. G. U. Gómez · A. E. Ulloa Pérez
IPN-CIIDIR Unidad Sinaloa, Guasave, México

M. E. González Vega
Instituto Nacional de Ciencias Agrícolas (INCA), San José de las Lajas, Cuba



M. P. Shah (ed.), *Sustainable Industrial Wastewater Treatment and Pollution Control*,
https://doi.org/10.1007/978-981-99-2560-5_5

concentrations of contaminants are low and some properties of the soil (physico-chemical and biological) can change in an irreversible way (Ali et al. 2013; Shah 2020). Therefore, there are some cost-effective, environment-friendly and efficient heavy metal remediation technologies such as bio- and phytoremediation.

Bioremediation, according to Medhat and Alsohim (2019), is a technique that uses microorganisms or their enzymes to diminish pollutants in the environment, transforming poisonous forms of heavy metals into less toxic. Some examples of these microorganisms are bacteria [*Acinetobacter junii*-Pb(II); *Bacillus thuringiensis*-Cd(II); *Pseudomonas* sp.- Cd(II), Cu(II), Pb(II), As(III), Cd(II) and Co(II); *Staphylococcus* sp.-Cd(II), Cu(II)], fungi (*Aspergillus niger*-Pb(II), Cd(II); *Trichoderma* sp.-Cd(II); *Trichoderma viride*-Cr(IV), Pb(II); *Saccharomyces cerevisiae*-Cu(II)) and algae [*Chlamydomonas reinhardtii*-As(III); *Chlorella vulgaris*-Pb(II); *Cystoseira crinitophylla*-Cu(II); *Fucus vesiculosus*-Pb(II); *Sargassum* sp.-Cu(II); *Sargassum filipendula*-Cd(II); *Sargassum muticum*-Sb(III); *Spirogyra* sp.-Cu(II)] (Tayang and Songachan 2021).

Microorganisms are abundant; they can multiply quickly and grow in massive numbers when inoculated to polluted sites. When they grow constantly in polluted sites, they develop tolerance and demonstrate great levels of ability to transform contaminants, and also, they can genetically adapt to degrade the pollutants (Tayang and Songachan 2021; Shah 2021a, 2021b).

Phytoremediation is an ecofriendly method that uses plants for an effective revegetation of soils contaminated with heavy metals (Yan et al. 2020). Examples of these plants are *Nicotiana tabacum* L. for Cu extraction (Thomas et al. 2003); *Brassica juncea* L.—Cd and Zn (Bennett et al. 2003); *Arabidopsis thaliana* L.—Ni (Freeman et al. 2004); *Artemisia vulgaris* L., *Silene vulgaris* (Moench.) Garcke, *Alopecurus pratensis* L., *Galium mollugo* L., *Bromus inermis* Leyss., *Arrhenatherum elatius*, *Achillea millefolium* L. and *Stellaria holostea* L.—Cd and Zn (Antoniadis et al. 2021). To increment the effectiveness of this technique, a well comprehension of the essential processes implicated in heavy metal tolerance and accumulation in plants is indispensable (Yan et al. 2020).

This chapter presents a review addressing topics like generalities of plants-microorganisms and heavy metals; detoxification and accumulation of heavy metals by terrestrial plants; and the role of fungi, bacteria and aquatic microorganisms in detoxification and accumulation of heavy metals.

## 2 Plants and Heavy Metals

Many plant species can live with high concentrations of metals due to the restriction of the absorption-translocation of them to the leaves, so this lets crops to keep constant and relatively low concentration regardless of the metallic concentration of the soil in a wide range (exclusion approach according to Baker 1981). However, others can absorb metals and accumulate them in forms that are non-toxic in their leaves (accumulation approach). Some of these plants cannot complete their life cycles when grown in "normal" soils and are called hyperaccumulator plants. The term "hyperaccumulator" refers to plants grown in the field, being capable to accumulate >1000 mg $kg^{-1}$ Ni of dry matter in some tissue of their aerial biomass (Brooks et al. 1977). In general, hyperaccumulators reach metal concentrations in leaves between 10 and 100 times the "normal" concentrations. Currently, the term hyperaccumulator of metals is applied to plants that accumulate >10000 mg $kg^{-1}$ of Zn and Mn >1000 mg $kg^{-1}$ of Se, Cu, Co, Ni, Pb and As >100 mg $kg^{-1}$ of Cd (Padmavathiamma and Li 2007).

According to Baker et al. (2000), until that year more than 400 plant taxa had been identified worldwide from at least 45 metal accumulating families, which is less than 0.2% of all plants species known. These plants have the ability to hyperaccumulate certain substances, meaning carry out a natural phytoextraction (Kidd et al. 2007). The most representative families of these plants are the Asteraceae, Cyperaceae, Lamiaceae, Poaceae, Violaceae, Brassicaceae and Fabaceae. The cultivation of these species permits the extraction of heavy metal from the soil by plant absorption and the harvesting of its aerial biomass (Kidd et al. 2007).

Hyperaccumulating plants can eliminate pollutants that persist in the environment through various mechanisms, for example, phytotransformation, phytofiltration, phytovolatilization, phytostabilization and phytoextraction, in addition to the containment and degradation of metals, phenolic compounds and various colorants, as well as other inorganic and organic contaminants (Tahir et al. 2016). The mechanisms mentioned will be analyzed in the next topic.

## 2.1 Accumulation and Detoxification of Heavy Metals By Plants

When plants are exposed to heavy metals, they can activate a selective pressure with essential evolutionary and ecological consequences, like the development of several mechanisms to evade uptake of metals while others can accumulate those metals at different levels. Accumulation and tolerance are very different definitions when applied to heavy metals with great levels of inter- and intraspecific discrepancy in plants. Nevertheless, the actual understanding of these discrepancy is derived mainly from angiosperms. Bryophytes also can tolerate and accumulate great quantities of these contaminants (Reeves et al. 2017).

According to the tolerance mechanism that plants possess, they can be classified into the following categories (Fig. 1) (Lu et al. 2017).

a. Bioaccumulators. Also known as hyperaccumulators. Plants take the metals through the root and translocate them to the aerial part where they accumulate and reach concentrations greater than those in the soil, with a lack of toxicity symptoms. These plants can be used for bioremediation as they can remove metals from the soil.

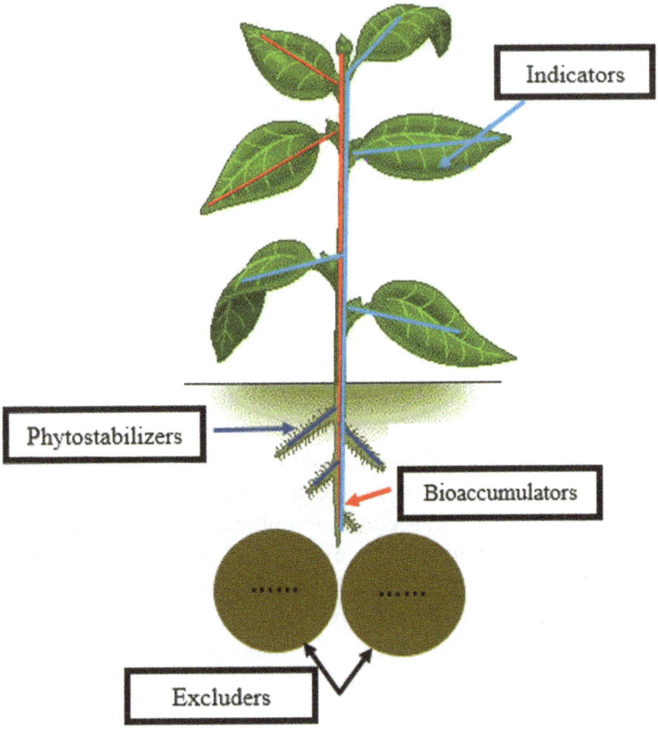


**Fig. 1** Classification of mechanisms of tolerance by plants

b. Phytostabilizers. These plants take and accumulate the metals in the root, but without translocating them to the aerial part. This mechanism is also useful for bioremediation as it fixes heavy metals preventing their leaching and subsequent arrival in water bodies.
c. Excluders. They do not allow the entry of metals to the root, this through the formation of complexes between organic compounds excreted by roots and the metals in the soil.
d. Indicators. They allow the metals to enter the root and go to the aerial parts, reaching concentrations in its tissues similar to those present in the soil. Plants usually present visible symptoms that are proportional to the amount of metals present in the soil.

Plants that can accumulate heavy metals in different parts like roots, leaves, stems, without visible indications of phytotoxicity are known as hyperaccumulators and they are very useful for soil decontamination (van der Ent et al. 2013). Some illustrations of these plants are presented in Table 1.

**Table 1** Examples of hyperaccumulators plants (modified from Yan et al. 2020)

| Plant species | Heavy metal | Concentration in plant (mg $kg^{-1}$) | References |
|---|---|---|---|
| *Alyssum murale* L<br>*Alyxia rubricaulis* (Baill.) Guillaumin | Ni<br>Mn | 4,730–20,100<br>14,000 | Bani et al. (2010),<br>Chaney et al. (2010) |
| *Brassica juncea* (L.) Vassiliĭ Matveievitch Czernajew (1796–1871)<br>*Brassica nigra* (L.) W.D.J. Koch | Pb<br>Pb | 10,300<br>9,400 | Koptsik (2014),<br>Koptsik (2014) |
| *Corrigiola telephiifolia* L<br>*Cicer arietinum* L | As<br>Hg | 2,110<br>0.2 | García-Salgado et al. (2012),<br>Wang et al. (2012) |
| *Hordeum* spp | Hg | 2.35 | Rodríguez et al. (2003) |
| *Medicago sativa* L. (Guillaumin) Virot<br>*Marrubium vulgare* L | Pb<br>Mn | 43,300<br>13.8 | Koptsik (2014),<br>Rodríguez et al. (2003) |

### 2.1.1 Detoxification Mechanism by Terrestrial Plants

Heavy metals enter the plants through several mechanisms (Dalvi and Bhalerao 2013), including:

- Metal mobilization by the releasing of roots exudates that increase the metal bioavailability.
- Root uptake: The metals are absorbed by roots through passive diffusion (apoplastic) and active transport (symplastic).
- Xylem loading: The metal ions move into the xylem torrent via the root symplastic (Thakur et al. 2016).
- Root-to-shoot transport: The metal ions go from the roots to the shoots through xylem tissue.
- Cellular compartmentation
- Sequestration: Ions accumulate in cell walls or vacuole but not into the cytosol (Tong et al. 2004).

The detoxification mechanism that plants use is mainly avoidance and tolerance, keeping in this way the cellular quantities of these metals lower than the harmfulness threshold ranks (Hall 2002).

Avoidance is the capacity of plants to diminish the uptake of heavy metals so they cannot move inside tissues. This is the first defense mechanism that plants use against heavy metal toxicity and includes synthesis of mycorrhizae, binding metals in the cell wall, callose, removal of excess metals and limitation of the metal assimilation by the root (Małachowska and Gnida 2015).

- Mycorrhizae: They are synergetic associations between fungi and plants, where the mycorrhizal fungi are capable to reduce the stress of heavy metals (Turnau and Mesjasz–Przybyłowicz 2003). Some examples of the application of mycorrhizae in plants to avoid heavy metals are: a RintPDX1, isolated from arbuscular mycorrhizal fungus (AMF) *Rhizophagus intraradices*, is involved in the reduction of Cu and hydrogen peroxide (Benabdellah et al. 2009); according to You et al. (2021), *R. irregularis* (FR717169) can reduce Zn and Cd on *Phragmites australis* by the increment of aerial part and by several regulatory patterns under diverse trace elements concentrations [five levels of Zn (2, 100, 300, 500 and 700 mg $L^{-1}$).
- Binding metals in the cell wall: The cell wall in plants is composed mainly by lignin, cellulose, hemicellulose and pectins; when these components are dissociated negatively charged groups appears, with calcium saturation these ions may be replaced by metals cations ($Cu^{2+}$, $Pb^{2+}$, $Cd^{2+}$).
- Synthesis of callose: This is the first action when heavy metal concentrations are high in the soil. Callose is a polysaccharide (β-1, 3 glucan) that is placed out of the cell membrane, so translocation of metal ions inside the cell is reduced (Miransari 2011).
- Removal of excess metals: Plants can remove excess of metals using leaves epidermis (salt glands) (e.g., *Armeria maritima* spp. may remove Cu, Ni, Zn) and hydathodes.
- Limitation of the metal assimilation by the roots: Organic acids and some other substances secreted by roots may bond to metal ions in a polluted environment, so the assimilation of these ions by plants is limited (Meier et al. 2012). Also, the pH may increase by roots in the soil, which reduces the availability of metals (Miransari 2011).

Tolerance occurs when metal ions pass the protection barriers of plants and go into the cells; in this case, several proteins can transport metals, after bind them, into the vacuoles. Examples of the must studied proteins are metallothioneins and phytochelatins (Miransari 2011).

- Metallothioneins (MTs) are proteins rich in cysteine with a low molecular weight that can link ions of metal with thiol groups, and they mostly detoxify metal ions of Cu and Cd (Hall 2002). These proteins in crops are part of metal homeostasis; some ecotypes of *Arabidopsis* show tolerance to heavy metals. Also, for ecological situations, they could be used as markers (Guo et al. 2008).
- Phytochelatins are enzymatically produced as small peptides because of the existence of several metals. These proteins can be obtained from reduced glutathione (GSH) and catalyzed by phytochelatin synthase (PCS), which is triggered by $Au^{+}$, $Cd^{2+}$, $Zn^{2+}$, $Bi^{3+}$, $Hg^{+}$, $Ag^{+}$, $Pb^{2+}$ and $Cu^{2+}$ (Grill et al. 1989).

### 2.1.2 Phytoremediation

The phytoremediation technologies include several techniques that are characterized by having particular mechanisms for the remediation of water, soil or sediment pollutants and have been used to clean bodies of water and soils. According to Maiti and Kumar (2016), the main five technologies are (Fig. 2):

- Rhizofiltration: Plants take metals from water and sequester them through the roots, which must have high biomass, wide surface area and tolerance to excess metals. This technique is easy to use and shows low maintenance cost and generates few secondary residues, serving to remedy various metals (Ni, Pb, Cr, Cd, Cu, V-vanadium) and radionucleotides (Sr-strontium, U-uranium and Cs-cesium), among others.

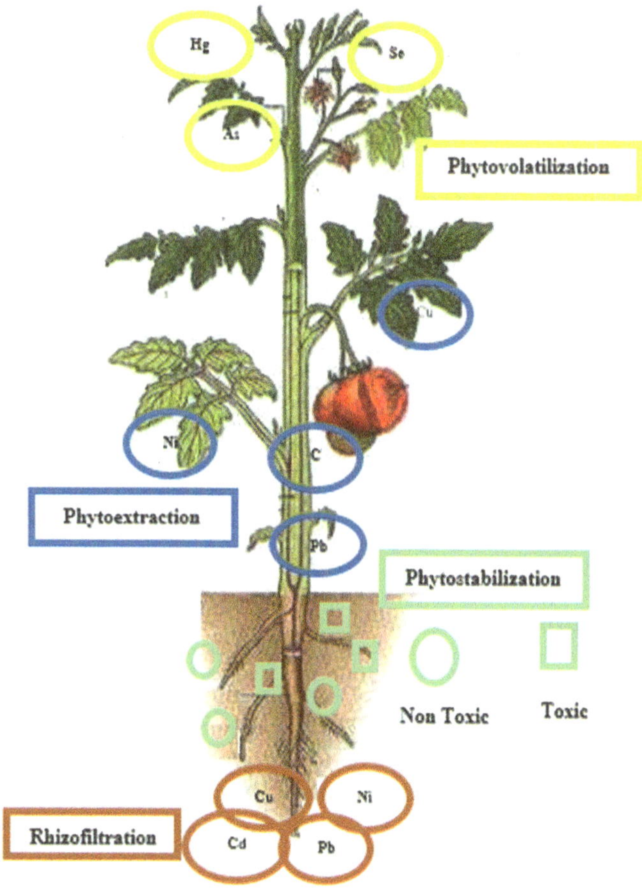


**Fig. 2** Illustration of phytoremediation technologies (Modified from Montenegro-Gómez et al. 2019)

- Phytostabilization: Plants may transform toxic moieties of metals into non-toxic forms, reducing the environmental risk. This technique can be useful to stabilize waste and prevent exposure pathways through water and wind erosion, also providing a hydraulic system that restricts vertical leaching of contaminants into groundwater, and at the root reduces mobility by adsorption and chemical fixation of physical and chemical pollutants. This method can be implemented as an indirect control when soil areas and water bodies are contaminated. For the success of this technique, selection of the correct species of plants is important because they must tolerate heavy metal concentrations (Marques et al. 2009).
- Phytovolatilization: Plants are used to absorb toxic metals and transform them into non-toxic volatile forms so they can be free in the air by plant transpiration process. Through this technique, metals like As, Se and Hg can be volatilized. For example, *Brassica juncea* can volatilize Se (Banuelos et al. 1993).
- Phytoextraction: This is the best-known technique. Here, plants take the contaminants by the roots, and they are translocated and accumulated in the harvested parts of the plants. After maturation, they are harvested, and this biomass is discharged into hermetic containers or landfills. With this process, metals like V, Cd, Ni, Cr, Pb and Cu are removed from polluted soils. The ideal plant species for this type of technology must have characteristics such as:

  - Tolerate concentration of pollutants that are accumulated;
  - High biomass and quick growth;
  - Accumulation of pollutants in the aerial and in harvested parts;
  - Easy to harvest; and
  - Possibility of appropriate cultivating the phytoextractor plant.

At present, around 45 angiosperm families with > 450 species of plants have been classified as metal hyperaccumulator plants (Suman et al. 2018), fluctuating from annual herbs, trees, to perennial bush, for example, Euphorbiaceae, Lamiaceae, Fabaceae, Brassicaceae, Scrophulariaceae and Asterraceae families. Species like *Sedum alfredii* can accumulate more than two elements (Cd, Zn and Pb) (Yang et al. 2004).

## 3 Microorganisms and Heavy Metals

Plants cannot degrade some kind of soil pollutants like organic toxins comprising carbon, for example, different fuels and hydrocarbons; these substances only become softer by microbial processes (Adesodun et al. 2010).

Microorganisms that form mutualism and association in the rhizosphere are very useful in the accumulation of metals. Once these microorganisms were treated with antibiotics, plants accumulated less heavy metals; in contrast if they grow axenically the accumulation of metals increased. Microorganisms are important in the phytoremediation process of cultivated soil because of their diversity and structure. They may play important role in edaphic processes like carbon and nitrogen mineralization and decomposition, and also, they facilitate the development of mechanisms by plants to diminish harmfulness of metals and toleration, and/or resistance to metal sequestration in a difficult contaminated ecosystem (Bruins et al. 2000). Nevertheless, metal concentrations are important for the alteration of species density, composition and biomass reduction of microorganisms. Metals like Pb, Cd, Mo Cr and Ni could increase the Gram-positive microorganisms (Acidobacteria, Proteobacteria, Chlorobi and Verrucomicrobia) in soils (Akerblom et al. 2007).

Microorganisms use several mechanisms to remove waste matter or metals. Examples of these mechanisms are formation of siderophores that might increase metal bioavailability (Visca et al. 2007); biosurfactants that may increase the solubility of organic compounds and consequently the quality of heavy metals (Zhang et al. 2012); biofilm (Ullah et al. 2015); biomethylation and oxidation–reduction processes (Ullah et al. 2015); secretion of plant hormones like cytokinins, gibberellins (GAs) and indole-3-acetic acid (IAA), that improve plant development (Kidd et al. 2009).

### *3.1 Accumulation and Detoxification of Heavy Metals in Soils By Fungi*

It has been documented that physicochemical strategies to remediate contaminated soils, such as solidification (Chen and Chiou 2008), extraction, in situ biopile, on-site land farming (Inoue and Katayama 2011) and soil washing (Day et al. 1997), are more expensive than biological techniques (Khalid et al. 2017). Thus, bioremediation is a low-cost, effective, simple and environmentally friendly remediation technology where biological agents, or their products, such as bacteria, algae, plants and fungi, are used for the decomposition, transformation and decontamination of materials and pollutants in products less dangerous or not dangerous (Mohammadian et al. 2017). In this technique, the degradation and removal of contaminants can be in situ and ex situ (Vidali 2001). Therefore, mycoremediation, or fungal bioremediation, is a type of technology that uses these organisms to diminish or reduce environmental contaminants into non-toxic compounds (Thakur 2014).

Fungi are eukaryotic organisms and omnipresent in any environmental condition. They exhibit peculiar characteristics such as colonization (their hyphae can penetrate the soil and expand further, compared to other microorganisms), tolerate high concentrations of toxins, are hyperaccumulators, secrete extracellular enzymes and bioactive compounds, and display exclusion by permeability barrier and translocation of nutrients, minerals and water (Thakur 2014; Mohammadian et al. 2017). Fungi have developed intracellular strategies such as the metal ion conjugation with organic ligands and proteins, and extracellular strategies such as chelation of metal and cell binding to prevent the entrance of the metal ions into the cell to resist the toxic effect of HM. Four main steps are observed in the defense mechanism: (1)

biosorption (Dhankhar and Hooda 2011); (2) bioaccumulation and compartmentalization (Ahalya et al. 2003); (3) detoxification or metal chelation (Tripathi et al. 2007); and (4) efflux transport for metal exclusion (Canovas et al. 2004).

To remediate HM-contaminated soils, fungi channel these metals and accumulate them in their fruiting bodies by virtue of their hyperaccumulating characteristic. Various species of fungi have phosphate, carboxylic, and amine or sulfhydryl groups in their cell walls, which bind to heavy metals due to differences in magnetic charge; others secrete ligninolytic and cellulolytic enzymes that are capable of degrading HM. Rhee et al. (2012) pointed out that *Paecilomyces javanicus* and *Metarhizium anisopliae* transform Pb into chloropyromorphite, while Huang et al. (2006) mentioned that *Phanerochaete chrysosporium* reduces the concentration of Pb. Similarly, various authors have pointed out that the genera *Acremonium*, *Alternaria*, *Aspergillus*, *Fusarium*, *Mucor*, *Penicillium*, *Rhizopus*, *Trichoderma*, among others, have the ability to eliminate HM (Mohammadian et al. 2017; Kumar and Dwivedi 2020).

Similarly, it has been confirmed that arbuscular mycorrhizal fungi (AMF) can bioremediate contaminated soils. These organisms supply nutrients to the roots of vascular plants from the soil and provide them protection against environmental stresses, including heavy metal toxicity. Remediation mechanisms by HMA include the production of organic compounds in fungal hyphae to make steady heavy metals, retain metals with the help of siderophores, metal binding to phytochelatins or metallothioneins within fungal and plant cells. Additionally, the cell wall of AMF binds to metals and deposits them in its vacuoles, reducing metal translocation within the plant (Kamal et al. 2010).

The efficiency of mycoremediation of soils contaminated with HM depends largely on concentration and bioavailability of the contaminant, physicochemical properties of the soil, carbon content, the redox potential, characteristics of the fungi used, availability of nutrients and environmental conditions (pH, temperature) (Pundir et al. 2018).

Fungi that are isolate form sites pollute with heavy metals are very useful for bioremediation technique, such as filamentous fungi like *Didymella glomerata*, *Cladosporium* sp., *Sarocladium kiliense*, *Phoma costaricensis* and *Fusarium oxysporum*. These fungi according to Văcar et al. (2021) show a MICs (Minimum Inhibitory Concentrations) of Hg of 140–200 mg $L^{-1}$, Zn of 2092–2353 mg $L^{-1}$, Pb of 1568 mg $L^{-1}$ or Cd 337 mg $L^{-1}$. Also those species were classified as highly resistant Hg concentrations ranging from 33.8 to 54.9 mg $g^{-1}$ dry weight, with a capacity of Hg removal from 47 to 97%.

In another study, Liaquat et al. (2020) found five fungi that showed resistant to heavy metals (*Komagataella phaffii*, *Aspergillus sclerotiorum*, *Trichoderma harzianum*, *Aspergillus niger* and *Aspergillus aculeatus*) and one of them, *K. phaffii*, showed 5500 ppm of Cd tolerance and resistance to Cr at 4000 ppm.

Bala et al. (2020) isolated several fungi from a soil that is a dumpsite. *Penicillium chrysogenum* and *Aspergillus niger* showed high tolerance for Zn, Pb and Cd (200 ppm, 400 ppm and 600 ppm). Related to Pb removal, *P. chrysogenum* removed at 5th day 1.07 ppm, at 10th 3.35 ppm and 4.19 ppm after 15th days compared with

*A. niger* that at 5th day removed 0.67 ppm, at 10th 3.11 ppm and 3.79 ppm at 15th day, respectively.

Alsabhan et al. (2022) in Saudi Arabia isolated several fungi such as Alternaria, Aspergillus, Fusarium, Rhizopus Penicillium, Botrytis Mucor and Trichoderma, and as a result, they found that the pH of the soil and the population of fungi had a robust negative correlation; same result was reported related to Cd and the population of fungi, and a positive correlation was observed between Zn and fungi.

### *3.2 Accumulation and Detoxification of Heavy Metals in Soils by Bacteria*

The soil is an ecosystem where microorganisms of diverse genera and species coexist. Each of them interacts with the soil in a particular way, which depends on the evolutionary processes that gave rise to its current form of existence; however, a few micrometers away there may be another microbe whose metabolism is completely different (Jackson et al. 2011). The macroscopic result of this joint activity that occurs at the microscopic level is a constant recycling of the nutrients that are necessary for plant life.

Heavy metals accumulate in the soil mainly as a result of mining, industry and agriculture, but almost all anthropic activities are potential contributors of these types of pollutants (Sarwar et al. 2017). In reducing heavy metal pollution, the role of bacteria is decisive; bacteria of the genera *Micrococcus, Bacillus, Penicillium, Pseudomonas, Enterobacter* and others have been shown to be capable of detoxifying lead, cadmium, copper, nickel, mercury, chromium, zinc and cobalt (Dhaliwal et al. 2020).

The capacity of microorganisms to modify the quantity of heavy metals in the soil lies in their tolerance to these metals, which according to Aly et al. (2018) is based on four mechanisms:

- Biosorption: Metals are captured on the surface of cells from the affinity of the biological compounds of their membranes.
- Biomineralization: Microbes synthesize compounds—such as sulfur—capable of immobilizing metals by combining with them to produce insoluble molecules.
- Biotransformation: It results from the oxidation, reduction, methylation or alkylation of metals, which turns them into compounds that can be more or less accessible to plants.
- Absorption: It involves the transport of metals into the microbial cells, which leads to their internal accumulation.

Certain species of plants have the capacity of extracting and accumulating heavy metals from the soil, which is why they are used for this in phytoremediation. However, although phytoremediation is relatively cheap, it is generally slow and not always effective, because it depends mainly on the growth of the plant; furthermore, in high concentrations of heavy metals, plant growth is inhibited, which means that phytoremediation ceases to take effect. On the other hand, the bacteria that live in the rhizosphere (rhizobacteria), and in particular the phosphorus-solubilizing bacteria (PSB), exhibit various metabolic mechanisms that directly contribute to the detoxification of heavy metals in the soil, or contribute to phytoremediation, enhancing the effect of extractor plants (Nadeem et al. 2020). These mechanisms are:

- Synthesis of organic acids that solubilize metal complexes, making metals accessible to extractor plants.
- Production of siderophores, which are not only capable of chelating iron but also other heavy metals like Ni, Cd, As and Pb.
- Indoleacetic acid synthesis, which stimulates root growth, allowing crops to access more water and nutrients to survive the stress caused by heavy metals.
- Production of the enzyme that hydrolyzes excessive 1-aminocyclopropane-1-carboxylate (ACC), an ethylene precursor that, when accumulated, prevents root growth.

The presence of microbial consortia increases the solubilization efficiency, as reported by Teymouri et al. (2016) as a result of a study that was based on the use of isolates of *Bacillus* sp., *Pseudomonas* sp. and *Acinetobacter* sp., separately and in combination. The possible explanation for this effect obtained when using consortia instead of isolates seems to lie in the sum of the actions of several of the mechanisms described.

The inclusion of plant growth-promoting bacteria (PGPR) in these consortiums is particularly favorable (Nadeem et al. 2014) because by favoring the growth of the plant, they increase its vigor and ability to resist different types of stress, including the one produced by heavy metals.

Isolates from *Bacillus subtilis*, *Paenibacillus jamilae* and *Pseudomonas aeruginosa* stimulate several processes like photosynthesis, transpiration and the stability of the membranes in *Spinacia olearacea* L., reducing the levels of cadmium and lead in the plant (Desoky et al. 2020), which is attributed to the suppression of heavy metals by the three bacterial genera.

The use of endophytic bacteria obtained from plants that accumulate heavy metals can enhance the bioremediation of soils, given the ability of these microbes to grow under these adverse conditions. Jing et al. (2014) isolated *Klebsiella* sp. and *Enterobacter* sp. of a heavy metal hyperaccumulating plant (*Polygonum pubescens*) and the isolates, once inoculated in *Brassica napus*, increased the capacity to accumulate lead, cadmium and zinc in the latter species. Siripan et al. (2018) found that isolates of *Paenibacillus* sp., *Bacillus* sp. and *Alcaligenes* sp. obtained from *Chromolaena odorata* (hyperaccumulating plant) increase the absorption of cadmium in *Helianthus annuus*. Jan et al. (2019) demonstrated that endophytic isolates of *Enterobacter ludwigii* and *Exiguobacterium indicum*, obtained from *Chenopodium fi cifolium, Artemisia princeps, Echinochloa crus-galli* and *Oenothera biennis,* (hyperaccumulating plants), are capable of absorbing cadmium and nickel, and thus protect *Oryza sativa* L. seedlings inoculated with these isolates from the toxic effects of the accumulation of these heavy metals. The bacteria improved the accumulation of Ni and Cd in the treated plants, but mainly in the roots, which allowed an adequate development of the aerial parts. These properties of the referred bacterial genera can be used to support the phytoremediation of polluted soils, and in this way, the combination between plants and bacteria would make the elimination of stress produced by heavy metals more efficient (Ullah et al. 2015).

The addition of certain nutrients together with bacteria can increase the capacity to extract heavy metals (Mishra et al. 2017). According to Franchi et al. (2017), the mobilization and absorption of As and Hg by *Brassica juncea* and *Lupinus albus* increase if the inoculation with growth-promoting bacteria (*Actinobacteria, Proteobacteria* and *Firmicutes*) is done with the manifestation of ammonium thiosulfate in the soil.

The use of plants and bacteria together for the remediation of soils with high concentration heavy metals has as main barrier the incomplete knowledge about the absorption, accumulation and transport of these elements (Mosa et al. 2016). The use of bacteria that have been genetically transformed becomes an alternative for the future of bioremediation, as can be the generation of hyperaccumulating plants via transgenesis, incorporating genes from detoxifying bacteria. The transgenic plants thus obtained can achieve the immobilization of heavy metals or increase their translocation to the leaves, stem and shoots to harvest them and extract the metals from the soil (Sarwar et al. 2017). However, for the moment, the results obtained have been limited to experiments carried out in laboratories (Tiwari and Lata 2018).

### 3.3 Accumulation and Detoxification of Heavy Metals in Water by Microorganisms

One of the problems that the first Industrial Revolution, in the eighteenth century, has bequeathed to humanity to this day is the contamination of water with heavy metals (HM). It is well recognized that some lead and tin mines exploited at that time continue to wash out, contaminating water sources. The transformation and use of natural resources have caused that, in the water, the load of various HM such as arsenic, cadmium, zinc, copper, iron, mercury, antimony, beryllium, nickel and silver accumulates in different substrates (organisms, soil, water) above permissible limits, putting at risk human health and that of all life forms that inhabit aquatic ecosystems (Briffa et al. 2020).

Besides the natural processes that produce heavy metals accumulation in the aquatic ecosystems (the weathering of minerals, volcanic eruptions and erosion), today, human activities like mining, manufacturing, urbanization, aquaculture and agriculture, among others, contribute to increase the HM levels in water (Vareda et al. 2019).

Since water is the most important resource for life, its care has also been a priority for research with immediate application. Conventional physical and chemical methods (air sparging/stripping, incineration, electron-beam irradiation, electroplating, radiocolloid treatment, removal by surprise to organo-oxides, among others) for the remediation of water due to HM accumulation are expensive, not very effective, non-specific, and generate a large amount of chemical waste (Anawar and Chowdhory 2020). The application of microorganisms and plants to eliminate heavy metals from water represents a cheap and environmentally friendly alternative.

Specifically, remediation by microbes consists of precipitating, absorbing, oxidizing or reducing the HM from the water, through microbial metabolic processes that use such pollutants to maintain the homeostasis or resistance of the heavy metals in the ecosystem (Ayangberno and Babalola 2017). The interaction between microorganisms/heavy metals can be identified as a process where the metal ion binds to proteins or polysaccharides of the cell membrane (biosorption), the metal ion is transported through the bi-lipidic layer of the membrane to the cell cytoplasm and sequestered (bioaccumulation), and the metal ion is solubilized by microorganism and purified by adsorption, ion exchange, membrane separation and selective precipitation (bioleaching), metabolic use of metal ions by microorganism (biomineralization), modification of the chemical structure involving the ion metal to modify it to a less toxic form (biotransformation), extracellular and intracellular sequestration of metallic ions, active transport and reduction of metal ions (Tayang and Songachan 2021).

The biosorption capacity of some microalgae (*Scenedesmus* sp., *Spirulina platensis, Dunaliella salina, Anabaena, Oscillatoria, Spirogyra*), fungi (*Trichoderma, Aspergillus, Penicillium*) and bacteria (*Bacillus, Acinetobacter, Pantoea, Rhodococcus*) for various metals is influenced by several factors like pH and temperature, as well as technical conditions such as contact time and metal concentration. The process of metals removal in water can be enhanced using mixed microorganism consortiums. Microorganisms offer several advantages that make them the most feasible option to water bioremediation of HM, since they are ubiquitous, fast-growth organisms and are easily produced in controlled environments and in large quantities; also, they become accustomed to heavy metal ions and, for that, they can develop tolerance and resistance against HM. In some studies, several microorganisms have been analyzed for the remotion of heavy metal from water such as:

- The bacterium *Pseudomonas putida* and the yeast *Saccharomyces cerevisiae* were used for the removal of Pb in water by using pure cultures, and the highest removals were achieved with the biosorption process, using *P. putida* inactivated in an autoclave (100% in 24 h), while with *S. cerevisiae* inactivated in a water bath, a removal of 90.23% was obtained. When using both microorganisms combined (consortium), a Pb removal of 52.11% was obtained (Espinoza et al. 2021).
- García-Torres et al. (2019) reported that *Pseudomonas koreensis* can remove efficiently Cr form water in a 97% at 32 h.
- Ijaz et al. (2015) combined the terrestrial plant *Brachiaria mutica* and several endophytic bacterial strains (*Bacillus licheniformis* strain BRSI58, *Bacillus cereus*

strain BRSI57 and *Acinetobacter* sp. strain BRSI56) for the detoxification of sewage effluent in floating treatment wetlands (FTWs), reporting that endophytic inoculation in FTWs increased the elimination of pollution efficiently; however, the best results were obtained with the combination of plant and bacteria.

The efficiency of microbial remediation could be increased through the genetic modification of MO, which would imply a careful review of ethical, legal and biological safety aspects in its experimental use and possible application in the environment (Tang et al. 2021).

## 4 Conclusions

The contamination of soil and water by heavy metals, resulting from natural sources and mainly from anthropogenic agricultural and industrial activity, has reached levels of concern for the health and life of many species on Earth, including human beings. Physicochemical procedures used for the reduction of heavy metals have exhibited high cost, low efficiency, and no specificity; this has led to studies on the ability of certain microorganisms to live in high concentrations of heavy metals and the use of these capacities for the detoxification of soils and water bodies.

Research has shown that fungi and bacteria of many genera, by themselves or in cooperation with plants, can contribute to the removal/reduction of high concentrations of toxic forms of heavy metals. The mechanisms by which these events occur are diverse in nature and depend on the way in which microorganisms and plants co-evolved. Microbial-mediated detoxification processes are cheaper, do not alter the physical and chemical properties of soils and are friendly to the environment, contributing to the action of hyperaccumulator plants, and in some cases stimulating plant growth in adverse conditions. The main limitations for the generalization of these technologies, including the projections of genetic engineering in plant using genes of remedial microbes, lie in the insufficient knowledge about the absorption, transport and accumulation of heavy metals, which continues to be a subject of study at present.

## References

Adesodun JK, Atayese MO, Agbaje TA, Osadiaye BA, Mafe OF, Soretire AA (2010) Phytoremediation potentials of sunflowers *(Tithonia diversifolia and Helianthus annuus)* for metals in soils contaminated with zinc and lead nitrates. Water Air Soil Pollut 207:195–201. https://doi.org/10.1007/s11270-009-0128-3

Ahalya N, Ramachandra TV, Kanamadi RD (2003) Biosorption of heavy metals. Res J Chem Environ 7:71–78

Akerblom S, Baath E, Bringmark L, Bringmark E (2007) Experimentally induced effects of heavy metal on microbial activity and community structure of forest mor layers. Biol Fertil Soils 44:79–91. https://doi.org/10.1007/s00374-007-0181-2

Ali H, Khan E, Sajad MA (2013) Phytoremediation of heavy metals-concepts and applications. Chemosphere 91:869–881. https://doi.org/10.1016/j.chemosphere.2013.01.075

Alsabhan AH, Perveen K, Alwadi AS (2022) Heavy metal content and microbial population in the soil of Riyadh Region, Saudi Arabia. J King Saud Univ Sci 34:1–7. https://doi.org/10.1016/j.jksus.2021.101671

Aly MM, Alzahrani N, Amashah RH, Jastaniah SD (2018) Bioremediation of hazardous heavy metals from solutions or soil using living or dead microbial biomass. IOSR-JPBS 13(6):75–80. https://doi.org/10.9790/3008-1306017580

Anawar HM, Chowdhury R (2020) Remediation of polluted river water by biological, chemical, ecological and engineering processes. Sustainability 12:7017. https://doi.org/10.3390/su12177017

Antoniadis V, Shaheen SM, Stärk HJ, Wennrich R, Levizou E, Merbach I, Rinklebe J (2021) Phytoremediation potential of twelve wild plant species for toxic elements in a contaminated soil. Environ Inter 146:1–14. https://doi.org/10.1016/j.envint.2020.106233

Ashraf S, Ali Q, Zahir ZA, Ashraf S, Asghar HN (2019) Phytoremediation: environmentally sustainable way for reclamation of heavy metal polluted soils. Ecotox Environ Safe 174:714–727. https://doi.org/10.1016/j.ecoenv2019.02.068

Ayangberno AS, Babalola OO (2017) A new strategy for heavy metal polluted environments: a review of microbial biosorbents. Int J Environ Res Public Health 14:94. https://doi.org/10.3390/ijerph14010094

Babu SMOF, Hossain MB, Rahman MS, Rahman M, Ahmed ASS, Hasan MM, Rakib A, Emran TB, Xiao J, Simal-Gandara J (2021) Phytoremediation of toxic metals: a sustainable green solution for clean environment. Appl Sci 11:1–34. https://doi.org/10.3390/app112110348

Baker AJ, McGrath SP, Reeves RD, Smith JA (2000) Plantas hiperacumuladoras de metal: una revisión de la ecología y la fisiología de un recurso biológico para la fitorremediación de suelos contaminados con metales. Phytoremediation Contam Soil Water 85–107

Baker AJM (1981) Accumulators and excluders strategies in the response of plants to heavy-metals. J Plant Nutr 3:643–654. https://doi.org/10.1080/01904168109362867

Bala JD, Kuta F, Nasiru A, Adedeji AS, Al-Gheethi AA, Fashola OH (2020) Biosorption potential of lead tolerant fungi isolated from refuse dumpsite soil in Nigeria. Acta Sci 42:e46753. https://doi.org/10.4025/actascibiolsci.v42i1.46753

Bani A, Pavlova D, Echevarria G, Mullaj A, Reeves RD, Mullaj JL, Reeves RD (2010) Nickel hyperaccumulation by the species of *Alyssum* and *Thlaspi* (Brassicaceae) from the ultramafic soils of the Balkans. Bot Serb 34(1):3–14

Banuelos G, Cardon G, Mackey B, Ben-Asher J, Wu L, Beuselinck P, Akohoue S, Zambrzuski S (1993) Boron and selenium removal in boron−laden soils by four sprinkler irrigated plant species. J Environ Qual 22:786–792. https://doi.org/10.2134/jeq1993.00472425002200040021x

Benabdellah K, Azcón-Aguilar C, Valderas A, Speziga D, Fitzpatrick TB, Ferrol N (2009) GintPDX1 encodes a protein involved in vitamin B6 biosynthesis that is up-regulated by oxidative stress in the arbuscular mycorrhizal fungus Glomus intraradices. New Phytol 184:682–693. https://doi.org/10.1111/j.1469-8137.2009.02978.x

Bennett LS, Burkhead JL, Hale KL, Terry N, Pilon M, Pilon-Smits EAH (2003) Analysis of transgenic indian mustard plants for phytoremediation of metal-contaminated mine tailings. J Environ Qual 32:432–440. https://doi.org/10.2134/jeq2003.4320

Briffa J, Sinagra E, Blundell R (2020) Heavy metal pollution in the environment and their toxicological effects on humans. Heliyon 6(9):e04691. https://doi.org/10.1016/j.heliyon.2020.e04691

Brooks RR, Lee J, Reeves RD Jaffre T (1977) Detection of nickeliferous rocks by analysis of herbarium specimens of indicator plants. J Geochem Exploration 7:49–57. https://doi.org/10.1016/0375-6742(77)90074-7

Bruins MR, Kapil S, Oehme FW (2000) Microbial resistance to metals in the environment. Ecotoxicol Environ Saf 45:198–207. https://doi.org/10.1006/eesa.1999.1860

Canovas D, Vooijs R, Schat H, De Lorenzo V (2004) Role of thiol species in the hypertolerance of *Aspergillus* sp. P37 to Arsenic. J Biol Chem 279:51234–51240. https://doi.org/10.1074/jbc.M408622200

Chaney RL, Broadhurst CL, Centofanti T (2010) Phytoremediation of soil trace elements. In: Hooda PS (ed) Trace elements in soils, Chichester: John Wiley & Sons, Inc., pp 311–352.

Chen C, Chiou I (2008) Remediation of heavy metal-contaminated farm soil using turnover and attenuation method guided with a sustainable management framework. Environ Eng Sci 25:11–32. https://doi.org/10.1089/ees.2006.0183

DalCorso G, Fasani E, Manara A, Visioli G Furini A (2019) Heavy metal pollutions: state of the art and innovation in phytoremediation. Int J Mol Sci 20: 1–17. https://doi.org/10.3390/ijms20143412

Dalvi AA, Bhalerao SA (2013) Response of plants towards heavy metal toxicity: an overview of avoidance, tolerance and uptake mechanism. Ann Plant Sci 2:362–368

Day SJ, Morse GK, Lester JN (1997) The cost effectiveness of contaminated land remediation strategies. Sci Total Environ 201:125–136. https://doi.org/10.1016/S0048-9697(97)00097-1

Desoky ESM, Merwad ARM, Semida WM, Ibrahim SA, El-Saadony MT, Rady MM (2020) Heavy metals-resistant bacteria (HM-RB): Potential bioremediators of heavy metals-stressed *Spinacia oleracea* plant. Ecotoxicol Environ Saf 198:110685. https://doi.org/10.1016/j.ecoenv.2020.110685

Dhaliwal SS, Singh J, Taneja PK, Mandal A (2020) Remediation techniques for removal of heavy metals from the soil contaminated through different sources: a review. Environ Sci Pollut Res 27:13191333. https://doi.org/10.1007/s11356-019-06967-1

Dhankhar R, Hooda A (2011) Fungal biosorption-an alternative to meet the challenges of heavy metal pollution in aqueous solutions. Environ Technol 32(5–6):467–491. https://doi.org/10.1080/09593330.2011.572922

Espinoza M, Selene-Gómez AV, Valladares MR, Flores E, Castañeda MT (2021) Bioremediation of lead in water by two pure microorganisms and bacteria-yeast consortium. Biotecnología 25(2):33–46

Franchi E, Rolli E, Marasco R, Agazzi G, Borin S, Cosmina P, Pedron F, Rosellini I, Barbafieri M, Petruzzelli G (2017) Phytoremediation of a multi contaminated soil: mercury and arsenic phytoextraction assisted by mobilizing agent and plant growth promoting bacteria. J Soils Sedim 17:1224–1236. https://doi.org/10.1007/s11368-015-1346-5

Freeman JL, Persans MW, Nieman K, Albrecht C, Peer W, Pickering IJ, Salt DE (2004) Increased glutathione biosynthesis plays a role in nickel tolerance in Thlaspi nickel hyperaccumulators. Plant Cell 16:2176–2191. https://doi.org/10.1105/tpc.104.023036

García-Salgado S, García-Casillas D, Quijano-Nieto MA, Bonilla-Simón MM (2012) Arsenic and heavy metal uptake and accumulation in native plant species from soils polluted by mining activities. Water Air Soil Pollut 223:559–572. https://doi.org/10.1007/s11270-011-0882-x

García-Torres SG, Hernández C, Jiménez DL (2019) Bioadsorption of Cr (vi) in aqueous solutions by *Pseudomonas koreensis* immobilized in alginate beads. Int J Environ Impacts 2(3):229–239

Grill E, Löffler S, Winnacker EL, Zenk MH (1989) Phytochelatins, the heavy metals-binding peptides of plants, are synthesized from glutathione by a specific $\gamma$-glutamylcysteine dipeptidyl transpeptidase (phytochelatin synthase). Proc Natl Acad Sci USA 86:6838–6842. https://doi.org/10.1073/pnas.86.18.6838

Guo WJ, Meetam M, Goldsbrough P (2008) Examining the specific contributions of individual Arabidopsis metallothioneins to copper distribution and metal tolerance. Plant Physiol 164(4):1697–1706. https://doi.org/10.1104/pp.108.115782

Hall J (2002) Cellular mechanisms for heavy metal detoxification and tolerance. J Exp Bot 53:1–11. https://doi.org/10.1093/jexbot/53.366.1

Huang DL, Zeng GM, Jiang XY, Feng CL, Yu HY, Huang GH, Liu HL (2006) Bioremediation of Pb-contaminated soil by incubating with *Phanerochaete chrysosporium* and straw. J Hazard Mater 134:268–276. https://doi.org/10.1016/j.jhazmat.2005.11.021

Ijaz A, Shabir G, Khan QM, Afzal M (2015) Enhanced remediation of sewage effluent by endophyte-assisted floating treatment wetlands. Ecol Eng 84:58–66. https://doi.org/10.1016/j.ecoleng.2015.07.025

Inoue Y, Katayama A (2011) Two-scale evaluation of remediation technologies for a contaminated site by applying economic input–output life cycle assessment: risk-cost, risk-energy consumption and risk-CO 2 emission. J Hazard Mater 192:1234–1242. https://doi.org/10.1016/j.jhazmat.2011.06.029

Jackson TA, West MM, Leppard GG (2011) Accumulation and partitioning of heavy metals by bacterial cells and associated colloidal minerals, with alteration, neoformation, and selective adsorption of minerals by bacteria, in metal-polluted lake sediment. Geomicrobiol J 28:23–55. https://doi.org/10.1080/01490451003739406

Jan R, Khan MA, Asaf S, Lubna LIJ, Kim KM (2019) Metal resistant endophytic bacteria reduces cadmium, nickel toxicity, and enhances expression of metal stress related genes with improved growth of *Oryza sativa*, via regulating its antioxidant machinery and endogenous hormones. Plants 8(10):363. https://doi.org/10.3390/plants8100363

Jing YX, Yan JL, He HD, Yang DJ, Xiao L, Zhong T, Yuan M, Cai XD, Li SB (2014) Characterization of bacteria in the rhizosphere soils of *Polygonum pubescens* and their potential in promoting growth and Cd, Pb, Zn uptake by *Brassica napus*. Int J Phytorem 16(4):321–333. https://doi.org/10.1080/15226514.2013.773283

Kamal S, Prasad R, Varma A (2010) Soil microbial diversity in relation to heavy metals. In: Sherameti I, Varma A (eds) Soil Heavy Metals, vol 19. Springer-Verlag, Berlin, pp. 31–64. https://doi.org/10.1007/978-3-642-02436-8_3

Khalid S, Shahid M, Niazi NK, Murtaza B, Bibi I, Dumat C (2017) A comparison of technologies for remediation of heavy metal contaminated soils. J Geochem Explor 182:247–268. https://doi.org/10.1016/j.gexplo.2016.11.021

Kidd P, Barcelob J, Bernal MP, Navari-Izzo F, Poschenriederb C, Shilev S, Clemente R, Monterroso C (2009) Trace element behaviour at the root-soil interface: implications in phytoremediation. Environ Exp Bot 67:243–259. https://doi.org/10.1016/j.envexpbot.2009.06.013

Kidd PS, Becerra C, García M, Monterroso C (2007) Aplicación de plantas hiperacumuladoras de níquel en la fitoextracción natural: el género *Alyssum* L. Ecosistemas 16(2):26–43. http://hdl.handle.net/10045/7666

Koptsik G (2014) Problems and prospects concerning the phytoremediation of heavy metal polluted soils: a review. Eurasian Soil Sci 47:923–939. https://doi.org/10.1134/S1064229314090075

Kumar V, Dwivedi SK (2020) Multimetal tolerant fungus *Aspergillus flavus* CR500 with remarkable stress response, simultaneous multiple metal/loid removal ability and bioremediation potential of wastewater. Environ Technol Innov 20:101075. https://doi.org/10.1016/j.eti.2020.101075

Liaquat F, Farooq M, Haroon U, Arif S, Saqib S, Zaman W, Khan AR, Shi J, Che Sh, Liu Q (2020) Evaluation of metal tolerance of fungal strains isolated from contaminated mining soil of Nanjing, China. Biol 9:1–12. https://doi.org/10.3390/biology9120469

Lu Y, Li X, He M, Zeng F, Li X (2017) Accumulation of heavy metals in native plants growing on mining-influenced sites in Jinchang: a typical industrial city (China). Environ Earth Sci 76:446. https://doi.org/10.1007/s12665-017-6779-2

Maiti S, Kumar A (2016) Energy plantations, medicinal and aromatic plants on contaminated soil. In: Prasad MNV (ed) Bioremediation and bioeconomy, Elsevier, pp 29–47. https://doi.org/10.1016/B978-0-12-802830-8.00002-2

Małachowska AJ, Gnida A (2015) Mechanisms of stress avoidance and tolerance by plants used in phytoremediation of heavy metals. Arch Environ Prot 41(4):96–103. https://doi.org/10.1515/aep2015-0045

Marques AP, Rangel AO, Castro PM (2009) Remediation of heavy metal contaminated soils: phytoremediation as a potentially promising clean-up technology. Crit Rev Env Sci. Technol 39:622–654. https://doi.org/10.1080/10643380701798272

Medhat R, Alsohim AS (2019) Bioremediation of heavy metals. In: Environmental chemistry and recent pollution control approaches. IntechOpen, pp 1–16. https://doi.org/10.5772/intechopen.88339
Meier S, Alvear M, Borie F, Aguilera P, Ginocchio R, Cornejo P (2012) Influence of copper on root exudate patterns in some metallophytes and agricultural plants. Ecotoxicol Environ Saf 75:8–15. https://doi.org/10.1016/j.ecoenv.2011.08.029
Miransari M (2011) Hyperaccumulators, arbuscular mycorrhizal fungi and stress of heavy metals. Biotechnol Adv 29:645–653. https://doi.org/10.1016/j.biotechadv.2011.04.006
Mishra J, Singh R, Arora NK (2017) Alleviation of heavy metal stress in plants and remediation of soil by rhizosphere microorganisms. Front Microbiol 8:1706. https://doi.org/10.3389/fmicb.2017.01706
Mohammadian E, Ahari AB, Arzanlou M, Oustan S, Khazaei SH (2017) Tolerance to heavy metals in filamentous fungi isolated from contaminated mining soils in the Zanjan Province, Iran. Chemosphere 185:290–296. https://doi.org/10.1016/j.chemosphere.2017.07.022
Montenegro-Gómez S, Pulido SY, Calderón-Vallejo LF (2019) Practicas de biorremediación en suelos y aguas. Escuela de Ciencias Agrícolas Pecuarias y del Medio Ambiente. Sello Editorial UNAD, Bogotá, DC, pp 1–49. https://doi.org/10.22490/notas.3451
Mosa KA, Saadoun I, Kumar K, Helmy M, Dhankher OP (2016) Potential biotechnological strategies for the cleanup of heavy metals and metalloids. Front Plant Sci: 7. https://doi.org/10.3389/fpls.2016.00303
Nadeem N, Asif R, Ayyub S, Salman S, Shafique F, Ali Q, Malik A (2020) Role of rhizobacteria in phytoremediation of heavy metals. Biol Clin Sci Res J: 35. https://doi.org/10.54112/bcsrj.v2020i1.35
Nadeem SF, Ahmad M, Zahir ZA, Javaid A, Ashraf M (2014) The role of mycorrhizae and plant growth promoting rhizobacteria (PGPR) in improving crop productivity under stressful environments. Biotech Adv 32(2):429–448. https://doi.org/10.1016/j.biotechadv.2013.12.005
Padmavathiamma PK, Li LY (2007) Phytoremediation technology: hyperaccumulation metals in plants. Water Air Soil Pollut 184:105–126. https://doi.org/10.1007/s11270-007-9401-5
Pundir R, Chary GHVC, Dastdar MG (2018) Application of Taguchi method for optimizing the process parameters for the removal of copper and nickel by growing *Aspergillus* sp. Water Resour Ind 20:83–92. https://doi.org/10.1016/j.wri.2016.05.001
Rafique N, Tariq SR (2016) Distribution and source apportionment studies of heavy metals in soil of cotton/wheat fields. Environ Monit Assess 188:309. https://doi.org/10.1007/s10661-016-5309-0
Reeves DR, Baker AJ, Jaffré T, Erskine PD, Echevarria G, van der Ent A (2017) A global database for plants that hyperaccumulate metal and metalloid trace elements. New Phytol 218:407–411. https://doi.org/10.1111/nph.14907
Rhee YJ, Hillier S, Gadd GM (2012) Lead transformation to pyromorphite by fungi. Curr Biol 22:237–241. https://doi.org/10.1016/j.cub.2011.12.017
Rodríguez L, López-Bellido F, Carnicer A, Alcalde-Morano V (2003) Phytoremediation of mercury-polluted soils using crop plants. Fresen Environ Bull 9:328–332. https://doi.org/10.1007/s11356-019-06563-3
Sarwar N, Imran M, Shaheen MR, Ishaque W, Kamran MA, Matloob A, Rehim A, Hussain S (2017) Phytoremediation strategies for soils contaminated with heavy metals: modifications and future perspectives. Chemosphere 171:710–721. https://doi.org/10.1016/j.chemosphere.2016.12.116
Shah MP (2020) Microbial bioremediation & biodegradation. Springer
Shah MP (2021a) Removal of refractory pollutants from wastewater treatment plants. CRC Press
Shah MP (2021b) Removal of emerging contaminants through microbial processes. Springer
Siripan O, Thamchaipenet A, Surat W (2018) Enhancement of the efficiency of Cd phytoextraction using bacterial endophytes isolated from *Chromolaena odorata*, a Cd hyperaccumulator. Int J Phytorem 20(11):1096–1105. https://doi.org/10.1080/15226514.2017.1365338
Suman J, Uhlik O, Viktorova J, Macek T (2018) Phytoextraction of heavy metals: a promising tool for clean-up of polluted environment? Front Plant Sci 9:1476. https://doi.org/10.3389/fpls.2018.01476

Tahir U, Yasmin A, Khan UH (2016) Phytoremediation: potential flora for synthetic dyestuff metabolism. J King Saud Univ Sci 28(2):119–130. https://doi.org/10.1016/j.jksus.2015.05.009
Tang X, Huang Y, Li Y, Wuang L, Pei X, Zhou D, He P, Hughes SS (2021) Study on detoxification and removal mechanisms of hexavalent chromium by microorganisms. Ecotol Environ Safety 208:111699. https://doi.org/10.1016/j.ecoenv.2020.111699
Tayang A, Songachan LS (2021) Microbial bioremediation of heavy metals. Curr Sci 120(6):1013. https://doi.org/10.18520/cs/v120/i6/1013-1025
Teymouri M, Akhtari J, Karkhane M, Marzban A (2016) Assessment of phosphate solubilization activity of rhizobacteria in mangrove forest. Biocatal Agric Biotechnol 5:168–172. https://doi.org/10.1016/j.bcab.2016.01.012
Thakur M (2014) Mycoremediation—a potential tool to control soil pollution. Asian J Environ Sci 9(1):21–31
Thakur S, Singh L, Wahid ZA, Siddiqui MF, Atnaw SM, Din MFM (2016) Plant-driven removal of heavy metals from soil: uptake, translocation, tolerance mechanism, challenges, and future perspectives. Environ Monit Assess 188:206. https://doi.org/10.1007/s10661-016-5211-9
Thomas JC, Davies EC, Malick FK, Endresz C, Williams CR, Abbas M, Petrella S, Swisher K, Perron M, Edwards R, Ostenkowski P, Urbanczyk N, Wiesend WN, Murray KS (2003) Yeast metallothionein in transgenic tobacco promotes copper uptake from contaminated soils. Biotechnol Prog 19:273–280. https://doi.org/10.1021/bp025623q
Tiwari S, Lata C (2018) Heavy metal stress, signaling, and tolerance due to plant-associated microbes: an overview. Fronts Plant Sci 452. https://doi.org/10.3389/fpls.2018.00452
Tong Y-P, Kneer R, Zhu Y-G (2004) Vacuolar compartmentalization: a second-generation approach to engineering plants for phytoremediation. Trends Plant Sci 9:7–9. https://doi.org/10.1016/j.tplants.2003.11.009
Tripathi AK, Harsh NSK, Gupta N (2007) Fungal treatment of industrial effluents, a mini-review. Life Sci J 4:78–81
Turnau K, Mesjasz-Przybyłowicz J (2003) Assessment of arbuscular mycorrhizal fungi diversity in the rhizosphere of Viola calaminaria and effect of these fungi on heavy metal uptake by clover. Mycorrhiza 10:161–168. https://doi.org/10.1007/s005720000072
Ullah A, Heng S, Munis MFH, Fahad S, Yang X (2015) Phytoremediation of heavy metals assisted by plant growth promoting (PGP) bacteria: a review. Env Exp Bot 117:28–40. https://doi.org/10.1016/j.envexpbot.2015.05.001
Văcar CL, Covaci E, Chakraborty S, Li B, Weindorf DC, Frent,iu T, Pârvu M, Podar D (2021) Heavy metal-resistant filamentous fungi as potential mercury bioremediators. J. Fungi 7:386. https://doi.org/10.3390/jof7050386
van der Ent A, Baker AJ, Reeves RD, Pollard AJ, Schat H (2013) Hyperaccumulators of metal and metalloid trace elements: facts and fiction. Plant Soil 362:319–334. https://doi.org/10.1007/s11104-012-1287-3
Vareda JP, Valente AJM, Durães L (2019) Assessment of heavy metal pollution from anthropogenic activities and remediation strategies: a review. J Environ Manage 246:101–118. https://doi.org/10.1016/j.jenvman.2019.05.126
Vidali M (2001) Bioremediation: an overview. Pure Appl Chem 73(7):1163–1172
Visca P, Imperi F, Lamont IL (2007) Pyoverdine siderophores: from biogenesis to biosignificance. Trends Microbiol 15:22–30. https://doi.org/10.1016/j.tim.2006.11.004
Wang J, Feng X, Anderson CW, Xing Y, Shang L (2012) Remediation of mercury contaminated sites—A review. J Hazard Mater 221:1–18. https://doi.org/10.1016/j.jhazmat.2012.04.035
Yan A, Wang Y, Tan SN, Mohd Yusof ML, Ghosh S, Chen Z (2020) Phytoremediation: a promising approach for revegetation of heavy metal-polluted land. Front Plant Sci 11:1–15. https://doi.org/10.3389/fpls.2020.00359
Yang XE, Long XX, Ye HB, He ZL, Calvert D, Stoffella P (2004) Cadmium tolerance and hyperaccumulation in a new Zn-hyperaccumulating plant species (*Sedum alfredii* Hance). Plant Soil 259:181–189. https://doi.org/10.1023/B:PLSO.0000020956.24027.f2

You Y, Wang L, Ju Ch, Wang G, Ma F, Wang Y, Yang D (2021) Effects of arbuscular mycorrhizal fungi on the growth and toxic element uptake of Phragmites australis (Cav.) Trin. ex Steud under zinc/cadmium stress. Ecotoxicol Environ Saf 213:1–10. https://doi.org/10.1016/j.ecoenv.2021.112023

Zhang X, Xu D, Zhu C, Lundaa T, Scherr KE (2012) Isolation and identification of biosurfactant producing and crude oil degrading *Pseudomonas aeruginosa* strains. Chem Eng J 209:138–146. https://doi.org/10.1016/j.cej.2012.07.110

# Hybrid Electrocoagulation and Ozonation Techniques for Industrial Wastewater Treatment

**Pranjal P. Das, Simons Dhara, and Mihir K. Purkait**

## 1 Introduction

Water is the most fundamental essential for life and is used for a variety of domestic and industrial activities. The rapid development of industries, viz. textile, pharmaceuticals, tanneries, distilleries, fertilizers, and mining industries, along with a rise in the living standards, urbanization, and human population, has resulted in the formation of toxic and recalcitrant effluent containing hazardous pollutants, responsible for the water crises around the world in recent years. Because of their low biodegradability index, these refractory compounds are resistant to biological treatment and have low biological oxygen demand and high chemical oxygen demand values (Ferrari et al. 2003). These non-biodegradable organic compounds are potential carcinogens, hormone disruptors, and very toxic in nature; thus, even their little concentration in wastewater poses a significant threat to the human health. Industries are the primary source of such non-biodegradable and toxic pollutants in the environment. As a result, prioritizing the treatment of industrial effluent is very crucial. In this context, primary treatment, secondary treatment, and tertiary treatment are the three main steps of conventional wastewater treatment process. Tertiary treatment is crucial as it produces water acceptable to various agricultural, industrial, and residential uses. Nevertheless, it was observed in the tertiary treatment that the membrane filtration process consists of several drawbacks, viz. membrane fouling and high energy usage. Also, the presence of refractory pollutants ($BOD_5$: COD < 0.2) in most of the industrial effluent makes the biological treatment inefficient, as the process is commonly used for wastewater having $BOD_5$: COD > 0.4 (Malik et al. 2017). Besides, adsorption is a pH-dependent process that necessitates a long treatment period in which the

P. P. Das (✉) · S. Dhara · M. K. Purkait
Department of Chemical Engineering, Indian Institute of Technology Guwahati, Guwahati, Assam, India
e-mail: pranj176107027@iitg.ac.in

 
M. P. Shah (ed.), *Sustainable Industrial Wastewater Treatment and Pollution Control*,
https://doi.org/10.1007/978-981-99-2560-5_6

adsorbents rapidly lose their potential as the number of cycle increases. Also, adsorbent regeneration necessitates the use of steam or vacuum. Thus, a more reliable, highly effective, and scalable technology which is also eco-friendly, sustainable, and cost-effective with high capacity throughput is required. Both electrocoagulation (EC) and ozonation ($O_3$) techniques have gained significant attention for their ability to drastically reduce the pollutant load from different industrial wastewater. The reason may be attributed to their ease of operation, low operating cost, rapid reaction rate, simple equipment design, strong oxidative ability, and low energy consumption. Nevertheless, it is well known that the individual technique has several constraints for the effective treatment of real industrial wastewater owing to its complex nature (Das et al. 2021a, 2021b). In this context, the focus is placed on the integration of electrocoagulation (EC) and ozonation ($O_3$) techniques with other water treatment processes to make it ideal for wastewater treatment, sustainable water desalination, and water reuse. Hybrid $O_3$ and EC are promising wastewater treatment technologies because it can minimize both the operational cost and treatment time while increasing the rate of mineralization and degradation of refractory organic compounds in the effluent. However, in order to improve the degradation efficiency and minimize the energy usage, integration of both $O_3$ and EC techniques with other treatment processes must be carefully chosen.

This chapter evaluates the efficacy of combining EC and $O_3$ with different water treatment processes, viz. electrodialysis (EC-ED), membrane distillation (EC-MD), biofiltration (EC-BF), membrane bioreactor (EC-MBR), and activated carbon adsorption (EC-AC), as well as sonolysis ($O_3$-US), hydrogen peroxide ($O_3/H_2O_2$), catalysts, and UV radiation ($O_3$/UV) for the mineralization and degradation of recalcitrant contaminants present in industrial effluents. The degrading efficiency and removal mechanism of the integrated processes are also discussed and summarized in this chapter, with an emphasis on improvements, drawbacks, future aspects, and a focus on sustainability and treatment of various industrial effluents.

## 2 Treatment Techniques for Wastewater Remediation

### 2.1 *Electrocoagulation Process (EC)*

Electrocoagulation as a water remediation process has attracted a lot of interest in pollutant removal from various effluents (domestic and industrial). It is a moderate-cost approach that uses polyelectrolytes or salt polymers to disrupt the stable emulsion and suspension. One of the critical processes that occur during electrocoagulation is the reduction of cathodic pollutants in water followed by electro-flotation. The electrodes play an important role in lifting the coagulated particles by the phenomenon of electro-flotation via generation of hydrogen or oxygen bubbles. Distance between electrodes, baseline pH, type of electrodes, applied current density, treatment duration, and electrolyte conductivity are the operational factors that impact the electrocoagulation process. Consequently, these operating parameters must be optimized in every electrocoagulation investigation. Furthermore, as compared to chemical coagulation, which uses additional chemicals, the electrocoagulation process produces less sludge (Rahman et al. 2021). In electrocoagulation, the electrolysis phenomenon is accountable for chemical and physiochemical phenomena. This means that electricity is necessary for the anodic breakdown of coagulants, the instability of particle and pollutant suspension, and the generation of flocs by aggregation of the destabilized phase. In the electrocoagulation chamber, ions move between the two electrodes due to the applied current density, with positive and negative ions moving to the cathode and anode surfaces, respectively. The electrocoagulation framework depends on the electrochemistry phenomena, in which oxidation or electron loss happens at the cathode and reduction or electron gain occurs at the anodic surface. During electrocoagulation, metal hydroxides, like $Fe(OH)_3$ or $Al(OH)_3$ (based on the utilized electrodes), are formed, which neutralizes the electrostatic charge existing between the colloidal particles, thereby causing coagulation or agglomeration of the suspended particles (Tahreen et al. 2020). The reactions are divided into three stages: (i) coagulant production through sacrificing anodes, (ii) emulsion dissolution and colloidal particle clustering, and (iii) adsorption or flocculation of coagulated metal hydroxide flocs. Electrocoagulation does not need additional chemicals since in situ coagulants are formed during the process. Three phenomena, viz. adsorption, coagulation, and flotation, are responsible for removing pollutants in the electrocoagulation method. The electrocoagulation chamber generates in situ coagulants as a result of anodic dissolution (Eq. 1), as well as the generation of $H_2$ gas at the cathode (Eq. 2) and hydroxyl ions at the anode (Eq. 3). These electrogenerated coagulants are responsible for producing a floc surrounded by metal hydroxides that act as an effective adsorbent (Nidheesh and Singh 2017).

$$\text{At anode: } M(s) \rightarrow M^{3+}_{aq} + ne^{-} \tag{1}$$

$$H_2O \rightarrow 4H^{+}_{aq} + O_2(g) + 4e^{-} \tag{2}$$

$$\text{At cathode: } nH_2O + ne^{-} \rightarrow \frac{n}{2}H_2(g)nOH^{-}_{aq} \tag{3}$$

## 2.2 *Ozonation Process ($O_3$)*

Ozonation is an intriguing and efficient water treatment approach since it generates no sludge and then dissipates the leftover ozone into $O_2$ and $H_2O$. With an oxidation potential of 2.07 V, ozone is regarded as a potent oxidizing agent. It is composed of one strong double bond and one weak single bond and interacts with the organic molecules present in water either directly through electrophiles or indirectly via an in situ-produced hydroxyl radical ($OH^{\cdot}$). The primary oxidants in aqueous solution under acidic conditions are molecular ozone and under alkaline conditions are hydroxyl radicals. The ozone formation is built on the premise of corona discharge technology that discharges extremely high voltage within a dried/cooled gaseous phase of oxygen. A rise in the ozone level of the solution enhances the oxidation rate for various organic molecules. However, it has no significant influence on the degradation of a few pollutants based on their molecular structure (Malik et al. 2020). The interaction of ozone with pollutants might be the reason as per the following categories: (i) The oxidation–reduction reaction of ozone with either superoxide or hydroperoxyl radicals is usually mediated by electron transfer, (ii) a cycloaddition process forms an ozonide structure when ozone interacts with organic molecules, (iii) ozone's electrophilic assault on the nucleophilic sites and groups in organic and aromatic compounds, such as –Cl, –OH, and $-NO_2^-$, has a significant influence on ozone's reactivity with aromatic rings, and (iv) ozone has nucleophilic properties, which means that compounds with carbonyl bonds undergo nucleophilic reactions (Abdurahman and Abdullah 2020). The various physicochemical characteristics of ozone gas are shown in Table 1.

Hypochlorite (0.9 V), oxygen (1.2 V), chlorine gas (1.4 V), chlorine dioxide (1.5 V), potassium permanganate (1.7 V), hydrogen peroxide (1.8 V), molecular ozone (2.1 V), and hydroxyl radical (2.8 V) have the most significant oxidation potential amid the many oxidizers utilized for water remediation. In water, ozone degradation is accompanied by a chain process that includes initiation, propagation, and termination. The following are the reactions that occur in the aqueous phase during ozonation (Das et al. 2021b; Guerra-Rodríguez et al. 2018):

**Table 1** Physicochemical properties of ozone (Obtained with permission from Abdurahman and Abdullah [2020] © Elsevier)

| Property | Value |
|---|---|
| Melting poling (°C) | −252 |
| Boiling point (°C) | −112 |
| Critical pressure (atm) | 54.62 |
| Critical temperature (°C) | −12.1 |
| Specific high gravity (g/cm$^3$) | 1.7 |
| Critical density (kg/m$^3$) | 436 |
| Heat of vaporization (cal/mol) | 2980 |
| Heat of formation (cal/mol) | 33,880 |
| Free energy of formation (cal/mol) | 38,860 |
| Oxidative potential (V) | 2.07 |

$$O_3 + 2H^+ + 2e^- \rightarrow O_2 + H_2O \quad (4)$$

$$O_3 + H_2O \rightarrow 2OH^{\cdot} + O_2 \quad (5)$$

$$O_3 + OH^- \rightarrow O_2^{\cdot-} + HO_2^{\cdot} \quad (6)$$

$$O_3 + OH^{\cdot} \rightarrow O_2^{\cdot-} + HO_2^{\cdot} \quad (7)$$

$$O_2 + HO_2^{\cdot} \rightleftharpoons 2O_2^{\cdot-} + H^+ \quad (8)$$

$$O_3 + HO_2^{\cdot} \rightarrow 2O_2 + HO^{\cdot} \quad (9)$$

$$2HO_2^{\cdot} \rightarrow O_2 + H_2O_2 \quad (10)$$

# 3 Hybrid Electrocoagulation Process for Wastewater Treatment

## *3.1 Electrocoagulation—Membrane Distillation (EC-MD)*

Membrane distillation (MD) is a separating technique that requires passing vapor across a permeable hydrophobic membrane. Direct contact membrane distillation (DCMD) is perhaps the utmost popular technique. The permeate water is considered as the best quality and appropriate for a variety of applications due to its high rate of salt rejection. Membrane wetting can be described as the water permeation through membrane pores, which causes a decline in permeate water quality, thereby affecting the MD process and resulting in membrane fouling. In most cases, many surfactants or pollutants with minimal surface tension in the inlet water promote membrane wetting. For surfactants, dissolved organic substances, and different contaminants with low surface tension, electrocoagulation has high removal efficiency. As a result, adopting EC as a pre-treatment procedure may minimize membrane wetting, potentially improving the treatment capacity (Deshmukh et al. 2018; Warsinger et al. 2015). Sardari et al. (2018) studied the effectiveness of a coupled EC-MD system for extremely salty hydraulic fracturing generated water (HPFW). The EC system is comprised of 5 aluminum electrodes set apart 0.5 cm from each other and coupled to a DC source. The influence of applied power on rates of removal was seen when the response time was set to 0.5 min and the pH was set to 6.4. As the current density was increased, the removal rate also increases. With a power intake of 1.41 kW h/m$^3$, the removal efficiency was found to be 96% for turbidity, 91% for TSS, and 61%

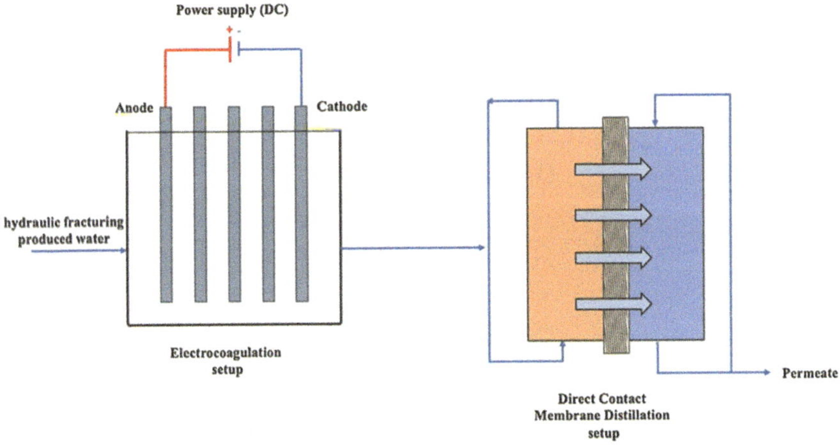


**Fig. 1** Schematic diagram for hybrid electrocoagulation–membrane distillation process (Reproduced with permission from Almukdad et al. [2021] © Elsevier)

for TOC. Before employing the MD process, the treated water was sedimented for 6 h. Ethylene chlorotrifluoroethylene (ECTFE) membrane was used during the MD procedure. After accumulating 2.45 L of permeate, the water flux rose from 18 L/m$^2$h to 70 L/m$^2$h at a flow velocity of 16.7 cm/s at temperatures of 50 °C and 70 °C, respectively. Moreover, at an FS temperature of 60 °C and flow velocities of 5.5 cm/s and 16.7 cm/s, the water flux rose from 24 L/m$^2$h to 47 L/m$^2$h. After collecting 2.45 L of permeate, it was observed that utilizing EC increased the water flow by about 50%. Recovery of water from the HFPW sample with 135 g/L dissolved solids was up to 57%. Long-term testing was also carried out with a flow velocity of 9.3 cm/s and an FS temperature of 60 °C, resulting in water flux of 30 L/m$^2$h after 434 h. When the EC-MD model was used to treat HFPW, the outcome was quite encouraging compared to the MD step alone. If properly fabricated, the EC pre-treatment process has a better opportunity of reducing the wetting and fouling problems throughout the MD process. However, very limited study has been conducted; additional exploration is needed to investigate the coupling of MD and EC techniques for wastewater remediation (Almukdad et al. 2021). The schematic representation of hybrid electrocoagulation–membrane distillation process is shown in Fig. 1.

## *3.2 Electrocoagulation—Membrane Bioreactor (EC-MBR)*

A membrane bioreactor (MBR) is a device that combines a membrane procedure, such as microfiltration, with biological wastewater treatment. Membrane fouling affects MBRs in three ways: organic, inorganic, and biofouling. Anions and cations such as $SO_4^{2-}$, $PO_4^{3-}$, $CO_3^{2-}$, $OH^-$ $Ca^{2+}$, $Mg^{2+}$, $Fe^{3+}$, and $Al^{3+}$ induce inorganic fouling. Membrane fouling can be prevented if the EC method is employed as a pre-treatment for MBR (Iorhemen et al. 2016; Ibeid et al. 2013). Al-Malack and Al-Nowaiser (2018) assessed the use of a coupled EC-MBR technique to treat hyper-saline oilfield-generated water. The study's feed solution was synthetically manufactured water. Two electrodes, either stainless steel or aluminum, were coupled to a DC energy source in the EC system. The effect of electrode material, current density, and electrolysis time was investigated. It was observed that an increase in the electrolysis time and current density significantly improves the COD and turbidity removal efficiency. The best removal efficiency was attained with 120 min of electrolysis, aluminum electrodes, and a current density of 30 $mA/cm^2$. COD was removed at a rate of 65%, while turbidity was removed at a rate of 52%. The bioreactor membrane was also submitted to testing utilizing artificially generated water as part of the EC process to identify the optimal conditions for that membrane. The optimal permeate flow was reported to be 12 $L/m^2h$, and as permeate flux rises, so does trans-membrane pressure till it touches the highest permissible pressure intended for that specific membrane. After a 90-day acclimatization period, the EC-MBR procedure began when 5,000 mg/L concentration was achieved by the volatile mixed liquor suspended solids (MLVSS). Further experiments were carried out in the EC-MBR with three phases varying oil contents of 100, 150, and 200 mg/L. After 60 days of operation, the COD removal efficiency was 97% with 48 mg/L residues, and the O&G exclusion efficiency was 95% with 5 mg/L residues via 100 mg/L of oil in the mixed setup. When oil concentration was raised to 200 mg/L, the O&G removal efficiency and COD were lowered to 80% with 40 mg/L residues and 91% with 162 mg/L residues, respectively. Additionally, it was claimed that using EC as a pre-treatment for the MBR method reduced fouling. Bani-Melhem and Smith (2012) studied the effect of treating greywater by an EC method and a submerged membrane bioreactor (SMBR). The dual setup was used to conduct the investigation, which lasted 24 days. The findings revealed that adopting the EC method decreased SMBR membrane fouling by 13%. The COD, turbidity, and color removal efficiency of the SMBR and hybrid system were almost identical. For example, using EC-SMBR, color, COD, and turbidity decreased by 93.7%, 88.9%, and 97.9%, respectively. SMBR (97.4%) had a 20% better ammonia nitrogen removal efficiency than EC-SMBR (77.8%). It was claimed that using less current and using electrodes made of iron might improve the hybrid system's removal effectiveness even further. The EC procedure used a DC power source (12 V), two aluminum electrodes positioned at 5.8 cm apart, and a 15 min of electrolysis time. In both setups, the transmembrane pressure was kept constant at 7.5 kPa. EC-SMBR and SMBR produced starting fluxes of 29 $L/m^2h$ and 27.5 $L/m^2h$, respectively. A 24-h brief trial was also carried out. The brief trial revealed that the EC-SMBR method enhanced membrane filtration capacity by 20% compared to the SMBR procedure. The schematic representation of hybrid electrocoagulation–membrane bioreactor process is shown in Fig. 2.

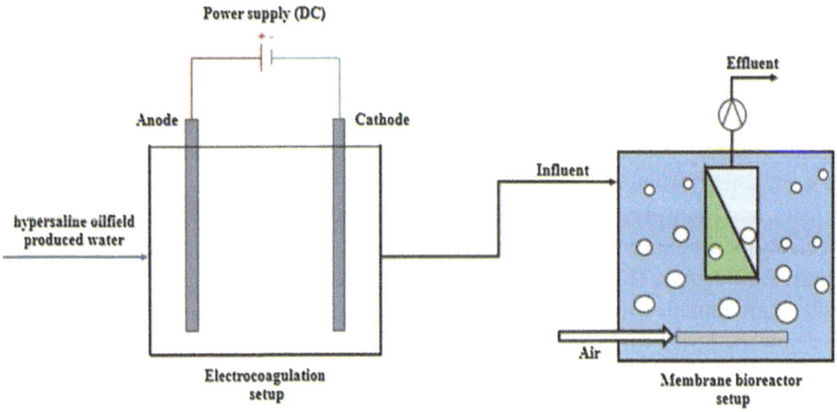


**Fig. 2** Schematic diagram for hybrid electrocoagulation–membrane bioreactor process (Reproduced with permission from Almukdad et al. [2021] © Elsevier)

## 3.3 Electrocoagulation—Electrodialysis (EC-ED)

Electrodialysis (ED) is a membrane technique that induces ions across the membrane using an electrical current rather than pressure. The system is a good way to purify water because of its maximum selectivity, better product recovery ratio, and capacity to eliminate the bulk of the elements in the water sample. However, one or two additional devices may be required to ensure effective wastewater processing (Kumar and Pan 2020; Zhang et al. 2021). To eliminate color, COD, $NH_3$-N, and Cr, from tannery effluent, Deghles and Kurt (2016) studied the efficacy of integrating an EC setup with an electrodialysis technique. The EC setup comprised of 5 pairs of electrodes in monopolar parallel (MP–P) configuration with a 0.7 cm gap between them and coupled to a DC power source. When the starting pH of 6 and 7 for aluminum and iron electrodes was established, respectively, the influence of electrode material (aluminum and iron) was investigated. The main detected factors in the EC procedure were electrolysis time and current density. At the optimal condition, the reported current density was 14 $mA/cm^2$ and the electrolysis time was 125 min. With a power intake of 6.75 $kWh/m^3$, the removal efficiency was observed to be 51% for $NH_3$-N, 73% for COD, 94% for color, and 100% for Cr. As a post-treatment for the EC process, bipolar membrane electrodialysis was used. The induced voltage was held consistent at 24 V. The electrodes were washed with 0.01 M HCl for all studies. Platinum-coated titanium electrodes were spaced by 0.05 cm in the anode and cathode. As aluminum electrodes were employed in the EC setup, the power intake of electrodialysis for 45 min was 14 $kWh/m^3$ at a current density of 190 $mA/cm^2$. The removal efficiencies of 92% for COD, 100% for Cr, 100% for NH3-N, 100% for ammonium, and 100% for color were achieved utilizing this blended system. According to the data, the hybridization of the EC process with the ED technique boosted the elimination effectiveness of the various contaminants contained in the effluent stream

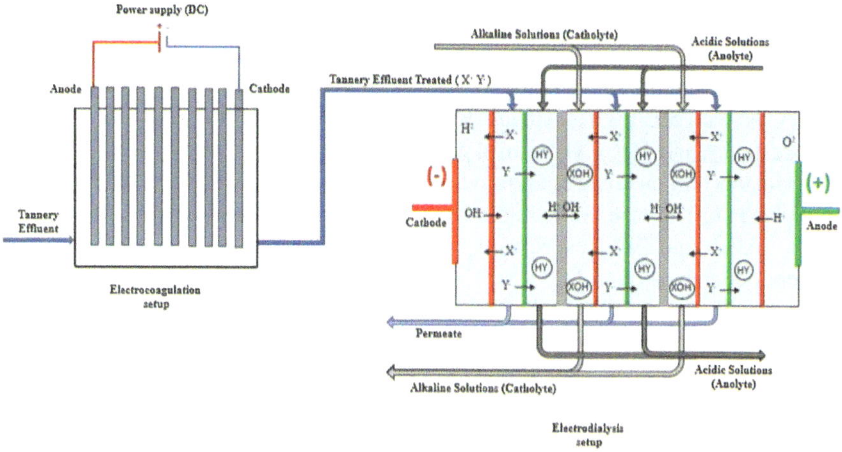


**Fig. 3** Schematic diagram for hybrid electrocoagulation–electrodialysis process (Reproduced with permission from Almukdad et al. [2021] © Elsevier)

while also drastically reducing the amount of waste emitted (Almukdad et al. 2021). The schematic representation of hybrid electrocoagulation–electrodialysis process is shown in Fig. 3.

## *3.4 Electrocoagulation—Biofiltration (EC-BF)*

Biofiltration has long been utilized to remediate landfill leachate because of its ease of use, cheap treatment cost, and great efficacy in removing biodegradable organic compounds and nitrogen. The method works by immobilizing bacteria on a medium that gives them a large surface area to absorb and biologically break down contaminants in wastewater (Kurniawan et al. 2006; Le et al. 2021). Dia et al. (2018) investigated the effect of adopting the EC method as a pre-treatment for biofiltration (BF) on landfill leachate treatment. The removal rates of COD, phosphate, $NH_4$, and metals were used to assess the efficacy of combining these two treatments. The EC setup employed a hollow stainless steel cylindrical rod as cathode and a complete cylindrical aluminum rod as the anode. Both were spaced at 1.55 cm and coupled to a DC electric source. At a baseline pH of 7.83, a current density of 8 mA/cm$^2$, and a reaction period of 20 min, the greatest COD elimination was 82%. An aerated column containing a combination of wood and peat scraps was employed to carry out the biofiltration process. Aeration is a sludge processing method that involves blowing the tank with air to encourage bacterial development in the effluent. The microorganisms subsequently consume organic matter, which causes the generated flocs to precipitate at the base of the container (Huang et al. 2020). Landfill leachate pre-treatment via EC method boosted the COD removal by 37%, TOC reduction by

15%, turbidity removal by 82%, color removal by 60%, phosphorus removal by 82%, zinc removal by 95%, and iron removal by 95%. The EC technique was shown to be ineffective in removing ammonia, with a clearance rate of just 6%. COD removal was 63.2% with 595 mg/L residues, $BOD_5$ removal was greater than 97% with five mg/L residues, and TOCHyl removal was 49% utilizing the hybrid system. Inorganic contaminants and metals were removed at a rate greater than 99%. Turbidity, color, zinc, phosphorous, iron, and aluminum removal rates were 95.5%, 47.5%, 69.9%, 78.9%, 96.5%, and 39%, respectively. The total price of operating the hybrid process was reported to be around 1.23 US/$m^3$ (Dia et al. 2017). The effectiveness of coupled biofiltration-EC on sanitary landfill leachate treatment was investigated by Oumar et al. (2016). A PVC column containing wood and peat mixes was utilized in the biofiltration system. For a 240-day filtering period, the airflow rate was fixed at 5 L/min. BF as a pre-treatment demonstrated favorable outcomes concerning phosphorous, ammonia, $BOD_5$, and turbidity with a removal rate of >98% (0.05 mg/L residues), 94% (20.5 mg/L residues), 94% (4.2 mg/L residues), and 95% (5.35 NTU residues), respectively. With an average COD removal rate of just 13%, the COD removal rate was shallow. When electrocoagulation was used as a post-treatment method with a cylindrical magnesium anode and a hollow stainless steel cathode for 30 min, the residual COD removal was decreased to 53%, and the color was reduced by 85%. Overall, the hybrid EC-BF method provides a cost-effective approach for landfill leachate treatment. The complementary characteristics of the two processes improve the removal rate of various contaminants, although additional study is needed to prevent electrode fouling from raw landfill leachate. The schematic representation of hybrid electrocoagulation–biofiltration process is shown in Fig. 4.

## 4 Hybrid Ozonation Process for Wastewater Treatment

### 4.1 Ozone—Hydrogen Peroxide ($O_3/H_2O_2$)

The peroxone process is defined as the reaction between $O_3$ and $H_2O_2$. It is one of the most extensively utilized hybrid AOPs for contaminant breakdown because it produces non-selective $OH^{\cdot}$ radicals at a reasonable cost. The interaction of ozone with peroxide anion ($HO_2^-$) during the peroxone process leads to the creation of $OH^{\cdot}$ precursors, which ultimately react with $OH^{\cdot}$. However, the remaining $H_2O_2$ in the treated water must be removed before being discharged into the atmosphere. The optimal molar ratio and standard ozone dosage ranges for the hybrid process are 0.5 mol/mol and 1–20 mg/$m^2$, respectively. The ozone and water matrix interaction can produce a tiny quantity of peroxide. Due to the ability of $H_2O_2$ to considerably boost the ozone breakdown, the peroxone process showed improved degradation of organic contaminants resulting in the generation of $OH^{\cdot}$ radicals through the electron transfer mechanism. Even when present in tiny amounts with ozone, $H_2O_2$ has a substantial contaminant breakdown capacity. Hydrogen peroxide speeds up

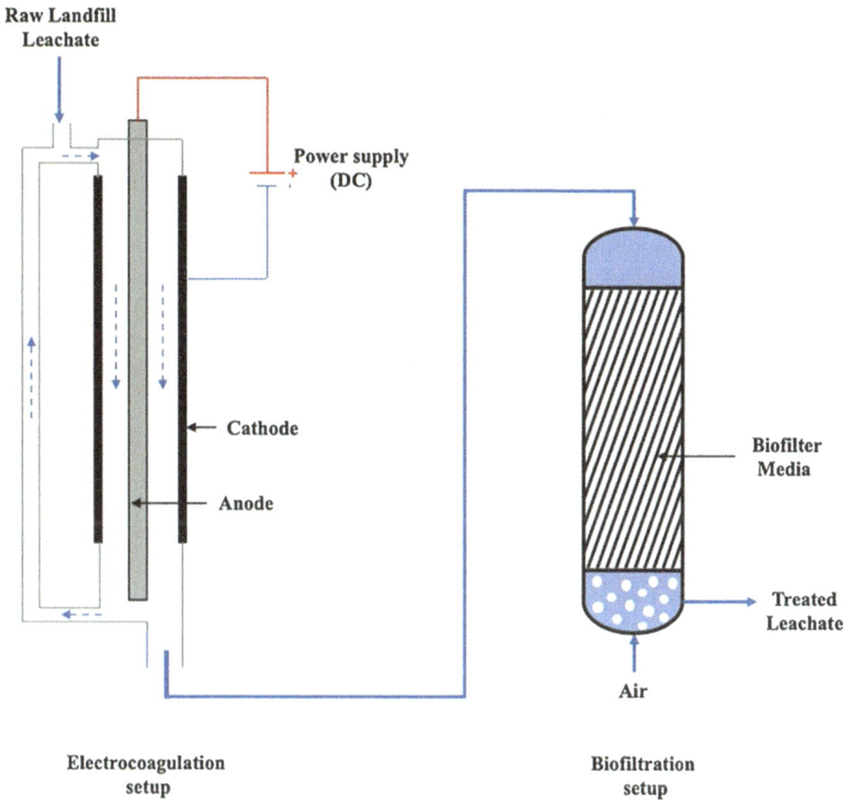


**Fig. 4** Schematic diagram for hybrid electrocoagulation–biofiltration process (Reproduced with permission from Almukdad et al. [2021] © Elsevier)

the process and enhances the grade of the wastewater, but leftover $H_2O_2$ is toxic to microorganisms and the surroundings (Malik et al. 2020; Fu et al. 2019). As a result, combining the ozonation step with a modest quantity of $H_2O_2$ may boost the efficiency levels by producing hydroxyl free radicals without toxic impact on the tertiary or biological remediation techniques. The coupled $O_3/H_2O_2$ process has a more effective reaction rate and method efficacy than an ozonation system on its own. However, the pH and $H_2O_2/O_3$ ratio of the remediation procedure should be appropriately considered for the optimum result of the coupled framework, as too much $H_2O_2$ begins to scavenge the hydroxyl radicals, lowering the procedure effectiveness. While the reaction rate and effectiveness of the peroxone method are significantly higher than those of the standalone ozonation techniques, the pH and $H_2O_2/O_3$ ratio of the coupled system must be carefully regulated, as an excess of $H_2O_2$ in the solution can excavate the $OH^{\cdot}$ radicals, leading to decreased process performance (Chen and Wang 2021). Lee et al. (2016, 2017) observed more than 95%

degradation of tetrachloro-1,3-butadienes (TCBDs) and 40% degradation of hexa-chloro-1,3-butadiene (HCBD) at ozone dosage of 0.5 mg/L and 4 mg/L, respectively. At the same time, $H_2O_2$ induces low abatement efficiency of the identified substances owing to the reduced ozone reactivity, caused by the quick conversion of ozone to hydroxyl ion by hydrogen peroxide. It is because the added $H_2O_2$ scavenging hydroxyl radicals moreover led to an increment of 8% in the scavenging rate of hydroxyl radical (($s^{-1}$) for $H_2O_2/O_3 = 2.0$ (w/w) and $O_3 = 2.0$ mg/L). However, when $H_2O_2$ was added more than 0.3 wt $H_2O_2$/wt $O_3$, the peroxone method was shown to be more successful in removing color than the ozone method alone. The optimal setting for color removal with ultrafiltration water was 0.7 wt $H_2O_2$/wt $O_3$ ratio due to OH's scavenging action, which was absorbed by elevated $H_2O_2$ levels. Jiao et al. (2016) used the peroxone procedure to treat nitrobenzene (NB) polluted wastewater. They observed that the effectiveness of NB degradation immediately rises with an increase in $H_2O_2$ concentration but after that decreases. The oxidative disintegration of organic molecules by the extremely reactive $OH^{\cdot}$ ions was entirely responsible for the fall in NB levels during the peroxone procedure. Shokri et al. (2016), on the other hand, demonstrated that increasing the preliminary intake of $H_2O_2$ enhanced the efficiency of OT (ortho-toluidine) degradation to a plateau, where adding more $H_2O_2$ had no impact since hydroxyl radicals recombine with extra portions of $H_2O_2$.

## 4.2 *Ozone—UV Radiation ($O_3$/UV)*

Ozone photolysis initiates the hybrid AOP, which consists of $O_3$ and UV light. The photodecomposition of ozone (at $\lambda < 310$ nm) generates $H_2O_2$ and $OH^{\cdot}$ radicals. UV light causes dissolved ozone to break, followed by an interaction between atomic oxygen and water to generate a thermally stimulated $H_2O_2$. The $H_2O_2$ is subsequently broken down into two $OH^{\cdot}$ radicals. Due to cage recombination, only a tiny quantity of $H_2O_2$ breaks down into $OH^{\cdot}$ radicals, resulting in a quantum yield of just 0.1. In addition, $H_2O_2$ experiences an acid–base equilibrium process, resulting in the formation of one $H_3O^+$ and one peroxyl anion ($HO_2^-$). Both $H_2O_2$ and ozone react with the hydroxyl radical in a chain propagation reaction to create the superoxide radicals ($O_2^-$) and peroxyl ($HO_2^-$) radicals with rate constants of $2.7 \times 10^7$ $M^{-1}$ $s^{-1}$ and $1.1 \times 10^8$ $M^{-1}$ $s^{-1}$, respectively. Subsequently, the superoxide radical combines with the hydroxyl radical to produce the hydroxyl ion $OH^-$. During the propagation process at basic pH with a rate constant of $9.4 \times 10^7$ $M^{-1}$ $s^{-1}$, the superoxide radical combines with ozone again to generate $O_3^-$ ($k = 1.6 \times 10^9$ $M^{-1}$ $s^{-1}$), which then releases hydroxyl ion and hydroxyl radical. Ozone combines with peroxyl ion in a chain reaction, producing superoxide and hydroxyl radicals ($k = 5.5 \times 10^6$ $M^{-1}$ $s^{-1}$). In a conclusion step, peroxyl, hydroxyl, and superoxide radicals recombine to create $H_2O_2$, which may again be subjected to photolysis (Malik et al. 2020). The scavenging process takes over when there is an excess of $H_2O_2$. The molar extinction coefficient ($\varepsilon$) of ozone at $\lambda = 254$ nm is 3300 $M^{-1}$ $cm^{-1}$, much greater than $H_2O_2$ at the same wavelength. Furthermore, the direct oxidation of organic micropollutants by the hybrid $O_3$/UV process may span a broad range of reactivity, which adds to the major advantage of this combination method (Gorito et al. 2021). In this context, to remove methyl methacrylate (MMA) from semiconductor effluent, Shang et al. (2007) used a hybrid $O_3$/UV approach. For MMA and COD, the research found a reduced efficiency of 96% and 22% after 30 min of therapy, respectively. After a 120-min response period, the COD degradation efficiency increases to 51%. The reason for this phenomenon is that, UV light causes the breakdown of dissolved ozone in the solution with simultaneous production of $OH^{\cdot}$ radicals. As a result of the increased reactivity of $OH^{\cdot}$ radicals toward organic molecules, mineralization efficiency improves dramatically. Furthermore, Pillai et al. (2009) treated terephthalic acid-contaminated wastewater via $O_3$/UV process. They found a slight decline in the solution pH from 8.5 to 7, which resulted from the implementation of $O_3$/UV process, contrary to the standalone ozonation system, thereby leading to a high acidic pH of the treated solution (pH 4). Guzmán et al. (2016), on the other hand, observed a decolorization efficiency of 68.4% at pH 7 after 60 min of reaction time while treating citrus wastewater. Nonetheless, the pH of the treated solution during the hybrid $O_3$/UV process was well within the desirable pH range of 6–9 in both instances, indicating appropriate involvement in the radical production pathway.

## *4.3 Ozone-Sonolysis ($O_3$-US)*

The sonolytic ozonation procedure is a mix of ultrasonic remediation and ozonation. Compared to a single ozonation method, combining sonolytic treatment with ozonation improves the extraction efficiency of resistive contaminants in wastewater. The cause for this is an increase in ozone disintegration productivity due to the system's bubble bursting, resulting in a considerable generation of $OH^{\cdot}$ radicals. Due to the ultrasonic dispersion created by the sonication wave, the breakdown of ozone in the solution is also accelerated, resulting in a synergistic impact in terms of removal efficiency. To create high-intensity bubbles, an ultrasonication process with a frequency range from 20 to 100 kHz is used, which is assisted through high- and low-pressure occurrence in alternate cycles (Anandan et al. 2020). Moreover, substantial shearing effects are created in the aqueous phase, resulting in efficient oxidation of the organic compounds present in the effluent because of the simultaneous formation of $H^{\cdot}$ and $OH^{\cdot}$ radicals. Also, the location of the ultrasonic transducer (either an ultrasonic bath or an ultrasonic probe) has a substantial impact on the batch sonolytic ozonation system (Zhang et al. 2006; Song et al. 2007). He et al. (2007) used a combination of ultrasonic treatment and ozonation to mineralize p-aminophenol (PAP) in the aqueous phase. The research found that increasing the treatment duration from 10 to 30 min increased the mineralization efficiency of PAP from 88% to 99%. Furthermore, Ziylan-Yavaş and Ince (2018) evaluated the effect of sonication on the catalytic ozonation treatment of paracetamol. Pt-supported $Al_2O_3$ nanocomposites were used as catalyst during the process. It was reported that addition of ultrasonic therapy to the procedure improved the mineralization rate significantly. The primary rationales for the hybrid process enhanced mineralization performance are the transport of reaction sites from the bulk solution to the solid surface, the formation of gassy hollow foams due to increased hydrophobicity, enhanced ozone decomposition in collapsing bubbles, and a high ozone mass transfer rate. Additionally, combining the sonication and ozonation processes helps to avoid the formation of harmful intermediary by-products during the ozonation step. Barbier and Pétrier (1996) found that the dynamic degradation pathway of 4-nitrophenol (4-NP) is a quick preferential reaction with molecular $O_3$ by OZ alone at pH 2, whereas combining US/OZ hinders 4-NP degradation at 20 kHz by ingesting a sizable portion of the $O_3$ transmitted to solution in the cavitation bubbles. The deterioration of 4-NP in the 500 kHz US/OZ setup was quicker than in the 20 kHz US/OZ and OZ separately. This means at 500 kHz, additional OH was created, and the disintegration rate was enhanced by both produced OH and molecular $O_3$.

## 4.4 *Ozone-Catalysts*

The catalytic ozonation process is a mix of catalyst and ozonation. It may be divided into two types: homogeneous and heterogeneous catalytic ozonation. Solid catalysts and transition metal ions are utilized extensively to speed up the ozone breakdown and $OH^{\cdot}$ radical formation. The two-phase catalytic cycle in homogeneous catalytic ozonation is (i) ozone breakdown via metal ions, which produces $OH^{\cdot}$ radicals, and (ii) the formation of complexes between organic molecules and the catalyst followed by its oxidation (Gracia et al. 1996; Beltrán et al. 2005). Cu(II), Mn(II), Fe(II), Fe(III), Cr(III), Co(II), and Zn(II) are some of the most widely used metal ions work as a powerful catalyst during the homogeneous catalytic ozonation process. Because metal ions act as an excellent catalyst, the generation of $OH^{\cdot}$ radicals rises dramatically, increasing the effectiveness of the ozonation method. Furthermore, a cheap metal ion like Fe(II) results in a cost-effective treatment, and as such, it may be considered the most favorable homogeneous metal catalyst (Pi et al. 2003). Pillai et al. (2009) used $Fe^{2+}$ dosages of 0.25, 0.50, and 0.75 mM to accelerate the ozonation process, attaining COD removal efficiencies of 73%, 78%, and 81%, respectively, after a reaction time of 240 min. Malik et al. (2019) used a catalytic ozonation technique to clean pharmaceutical wastewater. According to the study, the addition of $Fe^{2+}$ to the ozonation process enhanced the biodegradability index (BI) of the wastewater by 3.5 times its original value. However, a large portion of $Fe^{2+}$ dosage might lead to a significant number of $OH^{\cdot}$ radicals being consumed, which leads to a poor performance characteristics. Although the metal ions utilized in the ozonation method have high catalytic activity, their removal from the treated solution is critical. In addition, owing to the low concentration of metal ions in the solution, their catalytic activity may be inhibited. Due to these constraints, the focus has been shifted to the heterogeneous catalytic ozonation process.

The heterogeneous catalytic ozonation process is mediated by applying metal oxides ($MnO_2$, $TiO_2$, $CeO_2$, $Al_2O_3$, and FeOOH) on supports (zeolites modified with metals, activated carbon, $SiO_2$, and $CeO_2$). These are the most often employed catalysts since they result in more complicated reaction routes and are based on the phenomena of multiple-phase transport. The process's success is dependent on operational factors such as the solution pH and the catalyst's surface characteristics. Moreover, it may be categorized into numerous variables, viz. activated carbon, photo-catalyzed ozonation, metal oxides, and metals on supports (Yuan et al. 2022; Psaltou et al. 2021). Bai et al. (2017) treated carbamazepine polluted municipal secondary wastewater by utilizing $Fe_3O_4$-$CeO_2$/MWCNTs as an effective catalyst. The study observed a degeneration efficiency of 50.3% and 64.1% in standalone and heterogeneous catalytic ozonation method, respectively, after a reaction time of 4 min. The findings reveal that the integration of catalyst throughout the procedure stimulated the produced $OH^{\cdot}$ radicals to boost the elimination rate significantly. Thus, the amount of catalyst provided has an integral part in promoting the $OH^{\cdot}$ radical generation during the heterogeneous catalytic ozonation process. Moreover, during the use of catalytic ozonation method for tannery wastewater treatment, Huang et al. (2016) showed that an increase in Mn-Cu/$Al_2O_3$ catalyst loading significantly boosts the COD removal efficiency. The reason may be linked to the overall increase in the active surface area concerning the rise in the catalyst amount. Thus, the presence of supplemental active sites on the surface of catalysts increases the creation rate of $OH^{\cdot}$ radicals/units of ozone conversion. The study also showed that an increase in the quantity of catalysts substantially assists in the breakdown of organic pollutants. As such, a detailed understanding of the use of different catalysts during ozonation is highly vital to build an optimal catalyst for the effective mineralization of the organic contents. Table 2 represents the application of hybrid ozonation system ($O_3/H_2O_2$, $O_3$/UV, $O_3$/catalyst and $O_3$/US) for the treatment of different wastewater pollutants.

## 5 Conclusion and Future Perspectives

In recent years, there has been a growing interest in the utilization of hybrid EC and $O_3$ techniques for water and wastewater treatment. It was observed that combining the EC and $O_3$ processes with other treatment techniques might improve the degradation of refractory contaminants while lowering the total energy usage of the integrated process. This review chapter has demonstrated the principles, benefits, and limitations of hybrid EC and $O_3$ processes along with the potential of some of the most successful applications of these combined processes in the removal of inorganic and organic contaminants, disinfection and decolorization of various industrial effluents. The use of EC as a pre-treatment process offers several benefits, such as membrane fouling mitigation, high permeate water quality and low operating cost of the overall process. Moreover, the enhanced hydroxyl radical production caused by the combined effect of the hybrid $O_3$ process allows for excellent contaminant removal across a wide pH range, and the substantially improved ozone utilization associated with efficient

Table 2 Application of hybrid ozonation process for wastewater treatment

| Processes | Pollutants | Experimental conditions | Performance/Remarks | Reference |
|---|---|---|---|---|
| $O_3/H_2O_2$ | Ortho-Toluidine (OT) | Sample volume = 2 L<br>COD = 100 mg/L<br>$H_2O_2$ dose = 40 mM<br>Ozone dose = 2.5 g/h<br>Temperature = 25 °C<br>Reaction time = 120 min<br>pH = 9 | 65.5% of COD and 89.5% OT removed after 120 min and 40 min of reaction time respectively | (Shokri et al. 2016) |
| | Sulfamethoxazole (SMX) | Sample volume = 2 L<br>Ozone dose = 20 g $O_3/m^3$<br>COD = 287 mg/L<br>TOC = 116.3 mg/L<br>Temperature = 25 °C<br>Reaction time = 120 min<br>pH = 7.5 | 76% COD and 32% TOC depletion were reached when ozone was aided by 130 mg/L (0.46 mg $H_2O_2$/mg $O_3$) of added peroxide | (Martins et al. 2015) |
| $O_3$/UV | Disinfection by-product precursors (DBP) | Sample volume = 6.2 L<br>TOC = 1.8 mg/L<br>Ozone dose = 12 mg/min<br>UV lamp = $9.7 \times 10^{-3}$ W/cm$^2$<br>Temperature = 25 °C<br>Reaction time = 60 min<br>pH = 6.6 | 50% of TOC was removed after 60 min with reduced trihalomethanes and haloacetic acids formation potentials by around 80% and 70% respectively | (Chin and Bérubé 2005) |

(continued)

**Table 2** (continued)

| Processes | Pollutants | Experimental conditions | Performance/Remarks | Reference |
|---|---|---|---|---|
| | Azo dyes | Sample Volume = 500 $dm^3$<br>UV = 50 W $dm^{-3}$<br>Dye content = 20 mg/L<br>Ozone dose = 6 mg/min<br>Temperature = 20 °C<br>Reaction time = 11.5 min<br>pH = 5.8 | More than 95% of color can be removed in less than 11 min | (Shu and Chang 2005) |
| $O_3$/catalyst | 2,4-dichlorophenol (DCP) | Sample volume = 2 L<br>DCP content = 10.0 mg /L<br>Ozone dose = 8.4 mg /L<br>$Mn^{2+}$ catalyst = 0.1 mg/L<br>Temperature = 25 °C<br>Reaction time = 10 min<br>pH = 5.5 | 100% DCP degradation and more than 80% TOC removal | (Xiao et al. 2008) |
| | Bisphenol A (BPA) | Sample volume = 500 mL<br>BPA content = 10 mg/L<br>Ozone dose = 6 L/h<br>Ru/γ-$Al_2O_3$ = 200 mg/L<br>Temperature = 25 °C<br>Reaction time = 240 min<br>pH = 5.9 | 82% of TOC removal with 56% mineralization | (Cotman et al. 2016) |
| $O_3$/US | Azobenzene (AB) and methyl orange (MO) | Sample volume = 650 mL<br>AB and MO content = 10 μM<br>Frequency = 500 kHz<br>Power density = 2 W/$cm^2$<br>Ozone dose = 50 mL/min<br>Temperature = 15 °C<br>Reaction time = 40 min<br>pH = 5.5–6.5 | For both dyes, strong synergy was obtained for TOC removal: 20% for US, ~30% for $O_3$, and > 80% for $O_3$/US at various ozone doses | (Destaillats et al. 2000) |

(continued)

**Table 2** (continued)

| Processes | Pollutants | Experimental conditions | Performance/Remarks | Reference |
|---|---|---|---|---|
| | 4-Chloro 2-aminophenol (4C2AP) | Sample volume = 200 mL<br>4C2AP content = 20 mg/L<br>Frequency = 20 kHz<br>US power = 120 W<br>Ozone dose = 15 mg/h<br>Temperature = 30 °C<br>Reaction time = 20 min<br>pH = 6 | A synergy index of 1.82 was obtained by the combined $O_3$/US process. 99% of the compound was removed by US/$O_3$ process against 24.1% for US and 40.2% for $O_3$ alone | (Barik and Gogate 2016) |

mass transfer makes the overall process sustainable. The current applications and lab-scale studies on the kinetic reaction of the processes demonstrate the promising potential of these integrated techniques for industrial usage.

However, no study has been published on the actual amount of energy saved by the hybrid techniques against other traditional treatment processes. To obtain a perceptible comparison, studies investigating the total energy usage of each of these hybrid processes during the degradation of refractory pollutants should be conducted. Also, considerable attention should be given on the identification of scale-up parameters and reaction intermediates, criteria for cost effectiveness and development of rate expressions based on established reaction mechanisms. Further studies are needed to determine how the system parameters affect the individual and overall process efficiency and eventually the total operating cost. In addition, studies must be performed on reducing the electrode usage and overcoming the problem of electrode passivation. This might be accomplished by evaluating the potential of dielectrophilic enhanced electrodes, which are projected to lower the electrode usage while also improving the quality of the permeate water. Moreover, to simulate the commercial scenario, it is strongly advised to perform the studies at pilot scale in continuous flow operation.

## References


Abdurahman MH, Abdullah AZ (2020) Mechanism and reaction kinetic of hybrid ozonation-ultrasonication treatment for intensified degradation of emerging organic contaminants in water: a critical review. Chem Eng Process Process Intensif 154:108047

Al-Malack MH, Al-Nowaiser WK (2018) Treatment of synthetic hypersaline produced water employing electrocoagulation-membrane bioreactor (EC-MBR) process and halophilic bacteria. J Environ Chem Eng 6(2):2442–2453

Almukdad A, Hafiz M, Yasir AT, Alfahel R, Hawari AH (2021) Unlocking the application potential of electrocoagulation process through hybrid processes. J Water Process Eng 40:101956

Anandan S, Ponnusamy VK, Ashokkumar M (2020) A review on hybrid techniques for the degradation of organic pollutants in aqueous environment. Ultrason Sonochem 67:105130

Bai Z, Wang J, Yang Q (2017) Advanced treatment of municipal secondary effluent by catalytic ozonation using $Fe_3O_4$-$CeO_2$/MWCNTs as efficient catalyst. Environ Sci Pollut Res 24(10):9337–9349

Bani-Melhem K, Smith E (2012) Grey water treatment by a continuous process of an electrocoagulation unit and a submerged membrane bioreactor system. Chem Eng J 198:201–210

Barbier PF, Petrier C (1996) Study at 20 kHz and 500 kHz of the ultrasound-ozone advanced oxidation system: 4-nitrophenol degradation. J Adv Oxid Technol 1(2):154–159

Barik AJ, Gogate PR (2016) Degradation of 4-chloro 2-aminophenol using combined strategies based on ultrasound, photolysis and ozone. Ultrason Sonochem 28:90–99

Beltrán FJ, Rivas FJ, Montero-de-Espinosa R (2005) Iron type catalysts for the ozonation of oxalic acid in water. Water Res 39(15):3553–3564

Chen H, Wang J (2021) Degradation and mineralization of ofloxacin by ozonation and peroxone ($O_3$/$H_2O_2$) process. Chemosphere 269:128775

Chin A, Bérubé PR (2005) Removal of disinfection by-product precursors with ozone-UV advanced oxidation process. Water Res 39(10):2136–2144

Cotman M, Erjavec B, Djinović P, Pintar A (2016) Catalyst support materials for prominent mineralization of bisphenol A in catalytic ozonation process. Environ Sci Pollut Res 23(10):10223–10233

Das PP, Mondal P, Sinha A, Biswas P, Sarkar S, Purkait MK (2021a) Treatment of steel plant generated biological oxidation treated (BOT) wastewater by hybrid process. Sep Purif Technol 258:118013

Das PP, Anweshan A, Mondal P, Sinha A, Biswas P, Sarkar S, Purkait MK (2021b) Integrated ozonation assisted electrocoagulation process for the removal of cyanide from steel industry wastewater. Chemosphere 263:128370

Deghles A, Kurt U (2016) Treatment of tannery wastewater by a hybrid electrocoagulation/electrodialysis process. Chem Eng Process 104:43–50

Deshmukh A, Boo C, Karanikola V, Lin S, Straub AP, Tong T, Elimelech M (2018) Membrane distillation at the water-energy nexus: limits, opportunities, and challenges. Energy Environ Sci 11(5):1177–1196

Destaillats H, Colussi AJ, Joseph JM, Hoffmann MR (2000) Synergistic effects of sonolysis combined with ozonolysis for the oxidation of azobenzene and methyl orange. J Phys Chem A 104(39):8930–8935

Dia O, Drogui P, Buelna G, Dubé R (2017) Strategical approach to prevent ammonia formation during electrocoagulation of landfill leachate obtained from a biofiltration process. Sep Purif Technol 189:253–259

Dia O, Drogui P, Buelna G, Dubé R (2018) Hybrid process, electrocoagulation-biofiltration for landfill leachate treatment. Waste Manage 75:391–399

Ferrari B, Paxéus N, Giudice RL, Pollio A, Garric J (2003) Ecotoxicological impact of pharmaceuticals found in treated wastewaters: study of carbamazepine, clofibric acid, and diclofenac. Ecotoxicol Environ Saf 55(3):359–370

Fu L, Wu C, Zhou Y, Zuo J, Song G, Tan Y (2019) Ozonation reactivity characteristics of dissolved organic matter in secondary petrochemical wastewater by single ozone, ozone/$H_2O_2$, and ozone/catalyst. Chemosphere 233:34–43

Gorito AM, Pesqueira JF, Moreira NF, Ribeiro AR, Pereira MFR, Nunes OC, Silva AM (2021) Ozone-based water treatment ($O_3$, $O_3$/UV, $O_3/H_2O_2$) for removal of organic micropollutants, bacteria inactivation and regrowth prevention. J Environ Chem Eng 9(4):105315

Gracia R, Aragües JL, Ovelleiro JL (1996) Study of the catalytic ozonation of humic substances in water and their ozonation by-products. Ozone Sci Eng 18:195–208

Guerra-Rodríguez S, Rodríguez E, Singh DN, Rodríguez-Chueca J (2018) Assessment of sulfate radical-based advanced oxidation processes for water and wastewater treatment: a review. Water 10(12):1828

Guzmán J, Mosteo R, Sarasa J, Alba JA, Ovelleiro JL (2016) Evaluation of solar photo-Fenton and ozone based processes as citrus wastewater pre-treatments. Sep Purif Technol 164:155–162

He Z, Song S, Ying H, Xu L, Chen J (2007) p-aminophenol degradation by ozonation combined with sonolysis: operating conditions influence and mechanism. Ultrason Sonochem 14(5):568–574

Huang G, Pan F, Fan G, Liu G (2016) Application of heterogeneous catalytic ozonation as a tertiary treatment of effluent of biologically treated tannery wastewater. J Environ Sci Health Part A 51(8):626–633

Huang D, Yang L, Xu W, Chen Q, Ko JH, Xu Q (2020) Enhancement of the methane removal efficiency via aeration for biochar-amended landfill soil cover. Environ Pollut 263:114413

Ibeid S, Elektorowicz M, Oleszkiewicz JA (2013) Novel electrokinetic approach reduces membrane fouling. Water Res 47(16):6358–6366

Iorhemen OT, Hamza RA, Tay JH (2016) Membrane bioreactor (MBR) technology for wastewater treatment and reclamation: membrane fouling. Membranes 6(2):33

Jiao W, Yu L, Feng Z, Guo L, Wang Y, Liu Y (2016) Optimization of nitrobenzene wastewater treatment with $O_3/H_2O_2$ in a rotating packed bed using response surface methodology. Desalin Water Treat 57(42):19996–20004

Kumar A, Pan SY (2020) Opportunities and challenges of electrochemical water treatment integrated with renewable energy at the water-energy nexus. Water-Energy Nexus 3:110–116

Kurniawan TA, Lo WH, Chan GY (2006) Physico-chemical treatments for removal of recalcitrant contaminants from landfill leachate. J Hazard Mater 129(1–3):80–100

Le TS, Dang NM, Tran DT (2021) Performance of coupling electrocoagulation and biofiltration processes for the treatment of leachate from the largest landfill in Hanoi, Vietnam: impact of operating conditions. Sep Purif Technol 255:117677

Lee Y, Gerrity D, Lee M, Gamage S, Pisarenko A, Trenholm RA, Von Gunten U (2016) Organic contaminant abatement in reclaimed water by $UV/H_2O_2$ and a combined process consisting of $O_3/H_2O_2$ followed by $UV/H_2O_2$: prediction of abatement efficiency, energy consumption, and byproduct formation. Environ Sci Technol 50(7):3809–3819

Lee M, Merle T, Rentsch D, Canonica S, Von Gunten U (2017) Abatement of polychoro-1, 3-butadienes in aqueous solution by ozone, UV photolysis, and advanced oxidation processes ($O_3/H_2O_2$ and $UV/H_2O_2$). Environ Sci Technol 51(1):497–505

Malik SN, Ghosh PC, Vaidya AN, Waindeskar V, Das S, Mudliar SN (2017) Comparison of coagulation, ozone and ferrate treatment processes for color, COD and toxicity removal from complex textile wastewater. Water Sci Technol 76(5):1001–1010

Malik SN, Khan SM, Ghosh PC, Vaidya AN, Kanade G, Mudliar SN (2019) Treatment of pharmaceutical industrial wastewater by nano-catalyzed ozonation in a semi-batch reactor for improved biodegradability. Sci Total Environ 678:114–122

Malik SN, Ghosh PC, Vaidya AN, Mudliar SN (2020) Hybrid ozonation process for industrial wastewater treatment: principles and applications—a review. J Water Process Eng 35:101193

Martins RC, Dantas RF, Sans C, Esplugas S, Quinta-Ferreira RM (2015) Ozone/$H_2O_2$ performance on the degradation of sulfamethoxazole. Ozone Sci Eng 37(6):509–517

Nidheesh PV, Singh TA (2017) Arsenic removal by electrocoagulation process: recent trends and removal mechanism. Chemosphere 181:418–432

Oumar D, Patrick D, Gerardo B, Rino D, Ihsen BS (2016) Coupling biofiltration process and electrocoagulation using magnesium-based anode for the treatment of landfill leachate. J Environ Manage 181:477–483

Pi Y, Ernst M, Schrotter JC (2003) Effect of phosphate buffer upon $CuO/Al_2O_3$ and Cu (II) catalyzed ozonation of oxalic acid solution. Ozone Sci Eng 25(5):393–397

Pillai KC, Kwon TO, Moon IS (2009) Degradation of wastewater from terephthalic acid manufacturing process by ozonation catalyzed with Fe2+, $H_2O_2$ and UV light: direct versus indirect ozonation reactions. Appl Catal B 91(1–2):319–328

Psaltou S, Kaprara E, Triantafyllidis K, Mitrakas M, Zouboulis A (2021) Heterogeneous catalytic ozonation: the significant contribution of PZC value and wettability of the catalysts. J Environ Chem Eng 9(5):106173

Rahman NA, Jol CJ, Linus AA, Ismail V (2021) Emerging application of electrocoagulation for tropical peat water treatment: a review. Chem Eng Process Process Intensif 108449

Sardari K, Fyfe P, Lincicome D, Wickramasinghe SR (2018) Combined electrocoagulation and membrane distillation for treating high salinity produced waters. J Membr Sci 564:82–96

Shang NC, Chen YH, Ma HW, Lee CW, Chang CH, Yu YH, Lee CH (2007) Oxidation of methyl methacrylate from semiconductor wastewater by $O_3$ and $O_3/UV$ processes. J Hazard Mater 147(1–2):307–312

Shokri A, Mahanpoor K, Soodbar D (2016) Degradation of ortho-toluidine in petrochemical wastewater by ozonation, $UV/O_3$, $O_3/H_2O_2$ and $UV/O_3/H_2O_2$ processes. Desalin Water Treat 57(35):16473–16482

Shu HY, Chang MC (2005) Decolorization effects of six azo dyes by $O_3$, $UV/O_3$ and $UV/H_2O_2$ processes. Dyes Pigm 65(1):25–31

Song S, Xia M, He Z, Ying H, Lü B, Chen J (2007) Degradation of p-nitrotoluene in aqueous solution by ozonation combined with sonolysis. J Hazard Mater 144(1–2):532–537

Tahreen A, Jami MS, Ali F (2020) Role of electrocoagulation in wastewater treatment: a developmental review. J Water Process Eng 37:101440

Warsinger DM, Swaminathan J, Guillen-Burrieza E, Arafat HA (2015) Scaling and fouling in membrane distillation for desalination applications: a review. Desalination 356:294–313

Xiao H, Liu R, Zhao X, Qu J (2008) Effect of manganese ion on the mineralization of 2, 4-dichlorophenol by ozone. Chemosphere 72(7):1006–1012

Yuan Y, Garg S, Wang Y, Li W, Chen G, Gao M, Waite TD (2022) Influence of salinity on the heterogeneous catalytic ozonation process: implications to treatment of high salinity wastewater. J Hazard Mater 423:127255

Zhang H, Duan L, Zhang D (2006) Decolorization of methyl orange by ozonation in combination with ultrasonic irradiation. J Hazard Mater 138(1):53–59

Zhang W, Zhou Y, Hu C, Qu J (2021) Electricity generation from salinity gradient to remove chromium using reverse electrodialysis coupled with electrocoagulation. Electrochim Acta 379:138153

Ziylan-Yavaş A, Ince NH (2018) Catalytic ozonation of paracetamol using commercial and Pt-supported nanocomposites of $Al_2O_3$: the impact of ultrasound. Ultrason Sonochem 40:175–182

# The Advancement of Membrane Bioreactors (MBRs) in Industrial Effluent Treatment

**K. Viswanath Aswani, Ajay S. Kalamdhad, and Chandan Das**

## 1 Introduction

Wastewater, its generation, treatment, and disposal, is a significant problem faced by industries all over the globe. Health problems and diseases caused by the discharge of these untreated or inadequately treated wastewater from different industries are common. Disposals and accidental releases of wastewater are called water pollution and result in the spreading of diseases, destruction of aquatic life, and contaminating groundwater and soil. The most instantaneous effect of wastewater on the environment is the destruction and contamination of natural habitats and the flora and fauna that live in those habitats by exposing them to harmful pollutants. If only partial treatment is done for the sewage before it is disposed of, it can pollute water and damage vast wildlife. Leaking and flooding of untreated wastewater is also a cause of pollution in wetlands and waterbodies.

Industrial wastewater has different properties and quantities depending upon the product and procedure of the industry. It may be organic, highly biodegradable, and sometimes inorganic and hazardous. The industries consume a considerable volume of freshwater for different processes and the treatment of effluents and cleaning purposes which is becoming threatened in many parts of the world. Researchers worldwide work hard to find a better method that is expected to significantly impact the environment by treating industrial and municipal wastewater and minimising freshwater consumption. They are developing efficient processes for an indigenous

K. V. Aswani
Centre for the Environment, Indian Institute of Technology Guwahati, Guwahati, India

A. S. Kalamdhad (✉)
Department of Civil Engineering, Indian Institute of Technology Guwahati, Guwahati, India
e-mail: kajay@iitg.ac.in

C. Das (✉)
Department of Chemical Engineering, Indian Institute of Technology Guwahati, Guwahati, India
e-mail: cdas@iitg.ac.in

 
M. P. Shah (ed.), *Sustainable Industrial Wastewater Treatment and Pollution Control*,
https://doi.org/10.1007/978-981-99-2560-5_7

solution to fulfil zero liquid discharge. It will also enhance the industrial economy by converting waste to wealth, creating a knowledge base, and generating employment.

The wastewater is 99% of water and nearly 1% of pollutants. Various industries generate large volumes of liquid and solid waste that having large amounts of BOD, COD, high levels of dissolved and suspended solids, Pathogens, Nutrients, Contaminants of Emerging Concern (CEC), Endocrine-disrupting compounds (EDCs), phenolic compounds, heavy metal ions, oil, grease, chlorides, sulphates, surfactants, etc. Direct discharge of these waste streams into rivers and wetlands results in eutrophication and other contrary environmental effects. Therefore, the treatment of these effluents is essential to bring down the level of contaminants as per the standards for safe disposal and water recovery and at the same time production of gaseous biofuel, e.g. methane, particularly when subjected to anaerobic digestion.

Corresponding to the mission for a cleaner environment besides environmental compliance regulations, eliminating organic and inorganic pollutants from industrial effluent is a thought-provoking mission for attaining sustainable growth. Typical industrial effluent treatment plants comprise primary and secondary treatment, subsequently biological treatment, and tertiary treatment (if necessary). The schematic of the industrial wastewater treatment system is given in Fig. 1.

Industrial wastewater treatment mainly amalgamates different physicochemical and biological treatments followed by filtration. Various industries choose other methods according to the type of wastewater generated and continence regarding efficiency and economy of the procedure. The initial step of industrial wastewater treatment is mainly physical processes, for example: 'Gravity separation', which removes a significant part of floating and settleable materials from effluent. The equalisation system helps to minimise the fluctuations triggered because of either unexpected

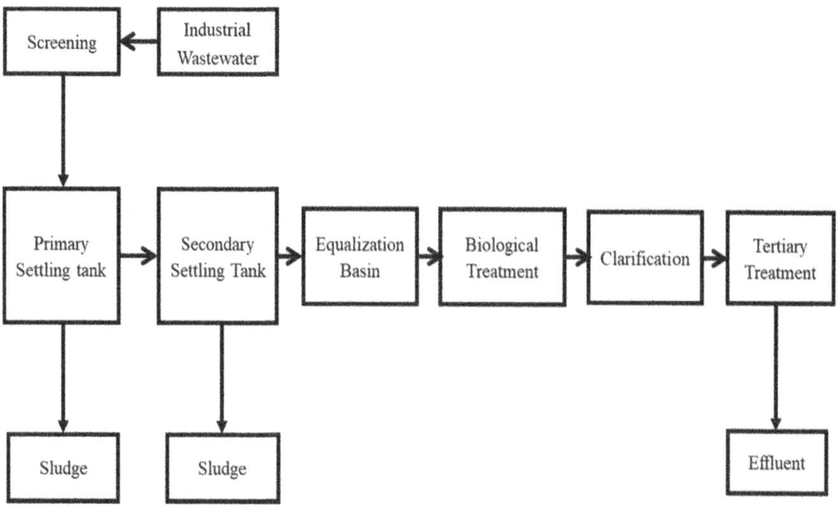


**Fig. 1** A typical refinery effluent treatment system

alteration of flow or composition in the effluent treatment plant. Biological treatment is the leading and extensively applied effluent treatment technology for removing liquefied organic compounds in wastewater. Tertiary treatment can be well-thought-out if the industry meets strict limits for diverse pollutants. The activated sludge treatment method is extensively applied in conventional effluent treatment plants. Traditional methods, such as the two-stage activated sludge process, anoxic/aerobic processes, and aerobic-settling-anoxic treatment, are also usually used to treat various industrial effluents.

MBR is an emerging technology with high efficiency and promising results. Studies are still going on to improve and commercialise the same. MBRs are suspended growth biological handling practices. It is a variant of the activated sludge system. MBRs are broadly distinct as the combination of conventional biological processes, for instance, activated sludge treatment through membrane clarification to deliver advanced reduction levels of organic matter and suspended solids. Most industries that installed and applied MBR technology in the effluent treatment plant are coming with impressive reviews. The shift from conventional effluent treatment methods to upcoming improved technologies is visible and welcoming.

## 2 Membrane Bioreactors (MBR)

An MBR combines membrane processes (suspended growth bioreactor), thus removing the necessity of secondary filtration applied in conventional activated sludge treatment. It combines a membrane procedure identical to the microfiltration or ultrafiltration with biological effluent treatment processes. It can be explained as biological degradation and membrane filtration (Farizoglu and Uzuner 2011). A schematic of an effluent treatment system with an MBR is given in Fig. 2.

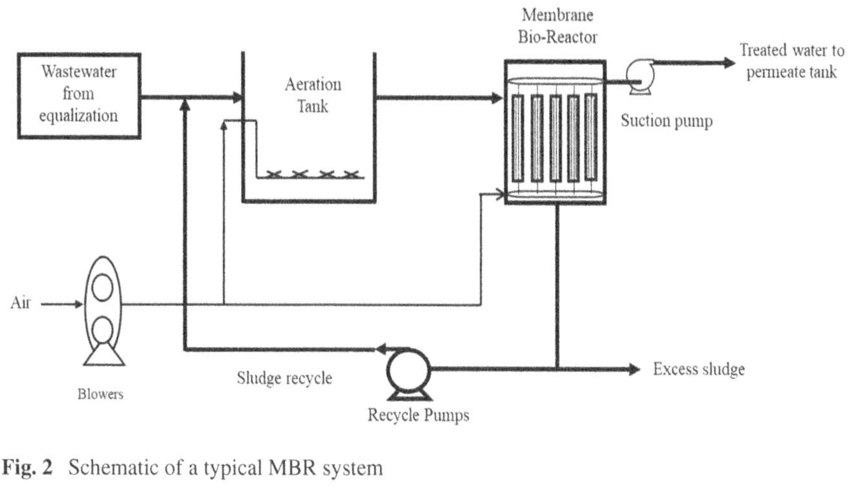


**Fig. 2** Schematic of a typical MBR system

In contrast to traditional biological effluent treatment systems, the amalgamation of a membrane into a digester offers numerous benefits, which include:

1. Less space utilisation due to smaller plant size.
2. Low energy consumption.
3. Whole retaining of particulate matter.
4. The reduced discharge rate of total suspended solids (TSS).
5. Automatic process.
6. Treated water is free from bacteria and pathogens.
7. Biological activity can be governed reasonably.
8. Bacteria and pathogens free high-quality effluent.
9. Higher organic loading rates, etc., (de Koning et al. 2008; Buntner et al. 2013).

These advantages make MBR striking for water reclamation and meeting strict effluent discharge requirements (Van den Broeck et al. 2012). The membranes can be outside the bioreactor and can function under pressure in solid–liquid separations. It is called an external membrane bioreactor (EMBR), or submerged in the bioreactor and operated under a vacuum; it is also known as a submerged membrane bioreactor (SMBR) (Busch et al. 2007; Tan and Ng 2008). For more clarification, it is explained separately below.

Generally, according to the Membrane alignment, two MBR configurations exist:

- Internal MBR:
  It is also called submerged MBR. In this type of MBR, the membranes are integral to the biological reactor immersed in them.

- External MBR:
  It is also known as side stream MBR. In this type of MBR, an external pumping setup is required. The membranes are a distinct unit procedure necessitating an intermediary pumping pace.

According to the flow type, there are two types of MBRs:

- Vacuum-driven MBR:
  It is also called gravity-driven MBR systems, which are immersed. They usually use hollow fibre or flat sheet membranes. Sometimes, it is fitted in either the bioreactor or a succeeding membrane tank.

- Pressure-driven MBR:
  These MBR setups are in-pipe cartridge systems positioned outside the bioreactor.

According to the method of biodegradation process, MBR can be again classified into two:

- Aerobic MBR:
  In aerobic MBR, the microbial reaction occurs in the availability of oxygen. The primary reaction products are water, carbon dioxide, and additional biomass.

- Anaerobic MBR:
  In anaerobic MBR, the microbial reaction occurs when oxygen is not available. The primary reaction outputs are methane, carbon dioxide, and excess biomass. Due to the inherent merits of less sludge production, bio-energy generation, and saving energy cost for aeration, anaerobic MBR systems are being used for treating an extensive range of industrial effluents.

  According to the materials used, there are:

- ***Ceramic MBR***:
  Membrane bioreactor (MBR) mainly using ceramic membrane is a relatively new technology and is considered a potentially viable, effective, and profitable process for industrial wastewater treatment. Though ceramic membranes are extra rigid in terms of resistance to chemical dosage and fouling, they continue limited to practical uses in MBR systems mainly because of their comparatively high initial expense. Recently, ceramic flat sheet configurations have been presented for submerged MBR systems. Nevertheless, marketable MBR membranes are more or less all polymeric.

- ***Polymeric MBR***:
  Over half of the membrane unit products for MBR on the market are built on polyvinylidene difluoride or PVDF. Additional membrane materials that are less common in the market comprise polyethersulfone (PES) and polyethylene (PE). The grouping of high chemical resistance, comparatively great mechanical strength, governable pore size, and decent flexibility has envisioned that the PVDF products lead. For example, the low-density polyethylene (LDPE), polyolefinic membranes, are among the bottommost in the introductory manufacture price of altogether MBR membrane supplies. The pores are produced by extruding that material under controlled circumstances. 'Dry spinning' is also applied to have slit-like pores. This slit-like pore structure is notable from the extra multifaceted thermal-induced phase separation (TIPS) process for PVDF, creating the additional classic pseudo-one-dimensional pores. The other materials, polysulphone, polyacrylonitrile (PAN), polytetrafluorethylene (PTFE), and polyvinyl alcohol (PVA), are much fewer in use.

## *2.1 Working Principle of MBR*

As the name implies, MBR combines a bioreactor and a membrane. The working principle is also as simple as that. MBRs combine dual primary units: the biological reactor and the membrane unit. The bioreactor is accountable for the digestion of the waste compounds. The organisms inside the bioreactor break down the organic contaminants into small fractions which are less harmful and disposable. The biodegradable pollutants in the wastewater are thus converted from complex form to simplex form. The membrane modules are for filtration of the treated effluent from

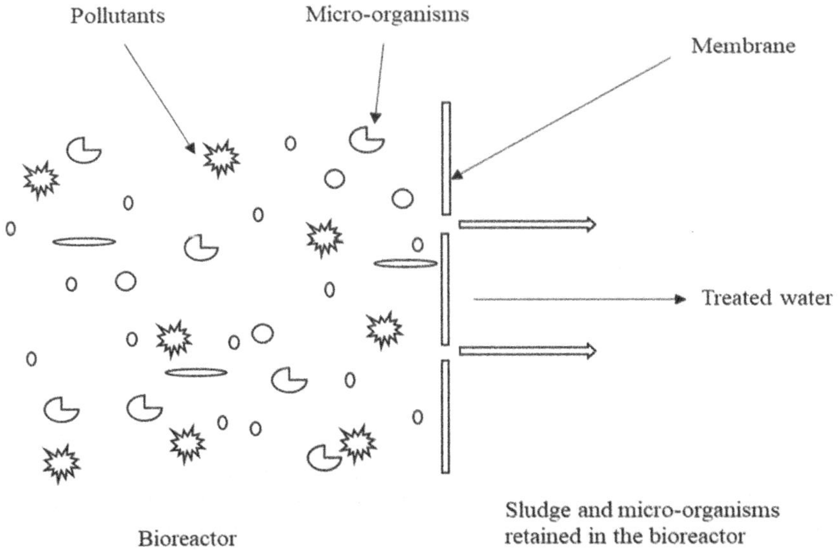


**Fig. 3** Working principle of Membrane Bioreactor

mixed liquor. It filtrates the treated water from the bioreactor, leaving the sludge and organisms inside the reactor. The working principle of MBR has shown as a schematic representation in Fig. 3.

## 2.2 *Role of Microorganism in MBR*

Microorganisms are the primary cause of the biodegradation of organic pollutants in industrial wastewater. It is essential to be familiar with the terms biodegradation and its mechanism before understanding the role of microorganisms in MBR.

### 2.2.1 Biodegradation

The term biodegradation is explained as the deterioration of organic materials by microorganisms. The microorganisms decay the organic substances into lesser toxicity by-products by their metabolic activities. Enzymes act as a catalytic part of the biodegradation procedure. The chemical compound is transformed from one state to another through numerous intermediates with the help of various enzymes. This conversion of organic substances is called mineralisation. Biodegradation is a very profitable method and an environmentally friendly selection that is habitually chosen, with the probability of complete mineralisation of the organic substance (Nair et al. 2008; Shah 2020; El-Naas et al. 2009). Since the aromatic structure of countless organic compounds like phenols is incredibly unchanging due to splitting the benzene ring, several microorganisms can utilise these chemical compounds for their various metabolic activities and biomass production. They use the compounds as energy and carbon sources. Biological conversion has been recognised as a critical solution to environmental contamination caused by countless tricky organic pollutants. By the way, the usage of pure and mixed cultures of microorganisms is well-thought-out as a promising and most excellent auspicious method (Shourian et al. 2009). Countless bacteria, algae, and fungi strains can destroy toxic organic compounds by degrading them. Bacterial cultures of the *Pseudomonas* genus are the maximum typically utilised biomass for the degradation of organic pollutants biologically, with particular attention paid to *Pseudomonas putida* for its excellent removal efficacy (El-Naas et al. 2010).

Nevertheless, the primary disadvantage in bioprocesses is inhibiting the enzymatic activity on heavy substrate concentrations of substrate. All organic substances can be disintegrated aerobically or anaerobically under factual circumstances (Wang et al. 2007). Conservatively, aerobic procedures are chosen. Aerobic microorganisms mature quicker and more effectively since they can mineralise toxic organic substances to inorganic components like carbon dioxide and water (Sahinkaya and Dilek 2007; Shah 2021a, b). This adds to low accompanying expenses (El-Naas et al. 2010). Conversely, the outcomes of biochemical reactions in anaerobic methods regularly produce visually unpleasant colours in addition to fouling odours in water (Diya'Uddeen et al. 2011). Consequently, there is a slight concentration in the use of anaerobic microorganisms to degrade organic waste.

### 2.2.2 Biodegradation Mechanism

Biodegradation is a multivariable phenomenon controlled by numerous biotic and abiotic features. The variables include microbial abundance, substrate concentration, oxygen content and availability, temperature, and pH (Nair et al. 2008; El-Naas et al. 2009). The chemical structure of all aromatic compounds is vital as represented by the type, degree of branching, and number and position of substituents on the aromatic ring. The more substituents in the structure, the more deadly and low degradable the compound remains.

Metabolic methods are conquered through the catalysis of enzymes. They are specific to particular kinds of reactions and biomass. A metabolic response is eventually a course of energy change. Significantly, fewer things are identified about the biodegradation mechanism by algae and fungi. Consequently, the subsequent is a short discussion of this mechanism by aerobic bacteria. Aerobic bacteria are classically signified by the biodegradation of aromatics.

Phenol is a primary structural component for various synthetic organic compounds. An enzymatic outbreak on the aromatic ring is originated from oxygen in aerobic biodegradation. A characteristic pathway for metabolising phenols is to hydroxylate the ring by the enzyme phenol hydroxylase, form catechol, and open the ring over *ortho-* or *meta*-oxidation. Therefore, catechol is a vital intermediary in the degradation pathways of numerous aromatic substances, and phenol hydroxylase is the primary enzyme. In the *meta*-pathway, the ring is cleaved by the enzyme catechol 2,3-dioxygenase (C23O). In the *ortho*-pathway, the aromatic ring is cleaved by the catechol 1,2-dioxygenase (C12O) enzyme. Therefore, the ring is opened and degraded (Nair et al. 2008). (C12O) and (C23O) label two types of alignments as to by what means the ring cleavage can ensue. Though the biodegradation of countless aromatics continues from side to side, the *ortho*-cleavage pathway afterwards the development of catechol since the *meta*-cleavage effects in the development of dead-end metabolites after catechol; the enzyme gets inactivated by the accretion of a toxic intermediate (Solyanikovak and Golovleva 2004). Most of the time, the change of catechol does not follow the *meta*-cleavage pathway, and usually, the *ortho*-cleavage pathway is essential to completely degrade various aromatic organics (Farrell and Quilty 2002). More details of biodegradation pathways and mechanisms are explained by many researchers later (Stoilova et al. 2007; Shah 2021a, b; Nair et al. 2008; Peng et al. 2008, Al-Khalid and El-Naas 2012).

As shown in Fig. 4, the microorganism utilises oxygen and nutrients in the biodegradation process to alter the organic matter to water and carbon dioxide in aerobic conditions. The biodegradation's excess cell biomass is retained in the bioreactor with sludge and periodically removed.

In anaerobic biodegradation, the microorganisms break down the pollutants into less harmful by-products without oxygen. The bacteria change organic matter to the anaerobic condition to carbon dioxide and methane, i.e. biogas. Figure 5 showcases the role of microorganisms in the anaerobic MBR. In this process, the organisms release the excess cell biomass as sludge. The leftover water is processed through a pressure-driven membrane filter with porosity 0.035– 0.4 μ for further treatment to generate reusable quality permeate water.

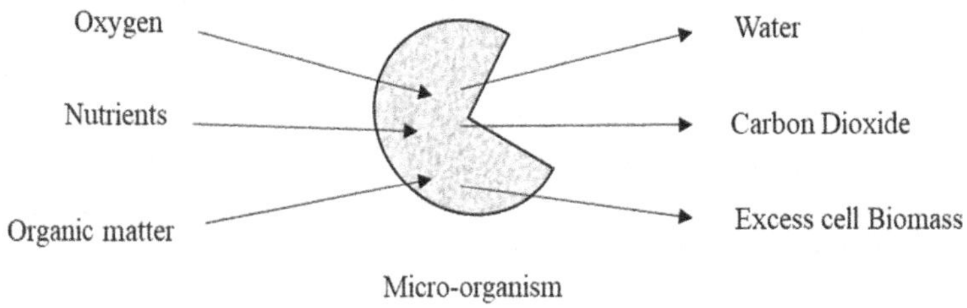


**Fig. 4** Role of Microorganisms in Aerobic MBR

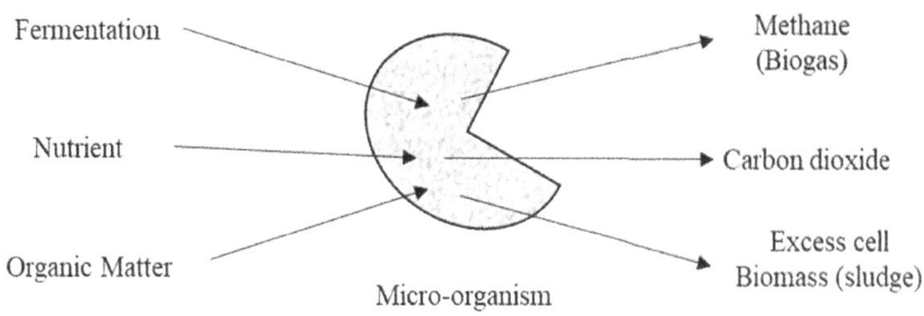


**Fig. 5** Role of Microorganisms in Anaerobic MBR

## 2.3 *MBR Setup and Modes of Operation*

The following parameters assess the performance of an MBR system:

- Activated sludge method:
  The biodegradation of sewage and industrial effluent utilising air and bacteria majorly happens in the activated sludge process.

- Substrate:
  The medium on which an organism matures or a surface on which it is attached. It is usually seen in the case of membrane fouling, where the microorganisms grow on the membrane by forming EPS (different polymeric substances).

- Soluble microbial products (SMP):
  SMP is any soluble material released from a biological system into the effluent not present in the influent.

- Mixed Liquor:
  It combines raw or settled effluent and activated sludge within an aeration chamber.

- Mixed liquor suspended solids (MLSS):
  Microbial suspension in the aeration chamber encompassing inert, biodegradable organic matter, alive and dead microorganisms. The operating concentration of this may fluctuate in the range of 1500–4000 mg/l.

- Mixed liquor volatile suspended solids (MLVSS):
  It is mainly well-distinct as the microbiological suspension in the aeration chamber of a conventional activated sludge system of the biological effluent treatment plant, proportional to the concentration of the microorganism in the aeration tank.

- Extracellular polymeric substances (EPS):

Extracellular polymeric substances as well recognised as exo-polysaccharide or EPS. They are high-molecular-weight compounds concealed by microorganisms into their environment. EPS are the construction material of bacterial settlements. It continues attached to the cell's outer surface or is secreted into its growth medium.

It depends upon the setup and modes of operation. MBR is classified into many.

### 2.3.1 Aerobic Treatment and Anaerobic Treatment

The process principle of aerobic treatment can be explained as the microbes degrade pollutants in the presence of oxygen. The primary application of this treatment is effluent with low to medium organic contaminations (COD < 1000 ppm) and for effluent that is hard to biodegrade, e.g. refinery wastewater, municipal sewage, etc. The reaction kinetics is comparatively fast compared with the anaerobic process, and the net sludge produced is relatively high. After treatment, the effluent typically discharges directly or sometimes diverts to filtration or disinfection. Figure 6 shows the MBR setup for aerobic modes of operation.

The process principle of anaerobic treatment can be explained as biodegradation without oxygen. The application of this treatment is effluent with medium to high organic contaminants (COD > 1000 ppm) and readily biodegradable effluent, e.g. beverage and food effluent rich in alcohol/sugar/starch. The reaction kinetics is comparatively deliberate compared with aerobic treatment. The net sludge yield is moderately low. It commonly produces one-fifth to one-tenth of sludge produced by aerobic treatment methods. The anaerobic process is invariably followed by aerobic treatment. Figure 7 shows the MBR setup for anaerobic modes of operation.

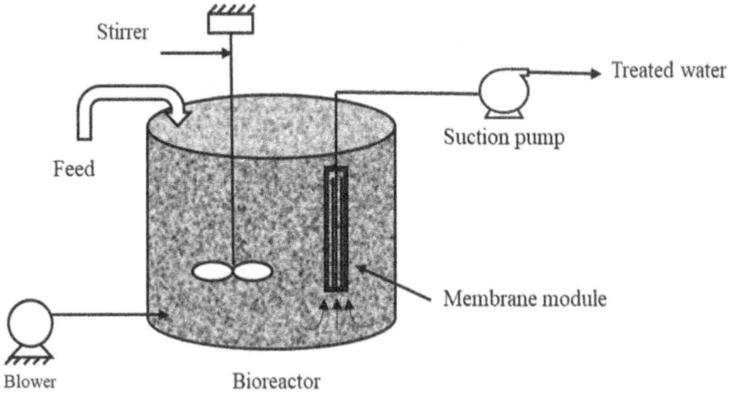


**Fig. 6** MBR setup for Aerobic mode of operation

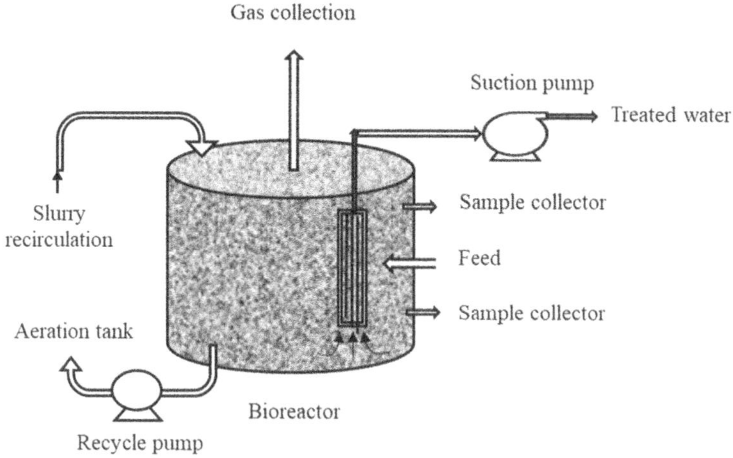


**Fig. 7** MBR setup for Anaerobic mode of operation

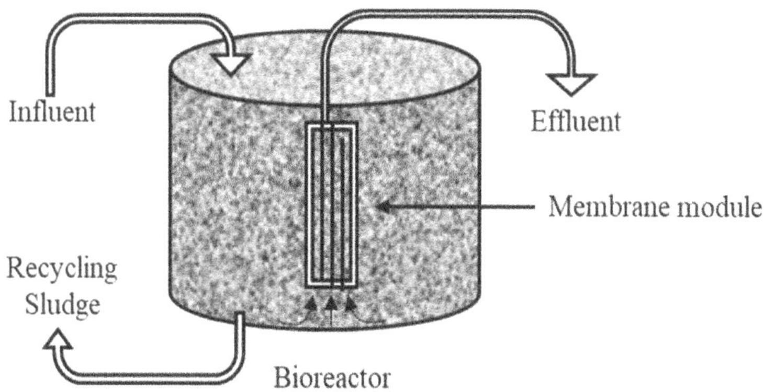


**Fig. 8** MBR operation in a submerged configuration

### 2.3.2 Submerged and Sidestream MBR

In an MBR system, the membrane filtration happens either within the bioreactor in submerged mode or externally through recirculation arranged in a side stream with the bioreactor. The choice between operating options relies on the application, as both systems have advantages and drawbacks. Figure 8 represents the MBR operation in submerged configuration. The information about the merits and demerits of these configurations are listed below. Submerged MBRs are available with meagre pumping costs, which require less operating costs. They didn't require frequent cleaning too. But the aeration cost of submerged MBR is high compared to the others. It has lower flux which means more giant footprints. One of the significant disadvantages of submerged MBR is the lower availability of the membrane area.

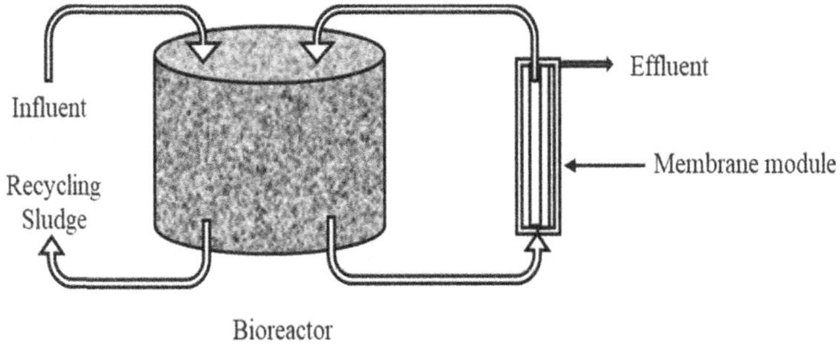


**Fig. 9** MBR operation in a sidestream configuration

Sidestream MBRs have higher availability of membrane area. Its aeration cost, as well as capital cost, is comparatively low. But the pumping and operating costs are high. Sidestream MBRs have higher flux which implies a smaller footprint. It requires more frequent cleaning. Figure 9 represents the MBR operation in the sidestream configuration.

## 3 Application of MBR in Industrial Effluent Treatment

MBR systems have been applied across numerous industrial areas. The industrial effluents most suitable for MBR treatment are those rich with biodegradable organic carbon content, which are readily available. MBR systems are generally selected as a prime treatment option when the space for effluent treatment plants is restricted or high-quality treated water is mandatory to satisfy the public regulations. Since the high initial expense compared to the conventional activated sludge system, MBRs were commonly not suggested for large scale effluent treatment plants in the past. But, later, the strict environmental legislation, together with lowering capital, majorly the money for membranes, operating expenses, etc., reduced the considerable part of limitations in MBR installation. After that, peak progress has been observed in the first 20 years of MBR implementation. Presently, many municipal MBR plants are working successfully with a capacity of more than 100 Megalitres/day (MLD) in a Peak Daily Flow (PDF).

Distillery spent wash and tannery effluents produce various waste, for example, solid waste and aqueous effluents that cover significant quantities of suspended solids, biochemical oxygen demand (BOD), high levels of dissolved oxygen, and chemical oxygen demand (COD). The produced effluents need treatment beforehand the disposal. Before they are discharged into rivers or wetlands, they require adequate treatment to reduce toxicity levels. This requirement is typically acquired by applying conventional treatment methods like two-stage-activated sludge, aerobic-settling-anoxic, and anoxic/aerobic treatments. The procedures mentioned above have become more and more costly to fulfil with more strict permitted discharge standards. The increased expenses related to conventional methods have directed importance in substitute processes that permit the effluent to be reused (de Koning et al. 2008; Farizoglu and Uzuner 2011; Buntner et al. 2013). MBR process is a method that has the capacity for much effective treatment of industrial effluents.

The MBR system is conceivably practical, effective, and economical with various chemicals, food, pharmaceutical, and biotechnology applications. Substantial paybacks of MBR over standard filtration methods comprise whole automated operation, retaining of particulate matter, reasonable governing of biological activity, high-quality bacteria and pathogens free effluent, the low discharge rate of total suspended solids (TSS), higher organic loading rates, and reduced plant size (Busch et al. 2007; Tan and Ng 2008; Van den Broeck et al. 2012). MBR syndicates biological processing of effluents to minimise toxicity and clarification by immersed low-pressure polymeric ultrafiltration (UF) membranes. Depending on the membrane separation process used, an MBR can be categorised. This can be completed by pressure-driven filtration in sidestream MBRs and vacuum-driven membranes submerged straight into the bioreactor in immersed MBRs. Immersed or sidestream types can function under aerobic or anaerobic circumstances (Poulton et al. 2002; Tan and Ng 2008). Abridged energy usage is a significant plus point accomplished using an immersed MBR compared to a sidestream MBR.

Heriot-Watt University, Edinburgh, used the MBR technology for drinking water de-nitrification. Department of Environmental Chemistry, IIQAB-CSIC, Spain, developed a novel membrane bioreactor for pharmaceuticals wastewater treatment (Van den Broeck et al. 2012). Tsinghua University, Beijing, is researching aerobic membrane bioreactors to treat domestic wastewater (Busch et al. 2007). IFA-Tulln, Austria, has developed MBR for oil-contaminated sewage (Tan and Ng 2008). Technical University of Berlin, Germany, used MBR technology for municipal wastewater treatment (Poulton et al. 2002). Rhodes University, South Africa, handling several research projects on membrane bioreactors using immobilised polyphenol oxidase to remove phenols from industrial effluents. Extensive research to treat textile wastewater using MBRs has been analysed by the University of Natural Resources and Applied Life Sciences, Vienna (Tian et al. 2009). The microfiltration membranes are placed inside a steel membrane tank. The membranes are exposed to a low-pressure vacuum that pulls water through the membranes and pumps the filtered water to keep solids inside the reactor. To scour the outside of the membranes, compressed air is injected into the system. The MBR process typically operates at higher MLSS concentrations (15,000–20,000 mg/l) than the activated sludge method. Refining industries have not promoted MBR systems because of their higher costs than the traditional activated sludge process. MBR is more cost-competitive for activated sludge systems requiring tertiary filtration since it corresponds to having an effluent filter. For special treatment systems where additional tertiary treatment for reverse osmosis will be used, MBR can be in striking contrast with the alternate choice of microfiltration and media filtration after biological treatment.

The application of both microfiltration (MF) and ultrafiltration (UF) membranes for separation is realised to be proliferating, and it outspreads to an extensive variety of industrial processes, including biochemical and biopharmaceutical applications

(Reif 2006; Gallucci et al. 2011). Using an MBR in such applications is a novel and accomplished method of treating diverse effluents. The MBR is getting more popularity and acceptance over time. It is becoming an operative process for industries like oil refineries. There is correspondingly short engagement in treating oil refinery effluents (Viero et al. 2008; Veronese et al. 2012). Veronese et al. (2012) applied COD to appraise the treatment of oil refinery effluent by applying a biological treatment collective with immersed UF membranes to determine the membrane capacity and efficiency for nine months of operation. The biodegradation of oil refinery effluent using immersed MBR was well understood by Viero et al. (2008). Their outcomes verified the potential of the MBR treatment method in treating high-strength feed.

Additionally, Rahman and Al-Malack explored using a cross-flow MBR in treating the discharged effluent from a petroleum refinery (Rahman and Al-Malack 2006). The performance of the MBR treatment was assessed at MLSS concentrations of 3000 and 5000 mg/l. It was proved that a COD removal competence of more than 93% was attained at both MLSS values. Scholz and Fuchs had tested the treatment of oily effluents by MBR process and testified that a high removal rate (about 99.99%) was attained for lubricating oil and fuel oil. Correspondingly, the average removal efficacy of COD through the MBR operation was 94–96 and 97% lubricating oil for and fuel oil.

# 4 Membrane Fouling

MBR technology has increased extraordinary acceptance in effluent treatment in the last decades. The extensive application of MBR is restricted due to the membrane fouling issue. It is one of the critical challenges of researchers. Severe loss of membrane permeability is majorly caused by fouling. This reduction of permeability leads to an upsurge in energy intake. Several academics have experimented on this problem, majorly, on the features disturbing the process performance. The significant areas of study include conventional elements, for example, reactor kinetic factors and biological but very rare on membrane performance factors, for example, pore size and distribution, nature of membrane material like whether they are hydrophobic or hydrophilic in nature, life span, impurity rejection and water flux, etc., (Gao et al. 2009; Tian et al. 2009).

## *4.1 Fouling and Control of Fouling*

The MBR's filtration ability unavoidably cuts through the filtration span. The primary reason behind the membrane fouling is the deposition of particulate and soluble materials onto and into the membrane recognised as the exchanges among the membrane and activated sludge components. This critical problem and process limitation have been studied in the early development of MBRs. After all these years of study, it

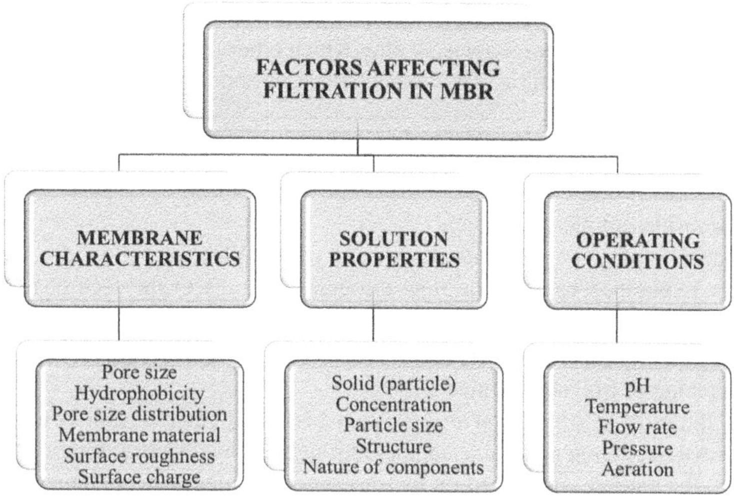


**Fig. 10** Factors influencing fouling (interactions in red)

remains one of the most vital challenges facing further advanced studies and applications on MBR. Figure 10 shows the various factors affecting fouling, thus the filtration in MBR.

It has been published in most of the recent reviews that cover membrane applications to bioreactors that fouling of membrane is the most severe complication that disturbs the system's overall performance. Fouling affects the entire membrane separation process. It primes to an essential upsurge in the hydraulic resistance of the membrane. Correspondingly, it is demonstrated as permeate flux weakening or transmembrane pressure (TMP) growth when the process functions under constant-flux or constant-TMP circumstances. The energy is obligatory to attain filtration rises where cumulative TMP upholds the flux. Otherwise, regular membrane washing is compulsory, provocatively swelling operating expenses due to production interruption and cleaning agents. Additional recurrent replacement of membrane is also required.

The interaction between the activated sludge liquor components and membrane material results in membrane fouling. It comprises biological flocs formed by colloidal and soluble compounds, live or dead microorganisms, etc. The suspended biomass or flocs has no fixed composition. It differs from the MBR operating conditions employed and the feed water composition. Therefore however numerous researches of fouling phenomenon of the membrane have been available on the different analytical methods used, various choice of operating conditions employed and, the feed water matrices provided, the limited evidence testified in maximum research on the suspended biomass composition has ended it problematic to find out any general trend about fouling in MBRs precisely.

In immersed MBR, the air-induced cross flow attained can eliminate or decrease the fouling layer formed on the surface of the membrane. A topical study reports the newest conclusions on uses of aeration in immersed membrane alignment and defines the improvement of gas bubbling performances. It was obtained using an optimum airflow rate that has been recognised, after which other upsurges in aeration do not affect the removal of fouling. Thus, calculating the optimal aeration rate is critical in designing MBR.

Several anti-fouling approaches can be helpful in MBR systems. Some of them are explained below:

- Backwashing of membranes:
  In backwashing, the permeate water is recycled. While recycling, the permeate water is pumped back to the flow through the pores to the feed channel and membrane. This helps in removing external and internal foulants.

- Slackening or irregular permeation:
  The filtration is stopped at a consistent period of intermission before being restarted. Non-stop aeration is applied throughout this inactive duration. At that point, the particles deposited on the surface of the membrane habitually diffuse back to the reactor.

- Patented anti-fouling products:
  Nalco's Membrane Performance Enhancer Technology is an example of this.

- Air backwashing:
  Pressurised air will build up in the permeate side of the membrane. This releases significant pressure within a limited duration. Consequently, membrane modules in a pressurised vessel are attached to an exhaust arrangement. Air typically does not pass through the membrane. If air passed, it would dry the pores and surface of the membrane. A re-wet pace would be required by pressurising the feed side of the membrane to make it in the previous condition.

Furthermore, various intensities of chemical washing or types may also be suggested:

- Chemically improved backwash (regular);
- Maintenance washing with higher chemical concentration (weekly); and
- Concentrated chemical washing (once or twice a year).

When additional filtration cannot be continued, the elevated TMP concentrated chemical cleaning is also performed. Each of the four leading MBR providers (Kubota, Evoqua, Mitsubishi and GE Water) has its chemical washing formulas. These formulas vary mainly in concentration and methods. Under normal circumstances, sodium hypochlorite (NaOCl) and citric acid remain the predominant washing agents. It is widespread for MBR dealers to accept definite protocols for chemical washings. These protocols majorly follow cleaning frequencies and chemical concentrations for separate conveniences.

## 5 Summary

There are four significant challenges facing operators of effluent treatment plants. These can be divided into four comprehensive groups: energy consumption, people skills and competence, sludge management, footprint, and facilities. Industrial effluents depreciate soil, crop, and environmental quality and are directly detrimental to human, animal, and aquatic lives. Accidental discharges of industrial effluents lower the rate of food crops. Wastes extremely high in suspended solids, dropping the efficacy of sewage treatment and growing the amount of sludge formed and upsetting its quality in connection with digestion and dewatering. Treatment of industrial and municipal effluent to accomplish the zero liquid release norms and reduce freshwater consumption is a significant concern in developing and, overall, the continents.

MBRs are the future of industrial wastewater treatment. It will be applied wherever high-quality effluent disposal is obligatory. In the future, high-quality effluent will be compulsory because of the reuse of water as process water or may be due to delicate receiving water bodies. MBRs are a faultless pre-treatment method for industrial effluents where additional handling using reverse osmosis or nanofiltration is well-thought-out. In all membrane processes, fouling of the membrane is generally triggered as a result of the precipitation and deposition of molecules or particulates on the membrane surface or pores. The significances of fouling which affects membrane's efficiency are amplified different membrane selectivity, membrane separation resistances, and reduced productivity. To avoid membrane fouling and extend the usage time through the choice of appropriate membrane materials; excellent configuration; pre-treatment of raw materials; optimisation of operating circumstances; control of inorganic salt solubility; cleaning the membrane frequently, etc., the MBRs have been well recognised for over three decades and are efficaciously commercialised for industrial and municipal effluent treatment. MBRs can be successfully applied to decrease the carbon footprint of the conventional activated sludge treatment method. This reduction is possible by eradicating some part of the liquid components of the mixed liquor. This produces a thick concentrated sludge treated through the activated sludge process. This status provides immense scope for developing cost-effective MBR technologies for different industrial and municipal effluents treatment.

**Acknowledgements** The author (KVA) would like to acknowledge the financial support provided by MoE, the Government of India, as part of the Ph.D. Scholarship Programme.

## References

Al-Khalid T, El-Naas MH (2012) Aerobic biodegradation of phenols: a comprehensive review. Crit Rev Environ Sci Technol 42:1631–1690. https://doi.org/10.1080/10643389.2011.569872

Buntner D, Sánchez A, Garrido JM (2013) Feasibility of combined UASB and MBR system in dairy wastewater treatment at ambient temperatures. Chem Eng J 230:475–481. https://doi.org/10.1016/J.CEJ.2013.06.043

Busch J, Cruse A, Marquardt W (2007) Modeling submerged hollow-fibre membrane filtration for wastewater treatment. J Memb Sci 288:94–111. https://doi.org/10.1016/J.MEMSCI.2006.11.008

de Koning J, Bixio D, Karabelas A et al (2008) Characterisation and assessment of water treatment technologies for reuse. Desalination 218:92–104. https://doi.org/10.1016/J.DESAL.2006.08.024

Diya'Uddeen BH, Daud WMAW, Abdul Aziz AR (2011) Treatment technologies for petroleum refinery effluents: a review. Process Saf Environ Prot 89:95–105. https://doi.org/10.1016/J.PSEP.2010.11.003

El-Naas MH, Al-Muhtaseb SA, Makhlouf S (2009) Biodegradation of phenol by Pseudomonas putida immobilised in polyvinyl alcohol (PVA) gel. J Hazard Mater 164:720–725. https://doi.org/10.1016/J.JHAZMAT.2008.08.059

El-Naas MH, Al-Zuhair S, Makhlouf S (2010) Continuous biodegradation of phenol in a spouted bed bioreactor (SBBR). Chem Eng J 160:565–570. https://doi.org/10.1016/J.CEJ.2010.03.068

Farizoglu B, Uzuner S (2011) The investigation of dairy industry wastewater treatment in a biological high-performance membrane system. Biochem Eng J 57:46–54. https://doi.org/10.1016/J.BEJ.2011.08.007

Farrell A, Quilty B (2002) The enhancement of 2-chlorophenol degradation by a mixed microbial community when augmented with Pseudomonas putida CP1. Water Res 36:2443–2450. https://doi.org/10.1016/S0043-1354(01)00481-X

Gallucci F, Basile A, Hai FI (2011) Introduction—A Review of membrane reactors. Membr Membr React Prep Optim Sel 1–61. https://doi.org/10.1002/9780470977569.ch

Gao DW, Fu Y, Tao Y et al (2009) Current research and development of controlling membrane fouling of MBR. African J Biotechnol 8:2993–2998. https://doi.org/10.4314/ajb.v8i13.60970

Nair CI, Jayachandran K, Shashidhar S (2008) Biodegradation of phenol. African J Biotechnol 7:4951–4958. https://doi.org/10.4018/978-1-5225-8903-7.ch045

Peng RH, Xiong AS, Xue Y et al (2008) Microbial biodegradation of polyaromatic hydrocarbons. FEMS Microbiol Rev 32:927–955. https://doi.org/10.1111/j.1574-6976.2008.00127.x

Poulton SW, Krom MD, Van Rijn J, Raiswell R (2002) The use of hydrous iron (III) oxides to remove hydrogen sulphide in aqueous systems. Water Res 36:825–834. https://doi.org/10.1016/S0043-1354(01)00314-1

Rahman MM, Al-Malack MH (2006) Performance of a crossflow membrane bioreactor (CF–MBR) when treating refinery wastewater. Desalination 191:16–26. https://doi.org/10.1016/J.DESAL.2005.05.022

Reif OWW (2006) Microfiltration Membranes: Characteristics and Manufacturing. In: Jornitz MW (ed) Sterile filtration. Springer, Berlin Heidelberg, Berlin, Heidelberg, pp 73–103

Sahinkaya E, Dilek FB (2007) Biodegradation kinetics of 2,4-dichlorophenol by acclimated mixed cultures. J Biotechnol 127:716–726. https://doi.org/10.1016/J.JBIOTEC.2006.08.009

Shah MP (2020) Microbial bioremediation & biodegradation. Springer

Shah MP (2021a) Removal of emerging contaminants through microbial processes. Springer

Shah MP (2021b) Removal of refractory pollutants from wastewater treatment plants. CRC Press

Shourian M, Noghabi KA, Zahiri HS et al (2009) Efficient phenol degradation by a newly characterised Pseudomonas sp. SA01 isolated from pharmaceutical wastewaters. Desalination 246:577–594. https://doi.org/10.1016/J.DESAL.2008.07.015

Solyanikovak IP, Golovleva LA (2004) Bacterial degradation of chlorophenols: pathways, biochemica, and genetic aspects. J Environ Sci Heal—Part B Pestic Food Contam Agric Wastes 39:333–351. https://doi.org/10.1081/PFC-120035921

Stoilova I, Krastanov A, Yanakieva I et al (2007) Biodegradation of mixed phenolic compounds by Aspergillus awamori NRRL 3112. Int Biodeterior Biodegradation 60:342–346. https://doi.org/10.1016/J.IBIOD.2007.05.011

Tan TW, Ng HY (2008) Influence of mixed liquor recycle ratio and dissolved oxygen on performance of pre-denitrification submerged membrane bioreactors. Water Res 42:1122–1132. https://doi.org/10.1016/J.WATRES.2007.08.028

Tian J yu, Liang H, Nan J, et al (2009) Submerged membrane bioreactor (sMBR) for the treatment of contaminated raw water. Chem Eng J 148:296–305. https://doi.org/10.1016/J.CEJ.2008.08.032
Van den Broeck R, Van Dierdonck J, Nijskens P et al (2012) The influence of solids retention time on activated sludge bioflocculation and membrane fouling in a membrane bioreactor (MBR). J Memb Sci 401–402:48–55. https://doi.org/10.1016/J.MEMSCI.2012.01.028
Veronese CG, Beal LL, Santiago VMJ et al (2012) Ultrafiltration hollow fiber membrane bioreactor (mbr) treating oil refinery wastewater. Procedia Eng 44:704–706. https://doi.org/10.1016/j.proeng.2012.08.538
Viero AF, de Melo TM, Torres APR et al (2008) The effects of long-term feeding of high organic loading in a submerged membrane bioreactor treating oil refinery wastewater. J Memb Sci 319:223–230. https://doi.org/10.1016/J.MEMSCI.2008.03.038
Wang SG, Liu XW, Zhang HY et al (2007) Aerobic granulation for 2,4-dichlorophenol biodegradation in a sequencing batch reactor. Chemosphere 69:769–775. https://doi.org/10.1016/J.CHEMOSPHERE.2007.05.026

# Nanofiltration Applications for Potable Water, Treatment, and Reuse

**Vandana Johnson, Caroline Biju Kurian, Diya Menon, Nilesh S. Wagh, and Jaya Lakkakula**

## Abbreviations

| | |
|---|---|
| AFM | Atomic Force Spectroscopy |
| BSA | Bovine serum albumin |
| COD | Chemical oxygen demand |
| DI | Deionised |
| DOM | Dissolved organic matter |
| DWTP | Drinking water treatment plant |
| FO | Forward osmosis |
| FRR | Flux recovery ratio |
| FSCF | Flat Sheet Cross Flow |
| GAC | Granular Activated Carbon |
| GO | Graphite oxide |
| HA | Humic acid |
| MAC | Maximum Allowable Concentration |
| MCL | Maximum contamination level |
| MWCO | Molecular Weight Cut-off |
| NF | Nanofiltration |

V. Johnson · C. B. Kurian · D. Menon · N. S. Wagh (✉) · J. Lakkakula
Amity Institute of Biotechnology, Amity University Maharashtra, Mumbai-Pune Expressway, Bhatan, Panvel, Mumbai, Maharashtra 410206, India
e-mail: waghnil@gmail.com

N. S. Wagh
Centre for Drug Discovery and Development, Amity University Maharashtra, Mumbai-Pune Expressway, Bhatan, Panvel, Mumbai, Maharashtra 410206, India

J. Lakkakula (✉)
Centre for Computational Biology and Translational Research, Amity University Maharashtra, Mumbai-Pune Expressway, Bhatan, Panvel, Mumbai, Maharashtra 410206, India
e-mail: spencerjaya@gmail.com



M. P. Shah (ed.), *Sustainable Industrial Wastewater Treatment and Pollution Control*,
https://doi.org/10.1007/978-981-99-2560-5_8

| | |
|---|---|
| PA | Polyamide |
| PAC | Powdered activated carbon |
| PEI | Polyethyleneimine |
| PES | Polyether sulfone |
| PIP | Piperazine |
| PNF | Positively charged NF |
| PSD | Pore size distribution |
| PSf | Polysulfone |
| RO | Reverse osmosis |
| TDS | Total Dissolved Salts |
| TFC | Thin film composite |
| THM | Trihalomethane |
| TMC | Trimesoyl Chloride |
| TMP | Transmembrane pressure |
| TOC | Total Organic Carbon |
| UF | Ultrafiltration |
| WHO | World Health Organisation |

## 1 Introduction

Pure drinking water is essential for our survival. The major sources of potable water include ground and surface water, rivers, and other freshwater sources (Vergili 2013). Increased pollution, climate changes, population growth, careless disposal of waste materials, etc., have all led to the deterioration of the water quality of these sources and resulted in severe water scarcity all over the world. More than 1 billion people in the world have limited access to safe drinking water. Furthermore, nearly 2.6 billion people have issues relating to poor sanitation that results in inferior water quality (Ahuja 2021). Hazardous wastewaters such as landfill leachate and industrial effluents discharged into the available potable water sources without proper pre-treatment pose a serious threat to human health globally since they contain high concentrations of toxic chemical compounds and pathogenic microorganisms (Singh et al. 2020). Conventional drinking water treatment methods have shown limited efficiency when it comes to meeting the standards of drinking water quality and curbing the concentration levels of emerging pollutants. Therefore, the development of advanced water purification technologies to tackle the ever-increasing contamination levels and to produce potable water has become an indispensable need in the current times (Lakhotia et al. 2018). In this context, membrane processes have proven to be an attractive alternative to achieve high quality permeate water which could be used as a source of direct or indirect potable water (Listiarini et al. 2010). Membrane filtration involves the use of membranes which act as selective filters to restrict the movement of pollutants. Depending on the pore size, membrane processes could be classified into four broad categories namely: microfiltration (MF), reverse osmosis (RO), nanofiltration (NF), and ultrafiltration (UF) (Karabelas and Plakas 2011).

Over the recent years, the development of NF technology has resulted in remarkable growth in its applications in numerous industries such as removal of pharmaceuticals compounds from fermentation broths, metal recovery and virus removal from wastewater, effluent treatment from textile industries, demineralisation in the dairy industry, etc. Commercial NF membranes usually have charged surfaces due to the dissociation of surface groups (Siddique et al. 2020). The characteristics of NF membranes fall between porous UF membranes and non-porous RO membranes which make it ideal for multivalent ions to be separated by a combination of size exclusion, electrostatic effects, and ion interactions (Hajibabania et al. 2011). The overall variables that influence filtration process include transmembrane pressure, temperature, fouling tendencies, permeability rate, etc., which in turn determine rejection rates of the target solutes (Uyak et al. 2008). NF membranes can remove a plethora of contaminants including organic matter, inorganic solutes, pathogens, heavy metals, hardness, turbidity, etc., owing to the pore size in NF membranes (nominally 1 nm) (Couto et al. 2020). Since NF technology incorporates several treatment goals in a single process, it is regarded as a suitable candidate for the purification of contaminated water such as in the treatment of tertiary wastewater to produce portable. However, the phenomenon of membrane fouling is one of the limiting factors hindering the successful operation of NF membranes since it leads to significant flux decline as well as altered surface morphology (Sarkar et al. 2007; Shah 2020).

Various techniques have been put forward to alleviate fouling like pre-treatment of the feed water (coagulation and adsorption), fabrication of novel nanocomposite membranes by introducing diverse nanofillers or nanoparticles (such as silver, graphene oxide, titanium oxide ($TiO_2$), carbon nanotubes, graphene quantum dot, molybdenum disulfide), etc. Besides fouling, low rejection rate of monovalent ions and inadequate management of membrane reject water (also referred to as retentate or concentrate) are major limitations of NF process (Franke et al. 2019).

This chapter primarily aims at describing the applications of NF membranes in potable water treatment and reuse by focusing on the removal of heavy metals, inorganic contaminants, and mixed contaminants from polluted sources such as wastewater, brackish groundwater, and contaminated surface water. A short note on customised NF process has also been highlighted towards the end of the chapter.

## 2 Removal of Contaminants by Nanofiltration

### *2.1 Removal of Heavy Metals*

Heavy metals are hazardous metals with a specific gravity, high density, or atomic weight that are normally found in the earth's crust. Human activities have disrupted their biogeochemical balance, causing them to invade our drinking water supply.

Cobalt, copper, zinc, and manganese are some of the necessary heavy metals that our bodies require in trace levels but can be harmful to our health if consumed in massive quantities. Other heavy metals, such as cadmium, lead, arsenic, mercury, selenium, chromium, and uranium, have no good impact on your body but can be dangerous if they accumulate in the body, creating health issues. Heavy metals are a serious concern in today's world as they are environmental contaminants, which are highly toxic and tend to accumulate in body tissues. In a study, the NF270 membrane was used to investigate the retention of five solutions (Cu (II), As (III), Mn (II), Pb (II), and Cd (II)) as well as the dependence of pressure, initial feed concentration, and solution Ph on membrane separation with a membrane area of 0.00076 $m^2$ and a pressure range of 1–5 bar.

NF270 is a polyamide membrane having a microporous polysulfone supporting layer and a thin semi-aromatic polyamide layer as its active layer. Arsenic was first dissolved in tap water to make a solution, and then Cu $(NO_3)$. $3H_2O$, Pb $(NO_3)_2$, $As_2O_3$, Mn $(NO_3)_2$, $4H_2O$, Cd $(NO_3)_2$, $4H_2O$, NaOH, and nitric acid (69–72%) were added in order after the former metal had fully dissolved. When the pH of the feed was lesser than the NF membrane's isoelectric point, all metals (excluding arsenic) displayed increased rejection due to electrostatic repulsion and metal adsorption on the membrane, revealing that Donnan exclusion or adsorption mechanisms are involved. The NF270 membrane rejected nearly 100% of Cu in a 1,000 mg/L copper solution at all pHs (1.50–5) and all pressures (3–5 bar), confirming its appropriateness for copper ejection. However, when the Cu concentration increased to 2,000 mg/L, its efficacy dropped. Utilising 1,000 mg/L level concentration at pH 1.50 and a pressure of 4 bar, NF270 eliminated nearly 99% of Cd, 89% of Mn, and 74% of Pb; nevertheless, being a loose membrane, it was unable to reject As (III). With all of the metals employed, the permeate flux rose when the pressure was increased. In the case of Cd (II) and As (III), the permeate flux was still linear with increasing pressure, denoting negligible concentration polarisation; however, the rise in flux was not linear with increasing pressure in the case of copper, lead, and manganese, indicating that a slight amount of concentration polarisation may have actually happened. Metals in the order $Cu^{2+} > Cd^{2+} > Mn^{2+} > Pb^{2+} > As^{3+}$ caused the flow reduction due to NF270 fouling. Apart from arsenic, which has a greater deposited amount and higher flux, the normalised flux decreased as the adsorbed concentration of metals onto NF270 increased. The difference in NF270 retention seen between single solution and the multicomponent solution was around 60% for Cu and Pb and 19% for Cd and Mn. The metal retentions were only moderately affected by the type of water used to manufacture the 5 components, with differences of just 4, 8, 5, and 6% for Cu, Mn, Cd, and Pb, respectively. The RMS roughness determined using Atomic Force Microscopy (AFM) revealed that membrane fouling reduced roughness (Al-Rashdi et al. 2013; Shah 2021a, b).

In this subsection, we'll look at some of the most common heavy metals present in drinking water, such as arsenic, chromium, selenium, and uranium, and how, based on previous research, nanofiltration could be used to eradicate these dangerous metals from water.

### 2.1.1 Removal of Arsenic

Arsenic is a natural compound of the Earth's crust which is highly toxic in its inorganic form. The groundwater in a number of countries including Argentina, Bangladesh, Chile, China, Mexico, USA, and India contains inorganic arsenic which is naturally present, at high levels. One of the major problems faced by today's world is arsenic-contaminated water which can even cause cancerous diseases (Some drinking water disinfectants and contaminants, including arsenic, 2004). Various studies have been conducted to purify water from arsenic which is a serious health concern.

In one of the studies, tap water was used to which arsenate and arsenite were added with the help of two types of nanofiltration membranes—NF90 and NF-200. The effective surface area of each membrane plate was 0.03412 $m^2$, and the NF90 had a water permeability of 8.6 L ($m^2$ h bar), while the NF200 had a water permeability of 9.8 L ($m^2$ h bar). The As(V) solution was made by diluting the As (V) standard solution in distilled water after the 1,000 mg/L stock solution had been diluted sufficiently. The As (III) stock solution was made by mixing $As_2O_3$ (Merck) in water, and the feed solution was made by diluting the stock solution with tap water. The amount of arsenic was detected using an Ag-DDTC spectrophotometric technique, and the conductivity and pH were measured using a conductivity metre and a pH metre, respectively. The influence of feed solution concentration, pressure, temperature, and pH on the elimination of As (V) and As(V) was examined in this experiment (III). Over the pressure ranges of 5–20 bar, the removal of As (V) was vastly higher than that of As (III). The removal of As (V) by NF90 was greater than that by NF200 over the pressure range examined, and the removal of As(V) by the membranes seems to be unaltered by the difference in applied pressure, reaching 98% in both cases. The removal of As(III) by the membranes increased with the increasing pressure, with NF200 removal ranging from 16% at 5 bar to 30% at 20 bar, and NF90 removal ranging from 50% at 5 bar to 63% at 2 bar. It did not depend on the feed concentration, with removal rates exceeding 98% for initial feed concentrations ranging of 100–1,000 g/L, while the expulsion of As(III) by NF90 was greater than that of NF-200, indicating that both types of membranes increased with increasing feed concentration. Over a pH range of 5–9, it was also discovered that the level of As(V) in the permeate was lesser than that of As(III) for both sorts of NF membranes. In addition, the elimination of arsenic for both membranes reduced as the temperature rose. It can be deduced from this study that source water that contains As(V) can be retrieved as drinkable water to EPA maximum contaminant level quality requirements by modifying the operating parameters, but source water having As(III) should undergo a pre-oxidation treatment before being passed through the NF membrane to retain water quality (Uddin et al. 2007; Shah 2021a, b).

Later in yet another experiment, NF90-4040 membrane was used to remove arsenate and arsenite samples which were added to the groundwater and four solutions (arsenite and arsenate) at concentrations of 100, 300, 500, and 1,000 μg/L were prepared to vary the laboratory conditions. Considering the above-stated input concentrations for both arsenate and arsenite, a steady pressure of 5 bar, at 28 °C, and pH 8 were employed. At 28 °C, the effect of pressure was investigated by administering pressures of 4, 5, 6, and 7 bar while maintaining a steady input concentration of 500 g/L and pH of 8. Temperature effects were also investigated at constant pressures of 5 bar and pH 8 with stable arsenate and arsenite ratios of 500 g/L at temperatures of 28, 31, 34, and 37 °C. According to the results of the aforementioned experimental analysis, raising arsenic concentration in the feed from 100 to 1,000 g/L resulted in lower arsenate and arsenite removal rates due to concentration build-up on the membrane surface, which clogged filtration passages through the membrane. When the pressure was increased from 4 to 7 bar, the removal efficiency of arsenate and arsenite elevated, most probably relationship existing between transmembrane pressure and water flux; even so, as the temperature increased, the removal efficiency for both reduced, most likely due to arsenic's diffusivity increasing with temperature, resulting in greater arsenic diffusive transport across the membrane. At a 500 g/L input concentration, the maximum arsenite removal was 90%, whereas arsenate removal was 93% at the same concentration at test conditions of 5 bar, pH 8, and 28 °C. Because of the negative charge on the surface of the membrane, arsenate has always had a better removal efficiency than arsenite (Harfoush et al. 2018).

Another experiment used GE, Osmonics' DK and HL membranes to decontaminate 3 real contaminated groundwater specimens from the Sila Massif (Calabria, Italy), renamed GW1, GW2, and GW3, with the goal of determining their effectiveness of arsenic exclusion and water production, as well as feed composition. The studies were conducted for each groundwater at varying transmembrane pressures (TMP) of 3 bar, 7 bar, 11 bar, and 15 bar using HL and DK membranes, which are made from same polymer and have comparable MWCO, with DK being a "High Rejection" type membrane and HL being a "Softening" type membrane. The results revealed that fluxes always increased with TMP due to the huge driving power used, and that the HL membrane had a stronger permeate flux than the DK membrane. Examine the influence of feed composition, the fluxes demonstrated by GW1 and GW2 groundwater resources were relatively similar; however, there was a considerable difference for the GW3 feed for both membranes, with the DK membrane showing the greatest fluxes and the HL membrane showing the lowest. At the conclusion of the tests, zero scaling or fouling was seen. At greater TMPs, As rejection rate increased noticeably, averaging between 94 and 98%, with the poorest rejection values recorded for the GW1 feed, and equivalent removals for the GW2 and GW3 waters. The DK membrane, on the other hand, demonstrated stronger As rejection at decreased TMP. The HL membrane performed better, combining good arsenic removal with increased water transmembrane fluxes. When the TMP was administered at 3 bar and 15 bar, the permeate recovery rate for the DK membrane was between 3 and 6% in steady-state conditions, and between 5 and 15% for the HL membrane (Figoli et al. 2020).

In a recent study conducted in 2021, arsenic was treated using nanofiltration with emphasis on water chemistry. The effectiveness of low and high salt rejection flat sheet membrane features was compared using NF270 and NF90 flat sheet membranes. Ionic strength was first examined by checking As(V) rejection at various salinities, and it was discovered that increasing NaCl concentrations influenced the pressure applied and, eventually, the requisite driving force for the separation with NF while operating at steady flux conditions, demonstrating the high rejection properties of NF90 contrasted to NF270. Increased NaCl concentration compresses the double layer on the membrane surface, lowering the Debye ratio, which can lead to decline in the amplitude of the membrane surface's zeta potential. When the bivalent $HAsO_4^{2-}$ is the dominating species at pH 7.9–8.2, size exclusion was determined to be the major As(V) rejection mechanism. Also, pH had no impact on As(V) rejection with NF90, whereas As(V) speciation was required for NF270 rejection. When As(V) rejection was examined in the presence of humic acid (HA) in the feed solution with elevated NaCl concentration, no major shift was observed with NF90; but even so, with NF270, the rejection improved by about 10%, resulting in a low As(V) concentration on the permeate of NF270 to values nearest to the WHO guideline at higher NaCl concentration. For example, at 20 g/L NaCl, As(V) rejection with NF270 was 94% when HA was present, but only 82% without HA, which can be interpreted by the elevation in the negative charge of the membrane surface. According to the findings, the enhancement mechanism on As(V) rejection via HA in NF is due to HA chemistry, and NF technology is an effective method for removing As(V) in aspects of salinity, pH, and the presence of organic matter (Boussouga et al. 2021).

### 2.1.2 Removal of Chromium

Chromium, an essential trace element for humans, is a metal found in natural deposits as ores containing other elements; however, it is poisonous in excess. Trivalent chromium (chromium-3), which is found in many vegetables, fruits, meats, cereals, and yeast, and hexavalent chromium (chromium-6), which occurs naturally in the environment through the weathering of natural chromium deposits, are the two most prevalent forms of chromium found in natural water. Hexavalent chromium is likely to be a carcinogen, if ingested as drinking water.

Nanofiltration membranes were applied to eliminate hexavalent chromium in a study as a rapidly evolving technology in the field of producing pure drinking water. In bench-scale and pilot-scale tests, genuine contaminated well water from the researched location near Turin, Italy was collected and treated employing two types of membranes, NF270 and NF90, with varied selectivity. At neutral pH conditions, zeta potential studies for both membranes revealed a negatively charged surface, indicating that NF270 is more permeable and less specific than NF90. Filtering well water with the membranes was done in the laboratory at a 100 psi pressure, a cross-flow velocity of 4.5 L/min, and a temperature of 22 °C; and 2 distinct pilot-scale experiments were done with full-fit fibreglass 4040 spiral wound elements for both NF270 and NF90 membranes at a flow rate of 2,000 L/h, a cross-flow rate of 26.7 L/min, and a recovery rate of 20. The facility operated at a continuous applied pressure of 5.25 bar for NF270 and 5.75 bar for NF90 for 42 days. The findings demonstrate that NF90 had maximum removal rates than NF270, with rejection rates over 80% with most elements or ions. The former separates compounds via size-dependent processes, whereas the latter separates them by charge exclusion. Cr removal was 78 and 98% in both NF270 and NF90 samples, and at standard conditions, the looser NF270 membrane had only marginally higher values than the firmer NF90 membrane. Additionally, the laboratory experiment's fluxes and rejection rates were found to be similar with those from the pilot-scale experiment. The favourable test findings led to the implementation of a full-scale NF plant for the particular water source explored in this study, where the membrane system was built to function with NF90 membrane modules especially, the NF90-400/34i, which has an area of 37.2 $m^2$ per module. According to a cost analysis, a plant with non-regenerable ion exchange resins can cost a minimum of 0.3 € per cubic metre of product water over a 10-year lifetime, which is significantly less than the price of water for a plant with non-regenerable ion exchange resins, which is approximated to be exceeding 1 €/$m^3$ due to the expense of resin replacement. In terms of the environment, the plant's power source might be the most significant component (Giagnorio et al. 2018b).

Different membranes: NF270, NF90, and HYDRACoRe70pHT (name is shortened to Hydra70) were used in a study to assess nanofiltration as a viable process to obtain lower chromium concentrations in drinking water, as well as to find the most suitable membrane based on specific treatment needs. Hydra70 is a selective sulfonated polyethersulfone on top of a polysulfone support layer. When water permeance was measured using water and permeate flux at various applied pressures, NF270 had the maximum permeance, whereas Hydra70 had the lowest. Three sets of tests were conducted using a cross-flow laboratory scale system, two of which used hexavalent chromium solution in DI water to change the ionic strength and pH, and the third set used chromium-contaminated well samples taken from three different sites in the Turin area, Italy. The membrane was compacted for 6 h at 300 psi prior to each experiment, with a constant cross-flow velocity of 4.5 L/min, an administered pressure of 100 psi, and feed water temperature of 22 °C. For oxidising the membranes, three oxidation baths were constructed, each comprising 2 L of oxidising solution, with the first containing 100 mg/L of NaOCl, the second containing 100 mg/L of $K_2Cr_2O_7$, and the third containing a mixture of both at a concentration of 100 mg/L, pH 7.6 in all cases. The surface zeta potentials of membranes submerged in various solutions were evaluated using an electrokinetic analyser, and the concentration of chromium in the obtained feed was quantified by inductively coupled plasma optical emission spectrometry. The flux through polyamide membranes fell sharply in the presence of $CaCl_2$, but the proportionate reduction of flux via Hydra70 membranes is minimal as the ionic strength increases. The looser NF270 membrane displayed greater negative zeta potential than the thicker NF90 membrane across the pH range tested, demonstrating that electrostatic effects are essential separation mechanisms, particularly for Hydra70 and NF270 membranes. When the pH of the feed solution was lowered, NF270 membranes had slightly greater water fluxes, although NF90 membranes had slightly lower fluxes at acidic pH, yet Hydra70 had a distinct response, with a minimal drop in permeate flux with declining pH. Calcium ions had the most negative impact on hexavalent chromium ion rejection by NF270 and Hydra70 membranes, although NF90 exhibited a high level of contaminant rejection even at high ionic strengths. The NF270 membranes had shown a 1.5-fold rise in permeance when exposed to oxidising agents alone; similarly, the NF90 membranes exhibited a 1.5-fold rise in permeance when exposed to both Cr(VI) and hypochlorite; even so, Hydra70 did not have any notable changes in permeance when exposed to oxidising agents, and only a small reduction in selectivity to Cr(VI) was observed after being in contact with a solution comprising both oxidising agents. The findings suggest that when treating sources contaminated by Cr(VI), all of the membranes investigated here might achieve new stringent limits for Cr(VI) in drinking water (Giagnorio et al. 2018a).

### 2.1.3 Removal of Selenium

Selenium is a naturally occurring element that can be dissolved through weathering of minerals in soil or rock, deposited from air sources, and found in groundwater. It is widely dispersed across the Earth's surface. Owing to its existence in ground water, that's used as a source of drinkable water, selenium has emerged as a severe health danger in today's environment of growing contaminants in potable water. Because selenium is extremely soluble in water and has a dangerous concentration range of 400–700 g/L, removing it from contaminated ground water is a difficult task. The needed quantity of selenium in water is 40 g/L. In this work, selenium was eliminated from living contaminated ground water retrieved from India's Punjab province, where the issue is particularly severe.

Three distinct commercial composite polyamide NF membranes, NF2, NF20, and NF1, were employed, each with a surface area of 100 $cm^2$. Studies were conducted in the flat sheet cross-flow module (FSCF) in steady state under response surface optimised settings to choose the right membrane from a specific lot. The research revealed that NF2 had the largest flow, followed by NF20 and NF1, signifying that NF1 is the tightest membrane and NF2 is the loosest. NF1 also had the maximum rejection, led by NF20 and NF2. Additionally, when the pH was raised from 2 to 12, Se rejection rose from 40 to 97.2% for the NF1 membrane, 26 to 66% for the NF2 membrane, and 32 to 78.5% for the NF20 membrane. Based on the results, the NF1 membrane, which had a greater rejection rate and permeate flux, was chosen for further module optimisation in order to build linear and combination correlations among independent individual variables and reliant parameters. Se contaminated water can be treated effectively with an enhanced flux of around 140 LMH and a rejection rate of >98% using a pressure of around 13–14 bar, an initial Se content of 1,600 ppb, an hourly fluid rate of 700 l, and a pH of 8. Despite a 240-h operation, the NF1 membrane showed just a 6.06% flow drop when evaluated for fouling propensity. Finally, this process reveals a fairly simple design with low surveillance and operational expenses, indicating that it could treat selenium-contaminated water for consumable use (Malhotra et al. 2020).

### 2.1.4 Removal of Uranium

Uranium is a naturally occurring, radioactive heavy metal which is a normal part of rocks, soil, air, and water. It occurs in the form of minerals in nature and enters water by leaching from soils and rocks or in releases from processing plants.

A strategy for selective extraction of uranium traces which are dissolved in drinking water was presented in this paper. Uranyl nitrate ($UO_2(NO_3)_{2.}$ $6H_2O$) was introduced to mineral water samples in plastic bottles collected through regular distribution channels. The investigations were done in three flat membranes, DK, DL, and G10, with a membrane surface area of 155 $cm^2$ and a polyamide selective layer supported on a polysulfone layer. When assessed for pure water permeability, the DK membrane is by far the most compact, whereas the G10 and DL membranes have the most open structures. Mineral water was filtered on the membrane at a fixed temperature of 25 °C with a transmembrane pressure of 1–4 bar, a tangential velocity of 145 mm/s, and a transmembrane pressure of 1–4 bar. The rejection of U(VI) with a reduced carbonate concentration in the sample water was 40% with G10, 95% with DL, and 100% with DK membrane, exhibiting the same pattern as the rejection of alkaline and alkaline-earth ions. DL and DK membranes displayed greater rejection regardless of the sample taken; however, G10 membrane reported reduced uranyl rejection. The charge effects between uranyl carbonate complexes and the charged membrane play a major role in uranyl rejection. G10 and DL membranes were employed in filtering tests to investigate the effect of operational conditions on U(VI) rejection. Transmembrane pressure was varied from 1 to 4 bar at a steady tangential velocity of 145 mm/s. When the feed composition remained constant and the rejection of every component grew linearly, the permeate water flux increased gradually as the pressure is increased. Further, using the DL membrane at a transmembrane pressure of 1 bar, a tangential velocity of 145 mm/s, and a temperature of 20 °C, a water sample containing up to 20 ppb of UV(I) may be reduced to 0.8 ppb, well below the WHO MAC. As an outcome, this technology for in situ water treatment is both practical and cost-effective (Favre-Réguillon et al. 2008).

## 2.2 *Removal of Inorganic Contaminants*

Potable water or groundwater is usually contaminated with a variety of inorganic contaminants like fluorides, sulphates, and other toxic chemicals. With modern age technology, we learn many methods to filter out these hazardous toxins from the drinking water being consumed, nanofiltration is one such technique which has proved to be highly effective in the removal of these inorganic toxins from drinking water. Nanofiltration uses membranes with pores of approximately 0.001 microns in size which can be coated with other chemicals to increase the efficiency of the process. The use of nanofiltration membranes has been studied through various investigations against different inorganic toxins in this subsection.

### 2.2.1 Removal of Fluorides

Groundwater is a great source of portable water that can contain increased fluoride concentration in the form of fluorine, biotite, fluoro-apatite, cryolite, and villiaumite. Fluorides can cause major health hazards like "blue baby" and "dental fluorosis". The acceptable contamination standards fixed by the WHO are 1.5 $mgL^{-1}$. The cost-effective method of nanofiltration has been suggested to defluorinate the water and makes it acceptable for using as it can provide desalination and remove divalent ions such as sulphates and calcium.

In a study, nanofiltration membranes such as NF5 and NF9, the NF1, NF2 and NF20, and spiral wound NF 4040 thin film composite membranes were used to investigate the efficiency of using nanofiltration as a method to filter out the fluoride ions in different concentrations in the water. Throughout the process the pressure and temperature were controlled to create the ideal environment for the study. The chemicals present after the study was determined using methods like ionic chromatography and scanning electron microscope. The results after conducting the experiment showed that the nanofiltration membranes could significantly reduce the presence of fluoride as well as salt in the water with high initial concentration up to 60 mg/L to a maximum fluoride level of 1.5 mg/L. Throughout the entire testing period, the NF90 membrane was most effective by rejecting 96–99% of fluoride ions in the water. It also showed that more hydrated ions were better retained by the membrane due to the energy generated by the hydrogen bonding of the molecules. The investigation showed positive economic value due to the cheap and easy installation. To conclude, the solar powered nanofiltration unit is low cost, easy to install as well efficiently removes fluoride from highly saline waters (Chakrabortty et al. 2013; Nasr et al. 2013; Bouhadjar et al. 2019).

### 2.2.2 Removal of Sulphates and Copper

Waters contaminated with sulphates and other inorganics which cause major health issues to the people who consume it. It was seen that the membranes ionic retention and its charge depends on the individual contribution of the ions absorbed on the membrane. The different pore size distribution (PSD) also affects the solute retention. A method known as interfacial polymerisation (IP) is a simple technique used to deposit polyamide active layer above a porous membrane through diffusion and rapid reaction of water phase monomer to oil phase monomer; the polyamide active layer is formed by the interfacial polymerisation is the part of NF membrane that plays a major role in separation, which is primarily responsible for permeance and selectivity. Thus, it is very important to regulate the structural parameters and chemical properties of the PA active layer. The process used to prepare the membrane will not only reduce the thickness of the active layer but also have an effective area larger than that of the unit support layer. All these features will facilitate water molecules to pass through the membrane matrix resulting in the enhanced permeation flux of NF membrane. The contaminated water was passed through the membrane and the concentration of the inorganic salts was determined by conductivity measurements and the organic marker concentration was checked using carbon analysis method. The growth of inorganic salt crystals on the membrane stretches the flexible PA layer. The inorganic salt crystals can dissolve themselves in water, resulting in the formation of a thin and a rough PA layer with a crumpled nanostructure. For NaCl, the membrane of the prepared NFM-1 was up to 24.1 L/m h bar while the rejection maintained above 96% for $Na_2SO_4$. The NF membrane showed possibilities of the removal of sulphates and chloride ions from contaminated water by the addition of an active PA layer on the membrane. To conclude for other soluble inorganic salts such as $Na_2SO_4$, KCl, $CaCl_2$, and $KNO_3$, the active PA layer improves the filtration performance of NF membrane, and it is ideal due to its simplicity and low cost (Košutić et al. 2004; Liu et al. 2021).

### 2.2.3 Removal of Nitrogen

Increased concentration of nitrates in drinking water is a huge problem due to industrial discharge, human activities, and excessive fertilisers used in intensive agriculture. The membrane is selected based on the size effect, charge effect, difference in diffusion rates, and solubility in solutes; the salt $NaNO_3$ was used in different concentrations to determine the retention rate of the nitrate ions when the concentration varies with difference in valances. The study was mainly based on three flat sheet organic polyamide membranes used in this experiment—NF90, OPMN—K and OPMN-P, as well as analytical grade salts. The procedure was done in batch mode in which the permeate solutions and retentate solutions were carried back to the feed tank. By keeping, velocity and the pressure constant the correct atmosphere was created. It was seen that OPMN-K and NF90 had higher flux and OPMN-P has smaller flux in comparison and for the NF90 membrane the retention remained constant which is the most important and the pores are smaller; the NF90 membrane works best with a simple $NaNO_3$ solution at 70 ppm. To conclude, the NF90 membrane is best suited for filtering the nitrate contaminated waters of France as it strongly reduces nitrate ion concentration ant high flux (Garcia et al. 2006).

### 2.2.4 Removal of Ammonia

High concentrations of ammonium in portable water are extremely hazardous to humans, it is converted into nitrate which is highly toxic for humans. Ammonium is adsorbed in the intestine through the microvilli and can be excreted directly through the kidneys in small quantities, thus, a standard ion concentration set by the European Community and Turkish Standard of 0.5 mg/l should be constantly maintained at all times. The use of membrane filtration has been used increasingly in the filtration of water as compared to other conventional treatment techniques. In this study, N30F, NF-PES-10 membranes, dual membrane module system with sheet type membranes were used to treat the water treatment plant effluent which has been contaminated with Ammonium ions, nitrate, and calcium ions. The results of the study proved to be effective in decontaminating the water of ammonium ions (Kurama et al. 2002).

The demand for clean hygienic portable water is on the rise and methods such as nanofiltration, reverse osmosis, and ultrafiltration have been investigated to find an effective and inexpensive way to achieve a solution for the problem. In the study, the membrane is modified by the aqueous sodium hypochlorite which chemically oxidising the NF membrane. This has seen to be highly effective in the removal of toxins from the surface water and groundwater. It was seen that the modification of the membrane by coating with sodium hypochlorite was not only efficient but also economical. Also, $ClO_2$ proved to be a safe disinfectant for the removal of contaminants without degrading the PA NF layer as it does not generate free chlorine in aqueous solutions. The process successfully produced safe drinking water with acceptable taste and mineral content (Govardhan et al. 2020).

Nanomaterials such as carbon nanoparticles have been incorporated into the matrix of the nanofiltration membrane have shown to be highly effective in the removal of inorganic ions such as sulphate and copper. In research conducted, nanoparticles were embedded in a mixed matrix of asymmetric polyethersulfone based on which nanofiltration units (I.e., membranes) were synthesised by casting a solution to increase the efficiency removal of sulphate and copper ions from the contaminated water. The nanofiltration units used in this study is an asymmetrical flat sheet which has a thickness of 150 $\mu$m and a pore size of 3–5 nm. The effect of concentration of the nanoparticles in the casting solution on the membrane morphology was studied, and the results were analysed by SEM (Search Engine Marketing) imaging, which showed that increasing the concentration of nanoparticles in the casting solution caused a decrease of the channel size in both the top layer and the sublayer of the membrane matrix and the SOM imaging showed that a uniform distribution of nanoparticles for the prepared membranes. Moreover, results showed that the flux decreased initially by utilising nanoparticles up to 0.05 wt% in the membrane matrix and then increased again for a nanoparticles concentration from 0.05 to 0.1 wt%. To conclude, the addition of super-activated carbon nanoparticles improved the rejection rate of sulphate and copper ions from the water (Hosseini et al. 2018).

## 2.3 *Removal of Mixed Contaminants*

NF has proven to be an excellent candidate for treating complex streams such as industrial effluents, polluted ground/surface water, and wastewater. Currently, a wide range of nanofiltration membranes are commercially available to target mass removals of specific or groups of compounds. Even though existing reverse osmosis (RO) technology can eliminate larger amounts of the same constituents than NF membranes, it might not be the best choice. RO membranes eliminate not just the contaminants of concern, but also hardness and dissolved minerals to a point where the product water is reactive towards concrete and metal conduits. Treatment systems must then reintroduce some of these hardness and minerals back. This results in "wasted" effort to overshoot the water quality objective which later must be brought to usable levels. Nanofiltration essentially solves this issue by eliminating the need to remineralise product water. The NF membrane process for advanced drinking water treatment is capable of removing a plethora of contaminants including organic, inorganic, and pathogens effectively under a low operating pressure (Shon et al. 2013).

Total dissolved solid (TDS) concentration of domestic wastewater effluents is generally in the range of 250–850 mg $L^{-1}$. While RO membranes are inevitable for desalination of extremely saline water, NF membrane can be considered as a feasible option for application in water reuse cases where the TDS of domestic wastewater is close to the secondary drinking water standard (i.e., 500 mg $L^{-1}$). The efficiency of NF in achieving the optimum TDS (100–500 ppm) in the product water was demonstrated in one study. With an inlet pressure of 10 bars, a thin film composite nanofiltration (TFC-NF) membrane incorporated within a spiral wound membrane module was used. TDS concentration in the feed water was measured at 451 mg/L. Following the treatment by NF, permeate water had only about 135 mg/L TDS concentration indicating 70% rejection (Naidu et al. 2015).

In terms of organics removal, NF can eliminate Total Organic Carbon (TOC) present in the feed water upto 90% effectively. TOC is basically composed of dissolved organic matter (DOM) and particulate matter. The major constituents of DOM in wastewater effluents comprise of fulvic substances, humic substances, and soluble microbial products (SMPs) (with molecular weight >1,000 Da). This is particularly relevant in case of NF since the pore size of most NF membrane are smaller than 1,000 Da, and therefore, it can achieve more than 90% of TOC rejection as well as about 60% of chemical oxygen demand (COD) removal. With respect to trihalomethanes (THMs) removal, it was observed that NF treatment could significantly reduce THM concentrations in the tap water from 55 μg/L to below 25 μg/L, which is much lower than the maximum contamination level (MCL) for potable water supply. The surface charges and fine pores of NF membranes are mainly responsible of the rejection of polarising molecules such as THMs. As a result, the stronger the polarisation and the greater the molecular weight, the easier it is to be rejected by the NF membrane. Similar trends are observed in the case of trihalomethane formation potential (THMFPs) when NF membranes such as ESNA1LF2 4040 (Hydranautics) and NF270 4040 (Dow Chemical) modules with membrane pore size of 0.49 and 0.5 nm, respectively, were used (Li et al. 2002).

In the case of uncharged organic molecules like pesticides, rejection occurs due to size exclusion mechanism but its role is affected by membrane/solvent/solute interactions, laying emphasis on the significance of the membrane pore size distribution (PSD) and material for achieving maximum rejection for uncharged organic solutes. In a particular study, four pesticides of different physiological properties and molecular structures were chosen and used in experiments pertaining to membrane rejection process. The rejections of all the pesticides were reasonably high supported by the rejection values, R, of pesticides by the NF270 which was found to be 0.407, 0.814, 0.998, and 0.931 for dichlorvos, atrazine, triadimefon, and diazinon, respectively (Košutić et al. 2005).

NF membranes can fairly remove inorganic feed components from various sources such as wastewater, brackish water, or seawater. Inorganic ions are generally rejected in the range of 50–90%. This NF product water would be healthier to drink as compared to distilled water since it can still retain certain amounts of inorganic ions like K, Na, and Zn. For chlorides, nitrates, sulphates, and phosphates approximately 60–90% rejection is observed. In drinking water treatment plants (DWTPs), NF membranes with lower salt ($Ca^{2+}$, $Mg^{2+}$, $Na^{+}$, $K^{+}$, Cl) rejection values have some beneficial implications since higher salt rejection leads to higher osmotic pressure differences on either side of the membrane implying higher feed pressures and energy requirements (Ribera et al. 2013).

Magnesium sulphate ($MgSO_4$) and sodium chloride (NaCl) are often used as model salts in salt rejection studies. In a particular study, 2,000 ppm $MgSO_4$ and NaCl aqueous solutions were used as feed solutions and passed perpendicularly through a NF membrane surface while maintaining 5 bar pressure and the results showed rejection of $25.43 \pm 3.12\%$ and $92.51 \pm 2.73\%$ for NaCl and $MgSO_4$, respectively (Yadav et al. 2022).

Heavy metal concentrations in ground water, as discussed earlier in the chapter, are generally below detection limits for potable water. However, industrial effluents found in wastewater can contain considerable concentrations of heavy metals. NF exhibit reliable performances in terms of heavy metal ions (e.g., As, Ni, Cd, Sb, Zn) removal with R values up-to 0.994. The removal efficiency of arsenic by NF shows good removal rate of about 80–96% (Zhu et al. 2017; Giagnorio et al. 2019).

Over the years, great advancements have been made in the development of high-performance nanofiltration membranes for advanced drinking water treatment. However, a proper pre-treatment prior to the NF process is needed to achieve certain feed water quality, minimise membrane fouling and to prolong the lifetime of the NF membrane used. One such setup is the compact NF purification system consisting of a granular-activated carbon (GAC) adsorption filter, dual media granular filter, a cartridge filter, and two NF membrane filters in parallel. A mixture of **ST** (a cationic polymer coagulant) and **PAC** (polymeric aluminium chloride) can be added to flocculate soluble and colloidal organics in the feed water, since this combined dosage can improve their individual flocculation efficiencies. The dosage of 3 and 1 mg/L for PAC and ST, respectively, demonstrated the highest silt density index (SDI) removal of 55%. Water sample analysis after the pre-treatment indicated that it could completely meet the requirements to feed the nanofilter. However, the application of GAC adsorption before NF process enhances biological growth which further leads to the occurrence of nitrification. To overcome this, the NF system can be operated using different operating modes like intermittent-operation and no bacteria would be detected in the NF permeate water confirming 100% rejection of all the bacteria and other pathogens. This developed NF purification system takes up little space and can work at low pressure making it a suitable candidate in water industries (Li et al. 2002).

Another approach towards customised nanofiltration technique is the integration of UF to create hybrid UF/NF process. This is done in order to avoid considerable fouling. In a particular study, hydrophilic polyethersulfone (PESH) UF membrane was used, with MWCO of 50 kDa and an effective area of 41.8 $cm^2$. NF device was equipped with a gear pump and a commercial cross-flow flat sheet membrane cell. The NF membrane (NF270) was used with an effective membrane area of 42 $cm^2$. Pre-coagulation and pre-adsorption were also employed to maximise efficiency of the process. The optimum dosage of the coagulant and powdered activated carbon (PAC) was found to be 35 mg/L and 3 g/L, respectively, later applied in the subsequent UF/NF process. The results indicated the removal rates of $Mg^{2+}$ and $Ca^{2+}$ in the coagulation/UF/NF to be enhanced to 98.9 and 99.1%, respectively. In case of organic matter removal, UF/NF offered a satisfactory rejection performance, especially for humic-like substances with removal up to 94.7%. After pre-coagulation, the removal of common metals (such as Mg, Ca and Na) also improved. While pre-coagulation failed to make significant reduction in membrane fouling, adsorption/UF/NF proved to be helpful in maintaining high and stable permeation (Wang et al. 2020).

NF membranes are also widely used in forward osmosis (FO) process. A typical **FO-NF** setup consists of a FO test cell (active area of 0.00114 $m^2$). The support layer faces draw solution, and the active layer of membrane faces feed solution (FS) (AL-FS or FO mode orientation). An initial detailed SEM micrograph study of the low-pressure NF membrane revealed porous PES top layer and highly porous support substrate facilitating water permeability essential for FO operation. The membrane rejected feed solutes efficiently and pure water was obtained as permeate when synthetic dye wastewater was fed through. Overall, the coupled FO-NF process could be employed to produce potable water from industrial wastewater. Furthermore, this setup could even be commercially viable if the energy efficiency of draw salute recovery is increased (Dutta et al. 2020).

Nowadays, there is also use of compounds such as **graphite oxide (GO)** and **vanillin** in NF membrane fabrication process where polyether sulfone (PSf) is used as base polymer. Vanillin and GO have antifouling and hydrophilicity properties which maintain the structural integrity of the PSf-based mixed matrix membranes. In a particular study, experimental work on the synergistic effect of 2D Graphite oxide (GO) nanolayers and vanillin on membrane permeability and selectivity using Polysulfone (PSf) as the base polymer for landfill leachate wastewater treatment was carried out. 5 membranes ($M_1$, $M_2$, $M_3$, $M_4$, $M_5$) were fabricated with varying concentrations of GO ranging from 0 to 200 mg along with 3.2 g PSf, 16 ml (N-Methyl-2- pyrrolidone) NMP, and 0.8 g vanillin in each membrane. Morphological and physicochemical analysis was carried out using various microscopic and spectroscopic techniques. It was noted that both the hydrophilicity of the membranes as well as surface zeta potential increased with increasing the GO concentration. The surface zeta potential increased from −20.52 mV to −25.07 mV for $M_1$ and $M_4$ membranes. However, a slight decrease in surface zeta potential was observed for $M_5$ membrane (−24.12 mV) due to the aggregation of GO flakes. This increase in negative zeta potential has beneficial implications such as the inhibition of foulant adsorption on the membrane surface and improved salt rejection. Divalent ions were rejected better as compared to monovalent ions in all membranes. Landfill leachate wastewater and BSA feed solutions were used to test the antifouling characteristics of PSf/GO-vanillin membranes. Antifouling results showed over 99% rejection for bovine serum albumin (BSA) and 93.57% flux recovery ratio (FRR). Moreover, the $PSf_{16}/GO_{0.15}$-vanillin$_{0.8}$ membrane outperformed the conventional UA60 NF membrane in parameters such as antifouling and salt rejection. Similar results were obtained for FRR wherein a significantly high FRR of 93.57% and 90.32% was recorded for BSA and landfill leachate wastewater, respectively. Therefore, such asymmetric membranes incorporating GO and vanillin offers a versatile, energy-efficient, robust, and cost-effective approach for preparing membranes with ideal permeability and fouling-resistant properties (Yadav et al. 2022).

## 3 Conclusion

In this developing world, clean drinking water is hard to obtain, for which nanofiltration provides a solution. It uses pressure to remove impurities from water streams, and it is one of the four membrane technologies. The advantage of NF over reverse osmosis is that it has a greater flux rate than RO, using fewer membrane elements and working at lower pump pressure, resulting in cheaper operational costs. NF is the most dependable membrane process available because of its unique features of maintaining optimal TDS (with critical minerals necessary by the human body), minimal energy consumption, and no chemical use.

We summarised the development of materials and the procedures used in this chapter to examine the performance of NF-based water filtration. Heavy metals, as well as organic and inorganic salts of substantial significance, were removed with care. Developing countries can use nanofiltration procedures, which are much less expensive than traditional methods, to enhance the availability of clean water. However, it is unclear how these economies will be able to integrate this modern technology into their economies without relying on international aid.

## References

Ahuja S (2021) Chapter 14 - select applications of nanomaterials for water purification. In: Ahuja S (ed) Handbook of water purity and quality, 2nd edn. Academic Press, Amsterdam, Second Edition, pp 339–357

Al-Rashdi BAM, Johnson DJ, Hilal N (2013) Removal of heavy metal ions by nanofiltration. Desalination 315:2–17. https://doi.org/10.1016/j.desal.2012.05.022

Bouhadjar SI, Kopp H, Britsch P et al (2019) Solar powered nanofiltration for drinking water production from fluoride-containing groundwater – a pilot study towards developing a sustainable and low-cost treatment plant. J Environ Manage 231:1263–1269. https://doi.org/10.1016/j.jenvman.2018.07.067

Boussouga YA, Frey H, Schäfer AI (2021) Removal of arsenic(V) by nanofiltration: impact of water salinity, pH and organic matter. J Membr Sci 618. https://doi.org/10.1016/j.memsci.2020.118631

Chakrabortty S, Roy M, Pal P (2013) Removal of fluoride from contaminated groundwater by cross flow nanofiltration: transport modeling and economic evaluation. Desalination 313:115–124. https://doi.org/10.1016/j.desal.2012.12.021

Couto CF, Santos AV, Amaral MCS et al (2020) Assessing potential of nanofiltration, reverse osmosis and membrane distillation drinking water treatment for pharmaceutically active compounds (PhACs) removal. J Water Process Eng 33. https://doi.org/10.1016/j.jwpe.2019.101029

Dutta S, Dave P, Nath K (2020) Performance of low pressure nanofiltration membrane in forward osmosis using magnesium chloride as draw solute. J Water Process Eng 33. https://doi.org/10.1016/j.jwpe.2019.101092

Favre-Réguillon A, Lebuzit G, Murat D et al (2008) Selective removal of dissolved uranium in drinking water by nanofiltration. Water Res 42:1160–1166. https://doi.org/10.1016/j.watres.2007.08.034

Figoli A, Fuoco I, Apollaro C et al (2020) Arsenic-contaminated groundwaters remediation by nanofiltration. Sep Purif Technol 238:116461. https://doi.org/10.1016/J.SEPPUR.2019.116461

Franke V, McCleaf P, Lindegren K, Ahrens L (2019) Efficient removal of per- and polyfluoroalkyl substances (PFASs) in drinking water treatment: nanofiltration combined with active carbon or anion exchange. Environ Sci Water Res Technol 5:1836–1843. https://doi.org/10.1039/c9ew00286c

Garcia F, Ciceron D, Saboni A, Alexandrova S (2006) Nitrate ions elimination from drinking water by nanofiltration: membrane choice. Sep Purif Technol 52:196–200. https://doi.org/10.1016/j.seppur.2006.03.023

Giagnorio M, Ricceri F, Tiraferri A (2019) Desalination of brackish groundwater and reuse of wastewater by forward osmosis coupled with nanofiltration for draw solution recovery. Water Res 153:134–143. https://doi.org/10.1016/j.watres.2019.01.014

Giagnorio M, Ruffino B, Grinic D et al (2018a) Achieving low concentrations of chromium in drinking water by nanofiltration: membrane performance and selection. Environ Sci Pollut Res 25:25294–25305. https://doi.org/10.1007/s11356-018-2627-5

Giagnorio M, Steffenino S, Meucci L et al (2018b) Design and performance of a nanofiltration plant for the removal of chromium aimed at the production of safe potable water. J Environ Chem Eng 6:4467–4475. https://doi.org/10.1016/j.jece.2018.06.055

Govardhan B, Fatima S, Madhumala M, Sridhar S (2020) Modification of used commercial reverse osmosis membranes to nanofiltration modules for the production of mineral-rich packaged drinking water. Appl Water Sci 10. https://doi.org/10.1007/s13201-020-01312-1

Hajibabania S, Verliefde A, Drewes JE et al (2011) Effect of fouling on removal of trace organic compounds by nanofiltration. Drinking Water Eng Sci 4:71–82. https://doi.org/10.5194/dwes-4-71-2011

Harfoush M, Mirbagheri SA, Ehteshami M, Nejati S (2018) Arsenic removal from drinking water using low-pressure nanofiltration under various operating conditions. Water Practice Technol 13:295–302. https://doi.org/10.2166/wpt.2018.042

Hosseini SM, Amini SH, Khodabakhshi AR et al (2018) Activated carbon nanoparticles entrapped mixed matrix polyethersulfone based nanofiltration membrane for sulfate and copper removal from water. J Taiwan Inst Chem Eng 82:169–178. https://doi.org/10.1016/j.jtice.2017.11.017

Karabelas A, Plakas K (2011) Membrane treatment of potable water for pesticides removal. In: Herbicides, theory and applications. InTech

Košutić K, Furač L, Sipos L, Kunst B (2005) Removal of arsenic and pesticides from drinking water by nanofiltration membranes. Sep Purif Technol 42:137–144. https://doi.org/10.1016/j.seppur.2004.07.003

Košutić K, Novak I, Sipos L, Kunst B (2004) Removal of sulfates and other inorganics from potable water by nanofiltration membranes of characterized porosity. Sep Purif Technol 37:177–185. https://doi.org/10.1016/S1383-5866(03)00206-5

Kurama H, Poetzschke J, Haseneder R (2002) The application of membrane filtration for the removal of ammonium ions from potable water

Lakhotia SR, Mukhopadhyay M, Kumari P (2018) Cerium oxide nanoparticles embedded thin-film nanocomposite nanofiltration membrane for water treatment. Sci Rep 8. https://doi.org/10.1038/s41598-018-23188-7

Li XZ, Chua H, Sun JM (2002) Advanced drinking water treatment by a NF membrane system. HKIE Transac Hong Kong Inst Eng 9:46–50. https://doi.org/10.1080/1023697X.2002.10667869

Listiarini K, Tor JT, Sun DD, Leckie JO (2010) Hybrid coagulation-nanofiltration membrane for removal of bromate and humic acid in water. J Membr Sci 365:154–159. https://doi.org/10.1016/j.memsci.2010.08.048

Liu Z, Wang T, Wang D, Mi Z (2021) Regulating the morphology of nanofiltration membrane by thermally induced inorganic salt crystals for efficient water purification. J Membr Sci 617. https://doi.org/10.1016/j.memsci.2020.118645

Malhotra M, Pal M, Pal P (2020) A response surface optimized nanofiltration-based system for efficient removal of selenium from drinking water. J Water Process Eng 33. https://doi.org/10.1016/j.jwpe.2019.101007

Naidu LD, Saravanan S, Chidambaram M et al (2015) Nanofiltration in transforming surface water into healthy water: comparison with reverse osmosis. J Chem 2015. https://doi.org/10.1155/2015/326869

Nasr A ben, Charcosset C, Amar R ben, Walha K (2013) Defluoridation of water by nanofiltration. J Fluorine Chem 150:92–97. https://doi.org/10.1016/j.jfluchem.2013.01.021

Ribera G, Llenas L, Rovira M et al (2013) Pilot plant comparison study of two commercial nanofiltration membranes in a drinking water treatment plant. Desalin Water Treat 51:448–457. https://doi.org/10.1080/19443994.2012.714528

Sarkar B, Venkateswralu N, Rao RN et al (2007) Treatment of pesticide contaminated surface water for production of potable water by a coagulation-adsorption-nanofiltration approach. Desalination 212:129–140. https://doi.org/10.1016/j.desal.2006.09.021

Shah MP (2020) Microbial bioremediation & biodegradation. Springer

Shah MP (2021a) Removal of emerging contaminants through microbial processes. Springer

Shah MP (2021b) Removal of refractory pollutants from wastewater treatment plants. CRC Press

Shon HK, Phuntsho S, Chaudhary DS et al (2013) Nanofiltration for water and wastewater treatment - a mini review. Drinking Water Eng Sci 6:47–53. https://doi.org/10.5194/dwes-6-47-2013

Siddique TA, Dutta NK, Choudhury NR (2020) Nanofiltration for arsenic removal: challenges, recent developments, and perspectives. Nanomaterials 10:1–37

Singh R, Bhadouria R, Singh P, et al (2020) Nanofiltration technology for removal of pathogens present in drinking water. In: Waterborne pathogens. Elsevier, pp 463–489

Uddin MT, Mozumder MSI, Islam MA et al (2007) Nanofiltration membrane process for the removal of arsepic from drinking water. Chem Eng Technol 30:1248–1254. https://doi.org/10.1002/ceat.200700169

Uyak V, Koyuncu I, Oktem I et al (2008) Removal of trihalomethanes from drinking water by nanofiltration membranes. J Hazard Mater 152:789–794. https://doi.org/10.1016/j.jhazmat.2007.07.082

Vergili I (2013) Application of nanofiltration for the removal of carbamazepine, diclofenac and ibuprofen from drinking water sources. J Environ Manage 127:177–187. https://doi.org/10.1016/j.jenvman.2013.04.036

Wang J, Tang X, Xu Y, et al (2020) Hybrid UF/NF process treating secondary effluent of wastewater treatment plants for potable water reuse: Adsorption vs. coagulation for removal improvements and membrane fouling alleviation. Environ Res 188. https://doi.org/10.1016/j.envres.2020.109833

Yadav S, Ibrar I, Samal AK et al (2022) Preparation of fouling resistant and highly perm-selective novel PSf/GO-vanillin nanofiltration membrane for efficient water purification. J Hazard Mater 421. https://doi.org/10.1016/j.jhazmat.2021.126744

Zhu H, Feng J, Teng Y (2017) The water quality monitoring of rural drinking water and analysis of improvement by nanofiltration membrane in Poyang Lake area. Integr Ferroelectr 178:138–143. https://doi.org/10.1080/10584587.2017.1325284

# Recent Advancements and Research Perspectives in Emerging and Advanced Wastewater Membrane Technologies

**Ramya Suresh, Rajivgandhi Subramaniyan, and Maheswari Chenniappan**

## 1 Introduction

Water is essential for the survival of all living beings. In India, most people and animals get their drinking water from underground sources. Groundwater contamination has risen as a result of an increase in population and the development of industry (Lutchmiah et al. 2014). Discharging untreated wastewater straight into the earth has a negative impact on the environment for most enterprises. So, industrial wastewater must be treated with the correct technologies. In general, biological, physicochemical, and tertiary approaches can all be used to treat wastewater (Lonsdale 1982). Every technique has its own advantages and disadvantages. However, the focus of this research is on membrane technology, which is one of the technologies used in the physic-chemical process to treat wastewater (Baker 2002).

Normally, membrane processes may be divided into two categories: organic processes and inorganic processes. To date, the most widely used types of organic membrane include polyvinylidene fluoride, polysulfone, polyamide-imide, polyethersulfone, polyacrylonitrile, and polytetrafluoroethylene. Ceramic, metallic, and zeolite membranes are the most prevalent inorganic membranes (Alzahrani and

R. Suresh (✉)
Department of Civil Engineering, Sanskrithi School of Engineering, Puttaparthi, Andhra Pradesh 515134, India
e-mail: ramya24591@gmail.com

R. Subramaniyan
Department of Mechanical Engineering, Sanskrithi School of Engineering, Puttaparthi, Andhra Pradesh 515134, India

M. Chenniappan
Department of Mechatronics Engineering, Kongu Engineering College, Perundurai, Erode, Tamil Nadu 638060, India

 
M. P. Shah (ed.), *Sustainable Industrial Wastewater Treatment and Pollution Control*,
https://doi.org/10.1007/978-981-99-2560-5_9

Mohammad 2014). Because they are more expensive to manufacture than polymeric membranes, yet inorganic membranes such as metals or ceramics have superior chemical, mechanical, and thermal capabilities when compared with polymeric membranes, they are not chosen (Politano et al. 2016).

Drinking water is mostly derived from safe water sources, such as groundwater, but as the world's population and economy have expanded, aquifers have been depleted and fresh water supplies have decreased. As a result, the unsustainable use of drinking water for reasons other than subsistence, such as industrial activities, is a source of considerable worry. Wastewater might be a viable option (San Román et al. 2010). High-quality water may be generated by a variety of methods, including microfiltration (MF), ultrafiltration (UF), nanofiltration (NF), and reverse osmosis (RO). The membrane technology, on the other hand, is falling behind. There are many applications where membranes are critical today, including hemodialysis, reverse osmosis saltwater desalination, micro- and ultrafiltration of surface water, the purification of foodstuffs and pharmaceuticals, fuel cells, and battery separators, to name a few. Organic and inorganic materials can both be used for membranes (de Morais Coutinho et al. 2009). When it comes to laboratory and industrial membranes, the most prevalent are organic membranes, such as Polysulfone, PES, PVDF, PAN, PTFE, and Polyamide-imide (PAI) polymeric membranes. In contrast, metallic, ceramic, and zeolite membranes are the most frequent inorganic membranes. Even though metals and ceramics have higher mechanical, chemical, and thermal characteristics compared to polymeric membranes, they are much more expensive to construct and so are not chosen over polymeric membranes (van Reis and Zydney 2007).

## 2 Membrane Separation Processes

During the last decade, membrane technology has emerged as an important method of separating different types of substances. Membrane technology has become an essential method of separating materials. There are several advantages to membrane technology: it operates without the use of chemicals, requires less energy, is simple to handle, and has well-organized process conductions (Ravanchi et al. 2009). It is already more efficient than conventional approaches, and the efficiency of a membrane is totally dependent on the material it is made of. The membrane works as a semi-permeable layer between two phases, and it regulates the flow of material between them. The filter lets water pass through the membrane while capturing suspended particles and other contaminants. Substances can pass through membranes in a variety of ways (Ravanchi et al. 2009). When creating process water from groundwater, surface water, or wastewater, membranes perform better than other methods. Membranes are becoming a viable alternative to traditional water filtration methods.

## 2.1 Pressure Driven Membrane Separation Process

Depending on the pore size and transmembrane pressure (TMP), there are four primary pressure-driven membrane processes: MF (0.15 μm, 110 bar), UF (500,100 Da, 1100 nm, 110 bar), NF (100, 500 Da, 0.510 nm, 1030 bar), and RO (0.5 nm, 35,100 bar). Classifies distinct membrane separation procedures according to their applicability in terms of particle or molecular size (Jhaveri and Murthy 2016). Aside from desalination, RO is commonly utilized in agricultural and bioprocessing because of its ability to purify water. When it comes to MF membranes, the diameter of the membrane's biggest pore is used to determine its absolute rating, whereas the nominal rating of the ultrafiltration and nanofiltration (UF/NF) membranes is based on its early use in purifying biological fluids. For example, the minimum rating is defined as the least molecular mass cut-off (MWCO) that the membrane has a greater than 90% rejection rate. Membrane nominal ratings tell us a lot about the separation process in MF/UF/NF: size exclusion (Shirazi et al. 2010). The other mode of separation is via electrostatic interactions among solute molecules and membranes, which are dependent on the interface and physiochemical characteristics of the solutes and membranes, respectively. The entire working cost of the system is determined by the applied pressure (Zhang et al. 2015).

## 2.2 Microfiltration Membranes

Microfiltration membranes (MF) is one of the earliest commercially utilized pressure-driven membrane applications. Based on the idea of physical separation, MF is effective in eliminating nanometer particles including such suspended particles, significant pathogens, big bacteria, enzymes, and fungal spores. Because of its capacity to reject a wide spectrum of large-scale pollutants, MF is a flexible filtration process (Warkiani et al. 2013). The pore sizes of MF membranes range from 0.1 to 5 μm. Particles larger than 0.1 m in diameter are often segregated by a relatively permeable cellular membrane. MF membranes be obtainable in spiral-wound and flat sheet forms (Jhaveri and Murthy 2016). Similarly, membrane units and filtering units could be modified to meet particular application demands. This brings up a plethora of opportunities for customized membrane systems in a variety of applications.

In the 1960s, the first marketable utilization of MF membranes had been in pharmaceutical and biological synthesis. MF membranes were primarily employed in sterile purification in the medical sector and final filtering of wash water in the electronics industry throughout the next 20 years. Although not as strict as in the biopharmaceutical industry, MF has also been utilized in the decontamination of beverages, as well as the clarifying of vinegar and other drinks, in a simple and cost-effective manner. Because of its economic effectiveness, MF had not been offered to the water treatment sector until the 1980s. Stricter pathogen elimination regulations for potable water have resulted in a substantial migration to low pressure filtration techniques (Caro and Noack 2008). With the introduction of extremely durable membranes like as polyethylene, polypropylene, and polysulphate, the MF process may now be employed on a big scale.

Frick described the first nitrocellulose MF membranes in 1855. The technology's first practical use was to cultivate pathogens in drinking water. MF, however, remained limited to small-scale industrial uses until the mid-1960s. In the 1970s, the invention of pleated membrane cartridges sparked large-scale commercial uses of MF in the medical, water, and microsystems industries. Several attempts were made to change the chemistry of the membranes or to incorporate new MF membrane elements (Wang and Chung 2015).

## 2.3 *Ultrafiltration Technology*

The ultrafiltration separation process employs membranes with membrane diameter varying from 2 to 100 nm and molecular mass thresholds ranging from 5–500 kDa. The working pressure is 2–8 bar that is frequently greater than the microfiltration pressure (Shi et al. 2014). Ultrafiltration typically eliminates all microbial species as well as humic substances. Ultrafiltration technique is receiving a lot of interest for heavy metal removal from contaminated water with the use of a right complexing agent. Based on the membrane properties, an ultrafiltration system may reach and over 90% removal with metal content of 9.5–11.8 mg/L. One of the drawbacks of ultrafiltration membranes is fouling, which has a variety of negative effects, including flux declination, increased transmembrane pressure, and biodegradation (Al Aani et al. 2020). The method will give an alternative to conventional water purification by chemicals, which can sometimes be regarded as a secondary barrier to pollution.

Polymeric membranes have their origins in commercialized viscose acetate filter media and have since evolved in tandem with membrane technology. Unlike RO membranes, no substantial osmotic pressure has been formed across ultrafiltration since the pore structure allows microsolutes to pass thru the membrane surface. In practice, UF membranes are utilized as a separator to isolate biomolecules, colloidal particles, and soluble compounds with molecular mass greater than 10,000 from entities with lower molecular mass. Even though these organisms may generate osmotic pressure, it really is usually just a few bars. As a result, the barometric pressure difference employed as the driving force in UF is in the 1–10 bar range (Hampu et al. 2020). Based on the size and components to be separated, the UF membranes can be selected. The majority of ultrafiltration have an asymmetrical pore structure and are frequently manufactured by the phase-inversion method. During the first era of UF, CA was the primary membrane material. CA, on the other hand, has limited chemical and thermal stabilities, a very restricted pH sensitivity range, and is readily biodegradable (Goosen et al. 2005).

The extraction of electrochemical pigments from wash water has been accomplished using UF membranes. The use of fibrous CA membranes without the need for thermo-chemical stability resulted in significant savings of paints and water, and the capillary electrophoresis pigment sector would have suffered without UF. The dairy sector is another key application for UF membranes. The membranes were not only utilized to extract nutrients, glucose, and lactic from whey, but they were also employed to condense nutritive whey or make cheese. Because of the dual effects of water contamination remediation and valuable product extraction, the dairy sector has become a big market region for UF membranes (Ren et al. 2021; Shah 2020).

## 2.4 Nanofiltration Technology

FilmTec coined the term "nanofiltration" in the mid-1980s to describe a "Reverse Osmosis operation" that intentionally and preferentially permits several anion substances in a brine to pass through. In contrast to RO membranes, which feature a solution-diffusion flow path and a nonporous construction, membrane processes operate at the junction between pores and non-porous membranes, using both diffusing transport processes and sieving (Mohammad et al. 2015). In the case of salt retention, ultrafiltration membrane provides the same function as nanofiltration membranes. As a result, it was discovered that nanofiltration has the same ability as "tight" ultrafiltration (sieving, porous) or "loose" reverse osmosis method (diffusion, nonporous). Because of their "loose" nature, nanofiltration membranes may operate at higher water flux with considerably lower pressure than reverse osmosis process membranes, resulting in considerable energy savings (Oatley-Radcliffe et al. 2017).

Furthermore, since many nanofiltration membranes were surface charged, electric chain is comprised to the selective rejection and transport of nanofiltration membranes and has a high permeability for monovalent salts, but they are also capable of removing multivalent salts and relatively tiny organic matter (Hilal et al. 2004).In the dairy sector, NF was employed as an alternative to RO for decalcification and concentrating of whey. Nanofiltration membranes have found increased use in the textile, seawater softening, food, and mining sectors. The nanofiltration 50 membranes would be the first reverse osmosis membrane to be capable of processing at ultra-filtration pressures (Agboola et al. 2014).

This membrane rejects MgSO4 at a rate of 84–90%, NaCl at a rate of 31–40%, raffinose at a rate of 99.98%, and sucrose at a rate of 98.99%. CA's NF (loose reverse osmosis) membranes, on the other hand, had poor chemical stability and biological performance in the first generation.

They did not permit application in organic solvents due to the membrane polymers' lack of chemical resistance to the solvents. The main drawback of NF membranes is indeed the difficulty in controlling the repeatability of the membranes pore diameter and pore size (Ji et al. 2017). Additionally, nanofiltration membranes are legally liable for fouling, which result in significant flux may reject. The nanofiltration parting technique has a pore size of 1–2 nm and a molecular weight cut-off of 0.1–5 kDa. This information typically necessitates higher pressure (6–12 bars) than ultrafiltration and microfiltration systems. It is capable of removing practically all viruses, germs, and humic compounds. Nanofiltration may also be used to remove alkalinity and hardness from water, which is known as softening membrane (Hołda and Vankelecom 2015). It is generally known that Nanofiltration may divide pollutants with molecular weights ranging from (0.1 to 500 kDa),which corresponds to the molecular weight of approximately of Reverse Osmosis membrane and ultrafiltration.

## 2.5 *Reverse Osmosis*

The most extensively utilized desalination technique nowadays is reverse osmosis membrane. Over the last several decades, amazing breakthroughs in the fabrication of reverse osmosis membranes from various materials have been accomplished. Profit incentive in RO technology is growing worldwide as a result of advancements in reverse osmosis technology in the area of membranes energy and material consumption, which has permitted a reduction in cost of regular water production (Glater 1998). Reverse osmosis is used in desalination businesses to generate clean water suitable for drinking, industries, and agriculture by eliminating minerals, salts, and irons from feed water. The effectiveness of the reverse osmosis process is determined by the qualities of the input water and the operating parameters. To counteract the natural osmotic pressure, a stream of water is subjected to external pressure(Petersen 1993) in the field of reverse osmosis desalination. In a naturally occurring osmosis, the solvent flows from a low to a high solute concentration region to equalization the concentration of solutes on the both sides of the membrane, resulting in osmotic pressure. To reverse the fluid flow of liquids in a reverse osmosis membrane, applied pressure is employed. So when applied pressure exceeds the osmotic pressure, the input water goes through to the membrane, leaving behind salt and other minerals, and emerges as pure water. Salt removal and water flux are the two factors that drive the functioning of the reverse osmosis membrane (Lee et al. 2011).

A key concern for reverse osmosis plants is the cost, which is determined by both membranes renewal costs and energy usage. Water manufacturing costs can be minimized by developing a hybrid system that combines one or more desalination processes. We use a variety of materials in the reverse osmosis membrane process, depending on what type of purpose about which it will be used. Polyamide, cellulose acetate, heterocyclic polymer, polymerizable monomer, polybenzimidazole, polyacrylonitrile, crosslinked water soluble polymers and polypiperazine-amide (PA) are the most often used reverse osmosis membrane materials (Joo and Tansel 2015).

Nonetheless, they are broadly categorized into following groups: asymmetrical membranes with a skinny-film and one polymer layer and composite membranes with two or more polymer layers. The most well-known anisotropic or asymmetrical membrane structure is the cellulose acetate membranes. It is made up of a very fine solute rejections layer on top of a coarser supporting layer. The supportive layer is frequently made of the same material as the selective layer (Saleem and Zaidi 2020); however, the thick of the selective transparent layer affects fluxes and rejection as well as the overall effectiveness of the membrane. The supporting layer simply serves to strengthen the membrane mechanically. Quasi phase inversion or polymers precipitation methods are commonly used to create these membranes.

A problem of asymmetrical membranes is that they have been exceedingly thin, ranging from 0.1 to 1.0 μm in thickness; as a result, they provide huge fluxes without performing well in dismissal. It was the era of straight aromatic polyamide membranes after cellulose acetate. It eliminates some of the drawbacks of cellulose acetate membrane (Zhao et al. 2021). Linear aromatic polyamide membranes gained popularity in the first generation with cellulose acetate membranes. Polyamide is among the most effective selective's predetermined to date. They have a high rejection performance and can be used for single-state saltwater desalination. However, these membranes were as tiny as cellulose acetate membrane (0.1 to 1.0 m) thick, and the problem of high fluxes persisted. It has a high rate of rejections. Nevertheless, fouling of thin-film composites membrane is a difficult problem in commercial processes (Tu et al. 2010). Various methods have been used to accurately determine the thin selective coating to reduce fouling. Until date, there have been a variety of additional challenges associated with reverse osmosis applications. To comprehend the challenges in a RO plant, we must first become acquainted with the operation of a conventional reverse osmosis plant. First, the high salinity feed water will flow to a pre-treatment stage. This procedure removes any big particles that may block the membrane and reduces its performance (Coutinho de Paula and Amaral 2017).

### *2.6 Forward Osmosis*

Forward osmosis is really a membrane technology that may clean wastewater and create high-quality water. Forward osmosis is a technical word that refers to the natural process of osmosis: the movement of water molecules over a semi-permeable membrane. In contrast to pressure-driven membrane activities, the dynamic force of water movement is the osmotic pressure differential (Yadav et al. 2020). Generally, forward osmosis benefits from naturally induced water diffusion through a quasi-membrane from a low concentration solution to a high concentration solution. The semi-permeable membrane should ideally act as a barrier, allowing water to pass through while rejecting salts or superfluous components. The raised concentration solution, which has a higher osmotic pressure than the feed water, functions as an attractive solution, drawing water from the feed solution over the membranes to itself (Chun et al. 2017). Thus, forward osmosis processes occur only when a semi-permeable membrane separates the feed solution first from draw solution so there is an osmotic pressure differential across the membrane. In comparison with typical pressure-driven membrane technologies such as reverse osmosis, forward osmosis uses substantially less power to produce a net flow of water over the membrane for water recycling and desalination. In contrast to reverse osmosis, the permeate of forward osmosis is a combination of draw water and draw solution rather than a finished product water (Coday et al. 2014). As just an outcome, a second separation process is required to generate clean water and replenish the draw solution.

If incorrect draw solutes and recycling procedures were used, the second phase of separation could be energy demanding. To provide a fair comparison of forward osmosis technologies with the other water production technologies, one must account both the prices of forward osmosis membranes and solute recycling (Lee and Hsieh 2019). Otherwise, the findings might be skewed and deceptive.

FO membranes are made from a variety of materials. The commonly used cellulose triacetate membranes are chlorine resistant and resistant to mineral and fatty oil adsorption, including petroleum. Although cellulose triacetate membranes is less susceptible to thermal, chemical, and biodegradation and hydrolysis in alkaline circumstances than cellulose, hydrolysis in severe settings, such as wastewater, may still be a challenge (Jafarinejad 2021).

Manufactured thin-film composites membranes for forward osmosis have been shown to be superior to cellulose triacetate membranes in terms of permeability and stability across a wider pH range (2.3–13 vs. 4–9 for cellulose triacetate membranes), while still withstanding the high pressures necessary in reverse osmosis. Durable and temperature resistant thermoplastic polymers were also employed. Polybenzimidazole has the ability to self-charge in aqueous systems , which results in increased salt rejections, relatively high hydrophilicity, and minimal fouling tendencies (Coday et al. 2014). Polyamide-imides offer both cation and anion repulsion to salt transport across the membrane and can form NF-like selective layers. Despite its hydrophobicity, nanoporous polyethersulfone with weak finger-like support systems and nano-sized pores is primarily used as the SL in TFC membranes; mixtures of polymers with polyacrylonitrile for membranes result in improvements over promotional forward osmosis membranes; and hydrophilic polydopamine tends to increase fouling resistance.

## 2.7 *Pervaporation*

These are some of the membrane technologies for liquid separation is pervaporation. Pervaporation is a form of unit process in which two components are separated by an inorganic membranes or a nonporous polymeric membrane based on their permeation rates (Shao and Huang 2007). Because an evaporative phase transition frequently occurs, vacuum operation is commonly provided to the downstream side of a membrane. When the ethanol concentration in solution ranges from 0.0 to 22%, the molar ratio of ethanol in vapor is substantially greater than its mass fraction in liquid, according the water vapor stability curve of ethanol–water (Ong et al. 2016). As a result, at low ethanol concentrations, pervaporation has an intrinsic preference for ethanol. The separating membrane is a crucial component of pervaporation apparatus. The membrane's pervaporation capabilities are commonly classified as selectivity and flux.

The membrane is an essential component of the pervaporation process. A typical membrane design roadmap begins with materials and progresses to preparation techniques, microenvironments, architectures, performance intensification, and mass transport mechanisms. Natural microenvironments, and also physical and chemical microenvironments, are explained by microenvironment (Liu and Jin 2021). The external microenvironment defined as the physical microscale structure of membranes, whereas the chemical microenvironment refers to the surface chemical microstructure (distribution, types of functional groups, quantity, and facilitated transport carriers) of membrane surface and cavities inside. By varying the materials and production procedures, membranes may be given with a variety of physical and chemical microenvironments, depending on the roadmap (Cheng et al. 2017).

### *2.8 Ion-Exchange-Membrane Process*

Membrane technology is presently applied in a variety of industries, including membrane-based sensors, medical devices, and so on. When compared to other membrane technologies, ion-exchange membranes are the most sophisticated separation procedures. The ion-exchange membrane operates on the basis of the Donnan membrane equilibrium principle (Nagarale et al. 2006; Shah 2021a). It is generally used to recover of ions and metals from by products, elimination of superfluous ionic species in wastewater notably harmful metals from effluent. It can also be divided into cation and anion membranes, which have high counter-ion mobility.

These membranes having more advantages such as to extraction of sodium chloride from sea water, extraction of nutrients from cheese and whey solutions, and demineralization of sugarcane juice and saline water. Sometimes, monovalent ion and proton permselective ion-exchange membranes are used to separate specific ions for industrial purposes. Aside from that, ion-exchange membranes are used in the field of transmitters for enzymatic reactions and vital functions, healthcare sensors, moisture levels, monoxide sensor, solid poly electrolytes, photocurrent, and voltage generation that are recent concept and power contribute to future implementations of ion-exchange membranes (Xu 2005).

### *2.9 Temperature-Driven Membrane Processes*

The temperature-driven membrane processes are divided into four categories: thermal pervaporation, pervaporation, membrane crystallization, and membrane distillation. When compared to traditional separation procedures, this technology features a more cost-effective approach (She et al. 2016). Unbalanced concentrations of solutions were separated using this procedure by evaporation on one side of the membrane and condensation of vapor on the other. The heavy force for each component is resolute by the biased pressure difference of the component due to the maintaining the process parameters in such a way that the first solution (feed) is operated at elevated temperatures, while the dense liquid (permeate) is controlled at minor temperatures (Yuan et al. 2021). Porous membranes are used to achieve membrane crystallization and membrane distillation. The selectivity of the MD process is only governed by the vapor–liquid equilibrium, which has led in the enhancement of the gaseous form by more volatile elements.

### *2.10 Membrane Distillation*

Membrane distillation is one of the desalination techniques used in the treatment of water and wastewater. It offers more benefits than other membrane approaches, such as easy membrane cleaning, no cleaning chemicals required, treated water may be reused, and less salinity influence. Direct contact and air gap membrane distillation are the two most common types of membranes utilized (Wang and Chung 2015; Shah 2021). The permeate and feed solutions have direct contact with the membrane in direct contact MD. In the case of air gap MD, the hot feed solution contacts the membrane's intake side, while the air layer covers the membrane's outlet side.

### *2.11 Liquid Membranes*

In the Liquid membrane (LM) process, among the two different concentrations of solutions, the thin organic liquid acts as a barrier and it do not require solid membranes. It having the features of high selectivity, single stage extraction, and non-equilibrium mass transfer. LM can be divided into with or without supports (Way et al. 1982). The liquid membrane can be classified into immobilized, supported and contained liquid membranes. Similarly, without liquid membrane can be parted into bulk liquid and emulsion liquid membranes. Supported LMs are made composed of an innocuous micro-porous substrate on which an aqueous solvent could be immobilized (Agarwal and Singh 2022). An inner aqueous layer lies between the two liquid phases in an LM mixture. Bulk LM utilizes a constrained dispersion route that is far from the fluid layers. The LM process's major applications include ionic species extraction from effluent, acid removal, extraction, and reduction, conversion of biomass, GS, and so on. The main disadvantage of LM is the unstable of the membrane interface, which may be caused by temperature and agitation differences inside the LM system. Diffusion is the mechanism of mass movement through the membrane (Lozano et al. 2011).

## 3 Application of Membrane Technology for Food Industry

Membrane technology is continually improving, with new uses in food production emerging all the time. Traditional procedures such as UF, MF, NF, and RO are now considered as somewhat typical process units used in a variety of processes. Emerging approaches, including such bipolar membrane separation or pervaporation, open up new ways, although they have been in the early stages of research. The current tendency in food processing membrane technology is to create highly customized membrane that have been committed with one operation, one item, or perhaps even merely enhancing the effectiveness of a current products (Daufin et al. 2001).The enhancement of food sensory attributes is another key reason for adopting membrane techniques in sterilization and clarifying. Nevertheless, given the complex nature of the formulations often involved, reliable extraction is challenging. As a result, extremely selective membranes are required. Aroma components, for example, may be eliminated together with the liquor during in the de-alcoholization of wine and beer (Shanmugasundaram and Pandit 2022). The extraction of flavor components from liquor by chilling and consequent reprocessing of the fragrances could be effective, but it increases the expense of the operation. In other way, the pervaporation/vapor permeation is an excellent substitute to reverse osmosis for extracting fragrance chemicals (Castro-Muñoz et al. 2022).When using the typical oil refining technique to extract the oil from the products, the following drawbacks occur: it consumes more energy, causes oil losses, necessitates the use of more water and chemical substances, and leads to the production of more wastewater. In contrast, membrane technology has been adopted because to the following advantages: low energy consumption, ability to function at lower temperatures, absence of chemical use, and ability to recover more valuable and desired products during extraction (Yadav et al. 2022).

Membrane technology is used in the dairy industry to separate and fractionate cream/fat from milk. Centrifugation is a process used in the dairy industry to extract cream/fat from milk. Ultrafiltration eliminates all lactose and minerals from dairy products while retaining all dairy proteins (Shrivastava et al. 2022). It is also used to distillate and reduce the required protein level in cheese milk. Ultrafiltered milk is not impacted by any intrinsic modifications of milk products driven by weather conditions, climates, state of lactation, nutrition, and breed due to protein standardization. Because heat evaporation is no longer necessary to standardize the protein, UF milk standardization opens up potential not just for cheese production but also for the production of curd, milkshakes, and drinking milk. This prevents gelatinization and any resulting changes in the functioning and nutritious content of milk proteins (Yi et al. 2022).

Membrane technologies such as NF, UF, RO, and MF can be employed as major or minor treatment processes in textile industry effluent treatment to improve unit overall treatment efficiency. When the UF procedure was used as a pre-treatment for textile operations such as washing, dyeing, and printing, the COD level was significantly lowered (Nasr and Ali 2022).The fouling effect of membranes is caused by the presence of organic and inorganic contaminants. When compared to conventional UF membrane modules, the spiral module UF is more effective at eliminating big particles, which significantly minimizes the fouling impact of NF. When NF was used as a subsequent treatment, it was possible to obtain a maximum of 92% COD and 95% color removal (Reddy et al. 2022). When the textile wastewater has a high concentration of contaminants; membrane treatment is sometimes utilized as a post-treatment approach following the biological process. The cotton industry effluent was treated using a combination of UF and electrodialysis methods (Manzoor et al. 2022). The UF process eliminates all micro and macromolecules in water, whereas the electrodialysis procedure removes or separates ionic particles in effluent.

## 4 Conclusion

Membrane technology was being utilized effectively and progressively for a variety of functions and applications in the cleaning and treatment of water and effluent. This technology was based on several criteria such as plant efficiency, low project duration, hygienic, simple, inexpensive, and protracted efficient performance and strong rejection rates for elements or pollutants. The requirement of water for lifeforms enhances the researcher's obligation to supply purified water at the lowest possible cost. When relative to other commercial approaches, the membrane separation procedure was more cost-effective. Furthermore, there has been a lot of research and development on membrane filtration in wastewater. Most of the processes have been performed on a laboratory scale, demonstrating that membrane filtration was appropriate for large use. Numerous ways were taken to increase membrane performance even more. Surface treatment with certain agents improves membrane performance and anti-fouling qualities. A few of the surface-modified membranes were tested in industrial applications and yielded good results. Nevertheless, operational settings have an essential role in improving removal efficiency. Various researches have been carried to investigate the influence of factors on operating parameters. Ceramic membranes, like polymer membranes, gained popularity in the divorce process. The continual and growing initiatives in the implementation of effective separators can provide sustainable and highly viable approach for the treatment of wastewater.

## References

Agarwal S, Singh AP (2022) Performance evaluation of textile wastewater treatment techniques using sustainability index: an integrated fuzzy approach of assessment. J Clean Prod 337:130384

Agboola O, Maree J, Mbaya R (2014) Characterization and performance of nanofiltration membranes. Environ Chem Lett 12(2):241–255

Al Aani S, Mustafa TN, Hilal N (2020) Ultrafiltration membranes for wastewater and water process engineering: a comprehensive statistical review over the past decade. J Water Process Eng 35:101241

Alzahrani S, Mohammad AW (2014) Challenges and trends in membrane technology implementation for produced water treatment: a review. J Water Process Eng 4:107–133

Baker RW (2002) Membrane technology. *Encyclopedia of Polymer Science, and Technology* 3.

Caro J, Noack M (2008) Zeolite membranes–recent developments and progress. Microporous Mesoporous Mater 115(3):215–233

Castro-Muñoz R, Correa-Delgado M, Córdova-Almeida R, Lara-Nava D, Chávez-Muñoz M, Velásquez-Chávez VF, Hernández-Torres CE, Gontarek-Castro E, Ahmad MZ (2022) Natural sweeteners: sources, extraction and current uses in foods and food industries. Food Chem 370:130991

Cheng X, Pan F, Wang M, Li W, Song Y, Liu G, Yang H, Gao B, Hong Wu, Jiang Z (2017) Hybrid membranes for pervaporation separations. J Membr Sci 541:329–346
Chun Y, Mulcahy D, Zou L, Kim IS (2017) A short review of membrane fouling in forward osmosis processes. Membranes 7(2):30
Coday BD, Yaffe BGM, Pei Xu, Cath TY (2014) Rejection of trace organic compounds by forward osmosis membranes: a literature review. Environ Sci Technol 48(7):3612–3624
Coutinho de Paula E, Amaral MCS (2017) Extending the life-cycle of reverse osmosis membranes: a review. Waste Manage Res 35(5):456–470
Daufin G, Escudier JP, Carrère H, Bérot S, Fillaudeau L, Decloux M (2001) Recent and emerging applications of membrane processes in the food and dairy industry. Food Bioprod Process 79(2):89–102
de Morais Coutinho C, Chiu MC, Basso RC, Ribeiro APB, Gonçalves LAG, Viotto LA (2009) State of art of the application of membrane technology to vegetable oils: a review. Food Res Int 42(5–6):536–550
Glater J (1998) The early history of reverse osmosis membrane development. Desalination 117(1–3):297–309
Goosen MFA, Sablani SS, Al-Hinai H, Al-Obeidani S, Al-Belushi R, Jackson A (2005) Fouling of reverse osmosis and ultrafiltration membranes: a critical review. Sep Sci Technol 39(10):2261–2297
Hampu N, Werber JR, Chan WY, Feinberg EC, Hillmyer MA (2020) Next-generation ultrafiltration membranes enabled by block polymers. ACS Nano 14(12):16446–16471
Hilal N, Al-Zoubi H, Darwish NA, Mohamma AW, Arabi MA (2004) A comprehensive review of nanofiltration membranes: treatment, pretreatment, modelling, and atomic force microscopy. Desalination 170(3):281–308
Hołda AK, Vankelecom IF (2015) Understanding and guiding the phase inversion process for synthesis of solvent resistant nanofiltration membranes. J Appl Polym Sci 132(27):43120
Jafarinejad S (2021) Forward osmosis membrane technology for nutrient removal/recovery from wastewater: recent advances, proposed designs, and future directions. Chemosphere 263:128116
Jhaveri JH, Murthy ZVP (2016) A comprehensive review on anti-fouling nanocomposite membranes for pressure driven membrane separation processes. Desalination 379:137–154
Ji Y, Qian W, Yu Y, An Q, Liu L, Zhou Y, Gao C (2017) Recent developments in nanofiltration membranes based on nanomaterials. Chin J Chem Eng 25(11):1639–1652
Joo SH, Tansel B (2015) Novel technologies for reverse osmosis concentrate treatment: a review. J Environ Manage 150:322–335
Lee DJ, Hsieh MH (2019) Forward osmosis membrane processes for wastewater bioremediation: research needs. Biores Technol 290:121795
Lee KP, Arnot TC, Mattia D (2011) A review of reverse osmosis membrane materials for desalination—development to date and future potential. J Membr Sci 370(1–2):1–22
Liu G, Jin W (2021) Pervaporation membrane materials: recent trends and perspectives. J Membr Sci 636:119557
Lonsdale HK (1982) The growth of membrane technology. J Membr Sci 10(2–3):81–181
Lozano LJ, Godínez C, de los Rios AP, Hernández-Fernández FJ, Sánchez-Segado S, Alguacil FJ (2011) Recent advances in supported ionic liquid membrane technology. J Membr Sci376(1–2):1–14
Lutchmiah K, Verliefde ARD, Roest K, Rietveld LC, Cornelissen ER (2014) Forward osmosis for application in wastewater treatment: a review. Water Res 58:179–197
Manzoor K, Khan SJ, Yasmeen M, Jamal Y, Arshad M (2022) Assessment of anaerobic membrane distillation bioreactor hybrid system at mesophilic and thermophilic temperatures treating textile wastewater. J Water Process Eng 46:102603
Mohammad AW, Teow YH, Ang WL, Chung YT, Oatley-Radcliffe DL, Hilal N (2015) Nanofiltration membranes review: recent advances and future prospects. Desalination 356:226–254
Nagarale RK, Gohil GS, Shahi VK (2006) Recent developments on ion-exchange membranes and electro-membrane processes. Adv Coll Interface Sci 119(2–3):97–130

Nasr RA, Ali EA (2022) Polyethersulfone/gelatin nano-membranes for the Rhodamine B dye removal and textile industry effluents treatment under cost effective condition. J Environ Chem Eng 10(2):107250

Oatley-Radcliffe DL, Walters M, Ainscough TJ, Williams PM, Mohammad AW, Hilal N (2017) Nanofiltration membranes and processes: a review of research trends over the past decade. J Water Process Eng 19:164–171

Ong YK, Shi GM, Le NL, Tang YP, Zuo J, Nunes SP, Chung TS (2016) Recent membrane development for pervaporation processes. Prog Polym Sci 57:1–31

Petersen RJ (1993) Composite reverse osmosis and nanofiltration membranes. J Membr Sci 83(1):81–150

Politano A, Cupolillo A, Di Profio G, Arafat HA, Chiarello G, Curcio E (2016) When plasmonics meets membrane technology. J Phys Condens Matter 28(36):363003

Ravanchi MT, Kaghazchi T, Kargari A (2009) Application of membrane separation processes in petrochemical industry: a review. Desalination 235(1–3):199–244

Reddy AS, Kalla S, Murthy ZVP (2022) Nano-particles enhanced hydrophobic membranes: high-performance study for dye wastewater treatment using membrane distillation. J Water Process Eng 46:102610

Ren Y, Ma Y, Min G, Zhang W, Lv L, Zhang W (2021) A mini review of multifunctional ultrafiltration membranes for wastewater decontamination: additional functions of adsorption and catalytic oxidation. Sci Total Environ 762:143083

Saleem H, Zaidi SJ (2020) Nanoparticles in reverse osmosis membranes for desalination: a state of the art review. Desalination 475:114171

San Román MF, Bringas E, Ibañez R, Ortiz I (2010) Liquid membrane technology: fundamentals and review of its applications. J Chem Technol Biotechnol 85(1):2–10

Shah MP (2020) Microbial bioremediation & biodegradation. Springer, Singapore

Shah MP (2021a) Removal of refractory pollutants from wastewater treatment plants. CRC Press

Shah MP (2021b) Removal of emerging contaminants through microbial processes. Springer

Shanmugasundaram OL, Pandit P (2022) Conclusion on biotechnology in textile production and future scope. In: Applications of biotechnology for sustainable textile production. Woodhead Publishing, pp 241–249

Shao P, Huang RYM (2007) Polymeric membrane pervaporation. J Membr Sci 287(2):162–179

She Q, Wang R, Fane AG, Tang CY (2016) Membrane fouling in osmotically driven membrane processes: a review. J Membr Sci 499:201–233

Shi X, Tal G, Hankins NP, Gitis V (2014) Fouling and cleaning of ultrafiltration membranes: a review. Journal of Water Process Engineering 1:121–138

Shirazi S, Lin C-J, Chen D (2010) Inorganic fouling of pressure-driven membrane processes—a critical review. Desalination 250(1):236–248

Shrivastava V, Ali I, Marjub MM, Rene ER, Soto AMF (2022) Wastewater in the food industry: treatment technologies and reuse potential. Chemosphere 293:133553

Tu KL, Nghiem LD, Chivas AR (2010) Boron removal by reverse osmosis membranes in seawater desalination applications. Sep Purif Technol 75(2):87–101

van Reis R, Zydney A (2007) Bioprocess membrane technology. J Membr Sci 297(1–2):16–50

Wang P, Chung T-S (2015) Recent advances in membrane distillation processes: membrane development, configuration design and application exploring. J Membr Sci 474:39–56

Warkiani ME, Bhagat AAS, Khoo BL, Han J, Lim CT, Gong HQ, Fane AG (2013) Isoporous micro/nanoengineered membranes. ACS Nano 7(3):1882–1904

Way JD, Noble RD, Flynn TM, Sloan ED (1982) Liquid membrane transport: a survey. J Membr Sci 12(2):239–259

Xu T (2005) Ion exchange membranes: state of their development and perspective. J Membr Sci 263(1–2):1–29

Yadav A, Patel RV, Singh CP, Labhasetwar PK, Shahi VK (2022) Experimental study and numerical optimization for removal of methyl orange using polytetrafluoroethylene membranes in vacuum membrane distillation process. Colloids Surf, A 635:128070

Yadav S, Saleem H, Ibrar I, Naji O, Hawari AA, Alanezi AA, Zaidi SJ, Altaee A, Zhou J (2020) Recent developments in forward osmosis membranes using carbon-based nanomaterials. Desalination 482:114375

Yi Y, Li P, Zhao F, Zhang T, Shan Y, Wang X, Liu B, Chen Y, Zhao X, Lü X (2022) Current status and potentiality of class II bacteriocins from lactic acid bacteria: structure, mode of action and applications in the food industry. Trends Food Sci Technol 120:387–401

Yuan Z, Yu Y, Sui X, Yao Y, Chen Y (2021) Carbon composite membranes for thermal-driven membrane processes. Carbon 179:600–626

Zhang W, Luo J, Ding L, Jaffrin MY (2015) A review on flux decline control strategies in pressure-driven membrane processes. Ind Eng Chem Res 54(11):2843–2861

Zhao S, Liao Z, Fane A, Li J, Tang C, Zheng C, Lin J, Kong L (2021) Engineering antifouling reverse osmosis membranes: a review. Desalination 499:114857

# Sequestration and Detoxification of Heavy Metals by Fungi

**Marwa Tamim A. Abdel-Wareth**

## 1 Introduction

Heavy metal pollution is considered one of the most important environmental issues. Mainly, two major sources introduce heavy metals into the environment: natural (deep-sea vents, volcanic emissions, forest fires, and geysers) and anthropogenic ones (Gadd 2009; Lata et al. 2019). Heavy metals toxicity is of great concern, because of their bioaccumulation and non-biodegradability (Gautam et al. 2015; Wai et al. 2012). Many metals including Ca, Cu, Na, Mg, Cr, and Ni are necessary for metabolic activities. Whereas Pb, Hg, Cd, Ag, Al, and Au are not needed for any biological process, moreover, they are toxic (Siddiquee et al. 2015; Lakherwal 2014; Turpeinen et al. 2002). As a consequence of technological applications and widespread industrial activities, an excessive increase of metals discharge into the environment is recorded. Many industries participate in maximizing this problem, e.g., pulp and wood processing, electroplating, leather tanning, steel manufacturing, milling, mining, and surface finishing (Alloway 1995; Congeevaram et al. 2007; Hemambika et al. 2011). This resulted in the contamination of surface water, groundwater, soil and sediments, and inevitably caused a real threat to human health (Choo et al. 2015). Heavy metals can accumulate in human body tissues, causing significant physiological disturbances, and irreversible destruction of the central nervous system and other vital organs (Çelebi and GökG 2020; Ifijen et al. 2020).

When agricultural and industrial wastewater containing heavy metals are discharged into surface waters without proper treatment, they disturb the balance of the ecosystem and might cause poisoning of various aquatic organisms, and consequently human who is on the top of the food chain (Baldrian 2003). Among the most

M. T. A. Abdel-Wareth (✉)
Aquatic Ecology, Environmental Research Department,
Theodor Bilharz Research Institute, Giza, Egypt
e-mail: marwatamim2001@hotmail.com

 
M. P. Shah (ed.), *Sustainable Industrial Wastewater Treatment and Pollution Control*,
https://doi.org/10.1007/978-981-99-2560-5_10

frequently encountered metals in industrial wastewaters are Pb, Cr, Cu, Hg Ni, Cd, and Zn (Panda et al. 2014).

When heavy metals reach soil systems, they deleteriously affect the biodiversity and composition of the soil ecosystem (Baath 1989). The most common heavy metals that contaminate soils are Ag, Pb, Fe, Cu, Cd, Zn, Hg, Cr, and As (Han et al. 2002). As persistent environmental contaminants, heavy metals may be deposited on the surfaces and then absorbed into the tissues of vegetables (Aishwarya et al. 2014). It was reported that the most problematic heavy metals because of their high toxicity are lead, chromium, and cadmium (Graz et al. 2011; Salinas et al. 2000). Even at low concentrations, heavy metals can cause chronic toxicity. So, the removal of such metals from industrial effluents is an urgent demand to protect different ecosystems (Ayele et al. 2021).

In naturally polluted environments, the response of a microorganism to heavy metals toxicity depends on the type, concentration and the availability of metals, the microbial species, and the nature of the medium that harbor them (Hassen et al. 1998). Filamentous fungi and yeasts are known to tolerate heavy metals (Gavrilesca 2004; Baldrian 2003), they possess many advantages such as their cell wall that shows the ability of metal binding (Gupta et al. 2000). Additionally, microorganisms have developed different mechanisms to tolerate heavy metals stress such as biosorption into cell walls, transportation via the cell membrane, sequestration in extracellular capsules, besides precipitation of metals (Malik 2004). Filamentous fungi and yeasts are versatile groups, as they can tolerate different conditions of pH, temperature, and nutrient scarcity, in addition to high concentrations of metal ions (Anand et al. 2006). It was reported that the microbial strains that were isolated from heavy metals contaminated ecosystems have a noticeable capability to remove considerable quantities of these metals (Malik 2004; Joshi et al. 2011; Oladipo et al. 2018). El-Morsy (2004) found that *Cunninghamella echinulata* biomass could be employed as a biosorbent of metal ions from wastewater. Similarly, Zafar et al. (2007) declared that *Aspergillus* sp. and *Rhizopus* sp. from metal-contaminated soil showed promising absorption of Cd and Cr. Moreover, it was found that the use of fungal biomass for removing heavy metals from industrial effluents is effective and cheap. Many fungal species have been investigated for heavy metal biosorption such as *Aspergillus niger* (Kapooret al. 1999; Srivastava and Thakur 2006; Park et al. 2005), *Saccharomyces cerevisiae* (Lin et al. 2005), *Mucor hiemalis* (Tewari et al. 2005), *Botrytis cinerea* (Akar and Tunali 2005a), *Neurospora crassa* (Tunali et al. 2005), *Lentinus sajor-caju* (Bayramoglu et al. 2005), and *Phanerochaete chrysosporium* (Iqbal and Saeed 2007).

## 2 Fungi from Different Ecological Niches Capable of Absorbing Heavy Metals

Filamentous fungi are widespread in the environment. The diameter of fungal hyphae is from 2 to 10 mm, while the fungal mycelium comprises an interconnected network of hyphae (from millimeters to centimeters long). Filamentous fungi are one of the most cheap and ecofriendly biosorbents (Shah et al. 2020). Fungal biomasses possess many advantages in comparison with bacteria, like their easy growth using available media constituents, high biomass yield, and abundance in industrial waste products after fermentation (Dhankhar and Hooda 2011; Cárdenas González et al. 2019). In addition to these advantages, the biosorption of different heavy metals by fungi can be investigated at an affordable cost (Drahansky et al. 2016). Moreover, the characteristic properties of the fungal cell wall that is composed of mineral ions and nitrogen, along with the filamentous growth pattern, allow fungi to efficiently decompose and absorb heavy metals (Naveena and Latha 2018). Thus, it is very promising to use fungal biomasses at a large scale for removing metal ions from contaminated industrial effluents. Fortunately, the majority of such biosorbents are non-toxic and hence, they can be easily and safely used (Lakshmi et al. 2018).

### *2.1 Fungi from Wastewater and Sediment*

Pal and Basumajumdar (2002) reported on the ability of *Aspergillus niger* to remove Cu (91%), Cd (59.67%), and Pb (45.54%) from swine wastewaters. In the same vein, Kapoor et al. (1999) found that the removal of Cd, Ni, Cu, and Pb from wastewaters was higher in case of using live *Aspergillus niger* biomass as compared to activated carbon. These two experiments demonstrated that *A. niger* is a good candidate for removing metals from the contaminated sources. Other fungal spp. such as *Fusarium oxysporum*, *Aspergillus terreus*, *Gliocladium roseum*, *Trichoderma koningii*, *Penicillium* spp., *Talaromyces helicus*, and *Cladosporium cladosporioides* were shown to be tolerant to 0.25–0.50 mg Cd/l, and that was attributed to their isolation from heavily polluted sediments located in an industrial area in Argentina (Massaccesi et al. 2002).

Additionally, *Trichoderma atroviride* isolated from sewage sludge was able to tolerate high Cd, Cu, and Zn concentrations (López Errasquín and Vázquez 2003), where the autolyzed mycelia showed the highest metal removal percentage. Also, Yazdani et al. (2009) demonstrated that *Trichoderma atroviride* could absorb 0.8 to 11.2 mg/g of Cu from the medium, on the basis of its isolation from a river containing copper-polluted sediments. So, using *T. atroviride* as an alternative to traditional chemical methodologies for wastewater treatment is very promising. Moreover, Simonovicova (2008) isolated a number of fungal strains from different water sources, where the highest removal of Al was observed by *Penicillium glabrum* (100%) followed by *A. niger* (43–57%) and *Aspergillus clavatus* (45–48%). The highest removal of Cd was recorded by *Trichoderma viride* and *A. niger*. On the other hand, *Neusartorya fischeri* highly removed each of Cd, Zn, Cu, and Cr.

A number of fungal species were isolated from water and sediments of heavy metal-contaminated sites in Tangier, Morocco (Ezzouhri et al. 2009). The isolated filamentous fungi belonged to the genera *Aspergillus, Penicillium, Fusarium, Alternaria,* and *Geotrichum.* The authors reported that most of the isolates were tolerant to Cu, Pb, Zn, and Cr, whereas only *Penicillium* sp. was able to grow in the presence of Cd. They demonstrated that *Penicillium* sp. was the most tolerant to each of Pb, Cr, and Cd, while *Alternaria alternata* was the most tolerant to Cu and Zn. The degree to which the isolate can tolerate the metal depended on both the type of the isolate and the isolation site. *Aspergillus* sp. and *Penicillium* sp. did not only show the highest tolerance to the heavy metals, but also grew well, even better than the control.

Ting and Choong (2009) demonstrated that *Trichoderma* SP2F1 from the contaminated sediment samples of Penchala River, Malaysia was successful in removing Cu from aqueous solutions. The bioaccumulation capacity of viable SP2F1 was 19.6 mg/g, whereas the biosorption capacity of non-viable SP2F1 cells was 28.75 mg/g. This indicated that *Trichoderma* sp. cells, whether viable or not, could be considered for removing Cu from sewage. The possible use of *Trichoderma* spp. for metals biosorption has been confirmed by Gupta et al. (2014) and Sharma et al. (2019). Similarly, Iskandar et al. (2011) showed that *Penicillium simplicissimum* had a strong potential to remove Cd, Co, Ni, Cu, and Cr on lab scale. These results matched the study of Mohammadian et al. (2017). They also highlighted the superior ability of *P. simplicissimum* to absorb Cu over other fungal species, such as *Seimatosporium pistaciae, Fusarium verticillioides, Alternaria chlamydosporigena,* and *Acremonium persicinum.*

Kumar et al. (2011) isolated *Aspergillus niger* from the sludge of an industrial effluent and trained it with many heavy metals of different concentrations. The strain was acclimatized to detoxify both Cd and Zn. The results emphasized the heavy metal detoxification abilities of *Aspergillus niger*, as its biomass reached a removal efficiency of 59% for Zn, and 51.05% for Cd. These data are significant for the environmental biotechnology as a simple bioremediation method that could allow the detoxification of these polluted sites to render them free from heavy metals. Moreover, Joshi et al. (2011) isolated 76 fungal species tolerant to Ni, Cr, Pb, and Cd from industrial effluents, sludge, and sewage contaminated by heavy metals. They found that for the metals Pb, Cd, Cr, and Ni, the maximum uptake was observed by *Aspergillus terreus* (59.67 mg/g), *Trichoderma viride* (16.25 mg/g), *Trichoderma longibrachiatum* (0.55 mg/g), and *A. niger* (0.55 mg/g), respectively. Similarly, Dwivedi et al. (2012) isolated various fungal spp. tolerant to heavy metals such as Ni, Pb, Cr, and Cd from industrial effluents, sludge, and sewage contaminated by heavy metals. They were *Phanerochaete chrysosporium, Aspergillus flavus, Aspergillus foetidus, Aspegillus awamori, Trichoderma viride,* and *Rhizopus* sp. Most of the fungal species were able to tolerate up to 400 ppm of Cu, Cd, Co, Pb, Ni, and Cr. *Aspergillus niger* and *Aspergillus flavus* succeeded in removal of varying amounts of heavy metals, e.g., Cr, Ni, and Pb, where the highest Pb uptake was that observed by *A. flavus* (17.35 mg/g), whereas the highest dry weight was that of *A. niger* (0.46 gm) exposed to 50 ppm of Pb. In the same vein, from effluent samples from Wise Park industrial area, India, *Aspergillus flavus* was investigated as a biosorbent of Cu (II), where the authors recorded 80% removal of copper (Jayaraman and Arumugam 2014). The findings indicated that *A. flavus* could be considered as a candidate for the clean-up of copper contaminated areas. Additionally, they showed that the biosorbents could be reused or regenerated again after the removal of copper or other heavy metals. Moreover, Rao et al. (2002) reported that *A. flavus* (DSF-8) could also be used in the uptake of copper. In another study, *Aspergillus terreus*, *Aspergillus niger*, *Aspergillus fumigatus*, *Aspergillus flavus*, *Aspergillus oryzae*, and *Aspergillus tamarii* were isolated from wastewater of three functional tanneries within Sokoto Metropolis, Nigeria (Abdullahi and Machido 2017). The researchers investigated the ability of these fungi—on a lab scale—to grow in the presence of various concentrations of Cd, Cr, and Pb, and found that the majority of the isolates were tolerant to Cr and Pb, while only a few were able to grow in the presence of Cd. The species that showed the highest level of tolerance to all the heavy metals, with a high biomass yield, were *A. terreus*, *A. niger* and *A. flavus*, where *A. terreus* showed the highest biomass when exposed to 15 μg/ml of Pb and Cr, while *A. tamarii* was the most tolerant to the same concentration of Cd. Therefore, these species of *Aspergillus* were shown to be capable of acclimatizing in this contaminated water through bioaccumulating heavy metals, and thus they can be considered in mycoremediation of the metals present in the tannery wastewater.

After searching for special microorganisms in microbiomes of some sulfur springs in Bavaria (Germany) over 200 $km^2$, three new strains of *Mucor hiemalis* that were capable of remediating many metals were found. These strains that were identified as EH8, EH10, and EH11showed high removal percentage (81–99%) of Ni, Al, Co, Zn, Cr, Cd, Cu, Hg, Pb, and U from water all within 48 h (Hoque and Fritscher 2019).

## 2.2 *Fungi from Soil*

Say et al. (2003) reported that the highest biosorption capacity of heavy metals onto the soil fungus; *Penicillium canescens* biomass under non-competitive conditions was 213.2 mg/g for Pb (II), 102.7 mg/g for Cd (II), 54.8 mg/g for Hg (II), and 26.4 mg/g for As (III). When investigating heavy metals biosorption in a mixture, the capacities were 32.1 mg/g for Pb (II), 11.7 mg/g for Cd (II), 5.8 mg/g for Hg (II), and 2.0 mg/g for As (III) at initial concentration of 50 mg/l. The equilibrium loading capacity of Pb (II) was higher than that of other metals, and the fungal biomass showed preference toward the binding of Pb over Cd, Hg, and As. They also showed that *Penicillium canescens* could be used for biosorption for more than six times. Moreover, Ahmad et al. (2005) isolated two fungi belonging to *Aspergillus* and *Rhizopus* from an agricultural field exposed to sewage and industrial effluents. They used the pretreated dead biomass of the two fungi for bioadsorption experiments and demonstrated that bioadsorption of Cr ranged from 6.20 to 9.5 mg/g of dry mass, whereas the bioadsorption of Cd ranged from 2.3 to 8.21 mg/g. They noticed that *Rhizopus* sp. could bioadsorb higher concentrations of both metals as compared to *Aspergillus* sp. They concluded that fungi from metal polluted areas showed higher metal tolerance and biosorption capacity of Cd and Cr. Also, Abd El-Ghany and Abd El-Mongy (2009) reported on the bioaccumulation of Cd, Ni, Co, and Pb by *Aspergillus tamarii* isolated from agricultural soil, as they found that the highest percentage of absorption was 88.48% for Cr, followed by 84.36% for Ni, and then 75.12 and 71.16% for Pb and Cd, respectively. They also mentioned that the addition of oxalic and citric acids improved the biosorption process.

Iram et al. (2009) isolated forty-one fungal species from soils contaminated with heavy metals. The most common fungal strains *Penicillium* sp., *Aspergillus* sp., *Aspergillus niger*, and *Fusarium* sp. were investigated for their tolerance to Cd, Zn, Ni, and Pb. The tolerance level was measured by minimum inhibitory concentration (MIC) in the presence of each metal. They demonstrated that among the isolated fungal strains of all locations, *Aspergillus niger* was the most tolerant. It showed strong radial growth even at 40 ppm of the all tested metals, followed by *Aspergillus* sp., *Pencillium* sp., and *Fusarium* sp. Furthermore, Shazia et al. (2013) conducted a study on highly tolerant filamentous fungal species; *Aspergillus fumigatus* isolated from a polluted soil at Kasur district, Pakistan, where the highest biosorption capacity (76.07 mg/g) was exhibited by *A. fumigatus* isolate K3 for Pb, followed by Cu (69.6 mg/g) and Cr (40.0 mg/g) at 800 ppm metal concentration.

Oladipo et al. (2018) investigated the tolerance of *Trichoderma ghanense* isolated from a gold and gemstone mine soil to each of Pb, As, Cd, Fe, and Cu. Their results showed that *T. ghanense* possessed high tolerance to high concentrations of heavy metals, which indicated that it can be used for removal of heavy metals from polluted sites. *T. ghanense* had also been found to produce antioxidant enzymes that can minimize oxidative stress caused by heavy metal pollution (Akhtar and Mannan 2020).

By the same token, Saad et al. (2019) isolated *Aspergillus tamarii* NRC3 from contaminated soil. They found that the live fungal biomass showed high Cu removal (90.94%), moderate Co, Ni, and Fe removal (60%, 40%, 34.5%, respectively), and low removal of both Pb (29.13%) and Cr (11.45%). Hassan et al. (2019), on the other hand, investigated filamentous fungi consortia in landfill metal-contaminated soil for their bioremediation ability. They divided 12 fungal species into two groups: highly tolerant fungi (*Aspergillus niger*, *Perenniporia subtephropora*, *Fusarium equiseti*, *Daldinia starbaeckii*, *Phanerochaete concrescens*, *Trametes versicolor*, *Cerrena aurantiopora*, *Polyporales* sp., and *Aspergillus fumigatus*) and moderately tolerant fungi(*Penicillium cataractum*, *Paecilomyces lilacinus*, and *Antrodia serialis*). They recorded the highest removal in the highly tolerant group as follows, As (62%) > Mn (59%) > Cu (49%) > Cr (42%) > Fe (38%). So, these data were important in highlighting the contribution of fungal consortia, besides showing how the synergistic effect of fungi can effectively reduce contamination of soils with heavy metals. Recently, Bala et al. (2020) isolated *Rhizomucor* sp., *Penicillium chrysogenum*, and *Aspergillus niger* from refuse dumpsite soil. *Aspergillus niger* and *Penicillium chrysogenum* showed high tolerance for the metals tested (Cd, Pb, and Zn) with concentrations of 200 ppm, 400 ppm, and 600 ppm. *Penicillium chrysogenum* showed higher lead removal of 1.07 ppm, 3.35 ppm, and 4.19 ppm as compared to *Aspergillus niger* that showed lead removal of 0.67 ppm, 3.11 ppm, and 3.79 ppm at the 5th, 10th, and 15th day, respectively. In the same vein, Imran et al. (2020) demonstrated that among 73 isolated fungal species from soil irrigated with industrial wastewater containing Cr, Ni, Co, Cu, and Cd, the highest tolerance was recorded for Cr followed by Cu, Co, Ni, and Cd. The minimum inhibitory concentration (MIC) values of all the metal tolerant strains were in the range of 200–2000 $\mu gml^{-1}$. As *Aspergillus* was the most frequent genus isolated from the contaminated soil, the authors found different MIC values among *A. terreus*, *A. niger*, *A. sydowii*, and *A. flavus*.

Vacar et al. (2021) assessed the heavy metals tolerance and biosorption capacity of Hg by fungi isolated from the rhizosphere of plants grown on highly Hg-contaminated soil, where a total of 32 heavy metal resistant fungal isolates were identified. They observed that Ascomycota members were the most abundant and diverse, besides showing a wide range of heavy metals tolerance, especially for the metals representing environmental threat. On the contrary, isolates belonging to Mucoromycota showed similar metal resistance patterns within phylogenetically related clades, despite observing isolate-specific tolerance. *Phoma costaricensis, Cladosporium* sp., *Fusarium oxysporum*, *Sarocladium kiliense,* and *Didymella glomerata* isolates exhibited high minimum inhibitory concentration (mg/L) for Zn (2092–2353), Pb (1568), Cu (381), Cd (337), and Hg (140–200). The biosorption capacity of Hg ranged from 33.8 to 54.9 mg/g dry weight, with a removal capacity from 47 to 97% as follows: *Didymella glomerata* live biomass showed 97% removal after 12 h of incubation, followed by *F. oxysporum* (62%), *Cladosporium* sp. (61%), *Phoma costaricensis* (56%), and finally *Sarocladium kiliense* (47%). Thus, these fungi could be considered excellent biosorbents of Hg from contaminated substrates.

## 2.3 Endophytic Fungi

It was found that the endophytes from the plants that are known as hyperaccumulators could be metal resistant (Idris et al. 2004). This was attributed to long-term adaptation to the high metal concentrations. Metal hyperaccumulating plants such as *Alnus firma*, *Alyssum bertoloni*, *Nicotiana tabacum*, *Brassica napus*, *Thlapsi caerulescens*, *Solanum nigrum,* and *T. goesingense* were found to harbor metal resistant endophytes. Nevertheless, metal resistant endophytes have been isolated from non -hyperaccumulating plants as well, for example, *Acacia decurrens*, *Symplocos paniculata,* and *Arabis hirusta* (Li et al. 2012). The metal resistant endophytic fungi were *Aspergillus*, *Peyronellaea*, *Phoma*, *Microspaeropsis*, *Mucor*, *Alternaria,* and *Steganosporium*. Xiao et al. (2010) isolated *Microsphaeropsis* sp. which is an endophytic fungus from cadmium hyperaccumulating *Solanum nigrum* L. and found that it showed a high biomass yield when cultured in vitro, in addition, it was a successful biosorbent of cadmium. Also, Deng et al. (2014) reported that the Cd, Pb, and Zn resistant endophytic strain *Lasiodiplodia* sp. MXSF31 which was isolated from metal accumulating *Portulaca oleracea* showed high biosorption capacity of Zn, Cd, and Pb from their solutions. Moreover, the strain showed high removal efficiency of various metals from rape soils. Possessing many advantages like their wide host range, the ability to live inside plant tissues and tolerance to many heavy metals, endophytic fungi from the plants that are able to accumulate various metals should be considered as biological tools for remediation of both contaminated water resources and soils (Aishwarya et al. 2014).

Choo et al. (2015) isolated endophytes from *Nypa fruticans* from Kuching Wetland National Park (KWNP), Malaysia. Ninety-three fungal isolates were recovered, where the most tolerant ones (able to grow at concentrations of 1000 ppm) were molecularly identified. All the endophytic fungi were found to be closely related to *Pestalotiopsis* sp., and this was the first study reporting the ability of *Pestalotiopsis* sp. to grow at high concentrations of Cu, Pb, Zn, and Cr. In the same vein, from the heavy metal hyperaccumulator plant *Vachellia farnesiana*, Salazar-Ramírez et al. (2020) isolated endophytic fungi. The plant environment was mine tailings that contained Cu, Zn, and Pb in high concentrations. Both phenotypic and molecular identifications demonstrated that the fungal strains were closely related to *Aspergillus* and *Neocosmospora* genera. The *Neocosmospora* isolate belonged to the *Fusarium solani* species complex (FSSC) that contains phytopathogenic species. Nevertheless, the host plants were healthy. *Neocosmospora* strain was found to tolerate Cu (500 ppm), Zn and Pb (700 ppm), with 55%, 86%, and 15% inhibition of growth, respectively. Meanwhile, when the lowest concentration was used (100 ppm for each of Cu and Pb, and 50 ppm for Zn), the best growth rate was observed, and it was very similar to that recorded in potato dextrose medium. On the other hand, the *Aspergillus* strain could tolerate Cu (1000 ppm), Pb (700 ppm), and Zn (350 ppm), with 88%, 89%, and 97%, growth inhibition, respectively. Also, the maximum growth rate was recorded when the lowest concentrations were applied, except for Zn, where the growth was the same for both control and in the presence of 50 ppm. Both fungal strains were very promising as bioremediating tools, *Neocosmospora* strain removed 34% of Cu, 48% of Zn, and 90% of Pb after 12 h, while *Aspergillus* strain removed 25% of Cu, 40% of Zn, and 85% of Pb. The authors also noticed that *Neocosmospora* strain secreted specific novel phenolic compounds in response to lead exposure. Regarding *Aspergillus* sp., it secreted glutamic acid, in addition to malic and succinic acids as a response to Pb exposure. This was recorded as a new mechanism that has not been reported before. The authors suggested that these acids were produced to cope with metal toxicity (Salazar-Ramírez et al. 2020). Recently, Abdel-Wareth et al. (2023) found that the dry biomass of *A. niger* isolated from seeds of *Jatropha curcas* plant was able to remove 97.8% of Mn, and 98.6% of Ni at 3.3 mg /L and 6.84 mg/L, respectively as initial concentrations.

## 2.4 Fungi from Other Sources

The dead biomass of *Fusarium flocciferum* fungus has the capacity of biosorption of copper, cadmium, and nickel. It was found that cadmium and nickel biosorption could fit in Langmuir isotherm equations, while copper had an irregular biosorption pattern. The highest biosorption capacities were 19.2 and 5.2 mg metal/100 mg biosorbent for Cd and Ni, respectively (Delgado et al. 1998). Additionally, Martino et al. (2000) reported that the better growth of the mycorrhizal fungus; *Oidiodendron maius* of *Vaccinium myrtillus* plant growing in highly contaminated soils was observed in the presence of higher concentration of Zn ions. Moreover, Ahmad et al. (2006) demonstrated that *Aspergillus niger* and *Pencillium* sp. showed high removal of Cd, Cr, and Ni from single and metal mixture solutions, and suggested their possible utilization in metal polluted habitats. By the same token, Kaoushik et al. (2013) found that the two isolated arsenic tolerant fungal strains *A. niger* (ASB3) and *A. flavus* (ASC1) showed high removal percentages (50–76%) of arsenic from different arsenic containing substrates, also they were tolerant to each of Cr, Cd, Hg, Pb, and Zn.

Simonescu and Ferdes (2012) investigated the use of fungal biomass for removal of Cu from a copper sulfide nanoparticles containing solution. They found that *Fusarium oxysporum* MUCL 791 showed the highest copper uptake capacity (7.52 mg/g). Regarding arbuscular mycorrhizal (AM) fungi, Vaishaly et al. (2015) demonstrated that they almost accumulate 100 mg/kg of Zn and 300 mg/kg of Cd present in the metal-contaminated soils, where 22 to 71% colonization, and 622 spore count for 100 mg of soil were recorded. For example, AM fungal colonization was up to 40% in high concentration of Cd and Pb. They also mentioned that AM fungal diversity was impeded with sewage sludge containing heavy metals such as Cu, Cd, and Pb. Long-term sludge application and increasing heavy metal concentrations generally lead to a decrease in the total number of fungal spores, on the contrary, AM fungal spores could withstand such environmental stress, and their spores were able to germinate well in naturally polluted soil when compared to clean soil.

From the polluted air in a fuel station, at San Luis Potosi, Mexico, Acosta-Rodriguez et al. (2017) isolated *Aspergillus niger*, where Acosta-Rodriguez et al. (2018) analyzed its tolerance to some heavy metals, as the fungus grew while exposed to 2000 ppm of Hg, Pb, and Zn, and 1200 and 1000 ppm of As (III) and (VI), 800 ppm of F and Co, and 400 ppm of Cd. The strain was able to remove 100% of Zn, 83.2% of Hg, 83% of F, 71.4% of Co, 48% of Ag, and 37% of Cu. Recently, Zhang et al. (2020) assessed the ability of an earthworm gut fungus *Trichoderma brevicompactum* QYCD-6 to tolerate and uptake individual and multi-metals. They found that MIC of Zn, Cu, Cd, and Cr ranged from 150 to 200 mg/L against the fungus on composite medium, while MIC of Pb was 1600 mg/L on potato dextrose medium. The highest metal removal rate was that of Pb (97.5%), which mainly was achieved by 80.0% bioaccumulation, and accompanied with a biomass of 6.13 g/L in potato dextrose medium. In case of the composite medium, the highest removal rate was observed for

Cu (64.5%). The authors also observed the cellular changes in the fungus by transmission electron microscopy analysis. Also, Fourier transform infrared and solid-state NMR analyses indicated the role of different functional groups such as hydroxyl, carbonyl, and amino groups in metals absorption. These findings indicated that the earthworm associated *T. brevicompactum* QYCD-6 could be considered a promising candidate for the removal of heavy metals from contaminated water.

## 3 Mechanisms of Heavy Metals Absorption

Filamentous fungi accumulate metal ions into their mycelium and spores via intracellular/biosorption and extracellular/adsorption mechanisms. These mechanisms mainly depend on the fungal cell wall that participates in the energetic metal uptake (Siddiquee et al. 2015; Igiri et al. 2018; Naveena and Latha 2018). This cell wall has certain properties, such as the chitin which contains chitosan (a polymer of N-acetyl D-glucosamine), that plays a major role in metal biosorption (Naveena and Latha 2018).

The response mechanism of a microorganism to a pollutant begins with the production of extracellular substances that participate in adsorption and precipitation of metal, and then, binding of metal ions to thiol-containing metabolites to form complexes stored in vacuoles or other compartments inside the cell (Ge et al. 2011). Metallothionein is one of the low molecular weight proteins that has been found to mediate hormones and redox signaling molecules in the metabolic responses of microbes to heavy metal toxicity (Verma and Kuila 2019). Additionally, the sulfur-containing compound, Glutathione (GSH), plays a role in metal detoxification (Ge et al. 2011).The remediation of water sources containing heavy metals is usually carried out as follows: (I) bioaccumulation, (II) surface complexation, (III) bioprecipitation, (IV) ion exchange, (V) electrostatic interactions, and (VI) cell surface adsorption (Ayangbenro and Babalola 2017; Hansda and Kumar 2016; Igiri et al. 2018).

Different researchers have defined biosorption; Abdi and Kazemi (2015) and Thakkar et al. (2016) described it as "the passive uptake of metal ions by dead/inactive biomass from an aqueous solution," whereas Gadd (2009) and Javanbakht et al. (2014) defined biosorption as "an activity of both living and dead biomass to uptake metal ions from an aqueous solution." Generally, the fungus has many advantages to be a candidate in biosorption processes, such as the multiple functional groups in its cell wall, its easy growth on a large scale, cheap growth media, and simple fermentation techniques (Kapoor and Viraraghavan 1995), besides being available as industrial waste product from organic acid and beverages industries (Dhankhar and Hooda 2011). The major dry weight of the fungus cell wall is due to the presence of 80 to 90% of polysaccharides, proteins, and lipids such as glucans, chitin, mannans, and phosphomannans (Javanbakht et al. 2014; Thakkar et al. 2016), also, the cell wall contains various metal-binding functional groups such as carboxyl, hydroxyl, acetamide, carbonyl, chitin, phosphonate, and sulfhydryls (Cai et al. 2016; Ayangbenro and Babalola 2017). Chemical bonds help in providing the ligand atoms needed for metal ions complexes by attracting and trapping metals in the biomass. The anion in the functional groups binds with the metal cation, and hence metals become attached to ligands formed by functional groups on the cell surface, and can be converted from one state of oxidation to another (Ayangbenro and Babalola 2017).

Biosorption can be achieved via passive adsorption in case of using inactivated biomass where cell surface binding is the only mechanism involved; or as an active process in case of using live biomass in which both internal and external cellular metabolism e.g., detoxification, chelation, volatilization, and bioaccumulation took place (Cai et al. 2016; Abdi and Kazemi 2015).

Li et al. (2010) and Iram et al. (2015) showed that the use of dead biomass for biosorption of heavy metals is better than living cells. On the other hand, Coelho et al. (2015) reported the opposite, as metal removal by inactive fungal biomass involves only physicochemical interaction. Both biosorption using non-living biomass and bioaccumulation using a living one involve interactions and concentration of toxic metals in the selected biomass (Chojnacka 2010).

Bioaccumulation is a gradual accumulation of a substance in an organism, where absorption occurs faster than metabolic processes. As an active process, it depends on the metabolic activity of a live organism. Biosorption, on the other hand, is a passive binding process via non-living microbial biomass or biologically derived materials. Many studies highlighted a higher efficiency of biosorption than bioaccumulation (Hansda and Kumar 2016). Generally, biosorption can be performed within few hours, whereas successful bioaccumulation might require many days or weeks (Salam 2019). To achieve the maximum microbial removal of heavy metals via either biosorption or bioaccumulation, there are three points that should be considered: the first one is that certain metals such as Cu, Co, Fe, Zn, and Mn are necessary for the proper functioning of microbial metabolism (Filote et al. 2020), while others such as Cd, Hg, and Pb are not (Rosca et al. 2015; Filote et al. 2020). The second one is that the development of microorganisms passes through the lag, exponential, stationary, and decline phases (Kurniati et al. 2014). In the lag phase, viable microorganisms adapt to the new environmental factors or stressors. This adaptation is carried out via the formation of growth enzymes and other substances that participate in cell development. Usually, the lag phase increases as the concentration of the metal to which the microorganism is exposed increases, and the maximum tolerance index in the stationary growth phase decreases (Ge et al. 2011). The third one is that biosorption potential differs according to metal properties, its bioavailability to the biomass, the presence of other pollutants or nutrients, temperature, redox potential, water characteristics, osmotic pressure, pH, and oxygen (Ayangbenro and Babalola 2017). As the fungal strains isolated from a metal polluted substrate are more efficient than isolates from non-polluted ones, The target areas to get promising isolates are industrial effluents, polluted soils, and waste disposal locations (Ayangbenro and Babalola 2017; Kapahi and Sachdeva 2017).

### 3.1 Metal Uptake by Dead Biomass

It is a biosorption process that is independent on metabolism. Simply, it is a physicochemical interaction between the metal and the functional groups of the fungal cell wall (Javanbakht et al. 2014). It was claimed that depending on living organisms as biosorbents of heavy metals is not recommended for many reasons (Dhankhar and Hooda 2011). On the contrary, dead biomasses showed many advantages, such as high environmental versatility, the possibility of regeneration and reuse, tolerance to many toxic metals, simplicity of absorption reactors modeling (Javanbakht et al. 2014), and nonnecessity for any culture media that the live organism requires (da Rocha Ferreira et al. 2019). Moreover, biosorbents can be stored for long time without altering their efficiency (da Rocha Ferreira et al. 2019; Javanbakht et al. 2014). Metal biosorption by dead fungal cells has been found to be higher than that by live fungi (Kapoor 1998). The mechanism of heavy metals biosorption by dead biomass includes physical adsorption, electrostatic interactions, ion exchange, and metal ion complexation (Thakkar et al. 2016). Physical adsorption is a surface sorption due to Van der Waals forces, it is a fast and reversible process. Electrostatic adsorption is also fast and reversible, it depends on the Coulombic attraction forces between the biosorbent and the metal (Javanbakht et al. 2014). Ion exchange is the substitution of an ion from a sorbent with another oppositely charged ion from cell wall (Gadd 2009). Actually, it is a costly method that cannot be applied on a large scale, especially when targeting wastewater contaminated with low concentrations of heavy metals (Dhankhar and Hooda 2011). Complexation is the formation of a complex between the metal and the biosorbent (Javanbakht et al. 2014). This complex consists of one or more positively charged central atoms bound to negatively charged ligands. Different functional groups located in the cell wall of the fungus provide these ligands (da Rocha Ferreira et al. 2019).

### 3.2 Metal Uptake by Live Biomass

It is an active process where the metabolic activity of a living organism is responsible for removal of heavy metals (Khan et al. 2019). This can be achieved through biotransformation, biomineralization, bioaccumulation, and bioprecipitation (Dhankhar and Hooda 2011). The living biomass cell wall carries a negative charge formed from functional groups ligands that are attached to the surrounding metal ions. The active metal sorption involves the transportation of metal ions across the cell membrane after binding to cell surface in a process known as biotransformation (Dhankhar and Hooda 2011). Bioprecipitation is another process in which specific metal-binding proteins such as metallothioneins and phytochelatins are synthesized (Hansda and Kumar 2016).

Xu et al. (2020) studied the mechanism by which Pb is removed from Pb-containing wastewater by *Penicillium polonicum.* They declared that *Penicillium polonicum*, within 12 days, tolerated 4 mmol/L of Pb (II), and showed 90.3% removal via extracellular immobilization, cell wall adsorption, and intracellular bioaccumulation. These three mechanisms were investigated by scanning and transmission electron microscopy, Fourier transform infrared spectroscopy, and X-ray diffraction analysis. It was demonstrated that Pb (II) was immobilized as lead oxalate outside the fungal cell, bound with halide, carboxyl, phosphate, hydroxyl, nitro, and amino groups on the cell wall, precipitated as pyromorphite [Pb5(PO4)3Cl] on the cell wall, and reduced to Pb (0) inside the cell. These findings provide more detailed information on removal mechanisms of Pb (II) by *P. polonicum* and suggest using this species for remediation of lead contaminated wastewaters.

## 4 Factors Controlling Heavy Metals Absorption by Fungi

Many factors can influence heavy metal absorption by fungi, such as pH, temperature, contact time, biomass dosage, and initial metal concentration. They are discussed below in details.

### *4.1 Hydrogen Ion Concentration (pH)*

The pH under which metal removal takes place directly affects the biosorption capacity. pH affects the chemistry of both metal ions and cell wall functional groups (Wei et al. 2016). Generally, the biosorption capacity increases as pH level increases until reaching the optimum pH where the maximum biosorption capacity takes place, beyond this, metals begin to precipitate where metal hydroxides or hydroxide anionic complexes are formed (Abdi and Kazemi 2015; Bilal et al. 2018). It was reported that biosorption of metal ions such as Ni, Zn, Co, Cd, and Cu is usually reduced at low pH, where metal cations compete with hydronium ions for the binding sites of the biosorbent (Gadd 2009). At low pH, the functional groups of the biosorbent prefer hydronium ions, and the binding between metal cations and functional groups does not occur due to the repulsive forces between them. Nevertheless, as the pH increases, some functional groups (phosphate, hydroxyl, and carboxyl) become negatively charged due to deprotonation, this would increase binding of metal cations and hence increases the biosorption rate and capacity (Abdi and Kazemi 2015; Bilal et al. 2018). pH might also affect the form of the metal ion in an aqueous solution and its concentration. This might lead to the formation of different species of a metal ion at different pHs (Gadd 2009).

## 4.2 Temperature

The effect of temperature on the biosorption process depends on the nature of the process. For an exothermic biosorption process, increasing the temperature would lead to a decrease in heavy metal removal, and for endothermic biosorption processes, the opposite is true (Tasar et al. 2014). Although it was reported that biosorption is not highly affected within temperatures of 20 °C to 35 °C (Sao 2014; Perpetuo et al. 2011), Jayaraman and Arumugam (2014) and Cárdenas González et al. (2019) demonstrated that higher temperatures enhanced the biosorption capacity of fungal biomasses, so in the tropics (characterized by high temperatures), using fungal biomasses for metals removal is suggested. They added that high temperatures would activate adsorbing surfaces and increase the mobility of metal ions.

## 4.3 Contact Time

The maximum biosorption capacity might be achieved upon increasing contact time, but when binding sites become fully saturated, extending the time become useless (Hajahmadi et al. 2015).

## 4.4 Biomass Dosage

The selective absorption of metal ions was found to increase at low biomass dosages (Sao 2014; Perpetuo et al. 2011). However, utilizing low biomass dose in complicated contaminated water would increase competition for the biosorbent's binding sites and would reduce its ability to absorb. Therefore, it is preferred to increase the biomass dosage when there are many metal ions to lessen competition for binding to the functional groups.

## 4.5 Initial Metal Concentration

It was reported that as the initial metal concentration rises, metal biosorption increases (Ashraf 2011). Similar to increasing contact time, increasing the initial metal concentration has an impact. On the one hand, utilizing a low metal concentration could not accurately represent the true biosorbent's maximum biosorption capacity because many of the metal-binding sites might not be occupied. On the other hand, increasing the metal concentration up to the point where the biosorbent's binding sites are fully saturated would show the maximum capability to remove the metal; however, after the biosorbent is fully saturated, further increasing the initial metal concentration will be ineffective.

## 4.6 Biosorbent Surface Area

The biosorbent material's surface area has an impact on its biosorption capacity since, under the same circumstances, a larger surface area would improve biosorption. Additionally, having more than one metal ion in a medium can affect how well another metal ion is removed since they may compete for binding sites due to one metal ion's stronger affinity for the biosorbent's binding sites (Sao 2014).

## 4.7 Examples on the Effect of Different Factors on Metals Biosorption

Many studies showed the effects of different factors on biosorption process. In their 2007 study, Ali and Hashem investigated how different pH and temperature ranges affected the ability of the zoosporic fungus *Saprolegnia delica* and the terrestrial fungus *Trichoderma viride* to remove Zn, Pb, and Cd from polluted water drainages in the Nile Delta in Egypt. They demonstrated that pH 8 resulted in the highest *S. delica* removal efficiency for Zn (II), Cd (II), and Pb (II), whereas pH 6 resulted in the highest removal efficiency for Pb (II). *T. viride*, on the other hand, required a pH of 6 to effectively remove the three tested heavy metals. Additionally, *S. delica* exhibited the highest Pb and Zn bioaccumulation potential at all pH levels other than pH 4, while Cd had the lowest removal percentage, and with *T. viride*, the opposite is true. *S. delica* produced the highest biomass dry weight when it was cultivated in a medium that had been treated with Zn and Cd at pH 8, and Pb at pH 6. When *T. viride* was treated with Zn, Pb, and Cd, the best biomass dry weight yield was recorded at pH 6 for all three heavy metals. Upon treating *T. viride* with Cd at all different pH values, followed by Pb, the largest yield of dry biomass weight was observed. Zn resulted in the lowest biomass, and the situation was reversed upon using *S. delica*. For the three heavy metals under evaluation, the greatest removal efficiency and biomass dry weight production were seen at incubation temperatures of 20 °C for *S. delica* and 25 °C for *T. viride*. The removal of Pb and Cd from *T. viride* was more effective at higher temperatures (30 °C and 35 °C) than at lower temperatures (15 °C and 20 °C), while the opposite was true for the removal of Zn. The largest yield of the two fungal biomasses was achieved in the case of Zn treatment at each of the tested incubation temperatures. The bioaccumulation potency of *S. delica* for Zn was higher than that for Pb at all incubation temperatures, and *T. viride* demonstrated the best removal efficiency of Pb followed by Cd.

In another study, the effect of pH on adsorption of Pb by the wet and dry biomass of *Penicillium* sp. isolated from a soil contaminated with heavy metals in Tangier (Morocco) was investigated by Ezzouhri et al. (2010). The biosorption capacity of this metal depended mainly on pH, as it increased with the increase in pH. The highest Pb uptake, 96.34% and 95.11% by wet and dry biomasses, respectively was recorded at pH 5.5. Below 3, insignificant Pb removal was observed due to competition between Pb ions and protons for biosorption sites (Wang and Chen 2006). As pH increases, the density of negative charges on the cell wall increases due to deprotonation of the metal-binding sites which enhance sorption of metals. Similarly, the removal% of lead ions by *Penicillium austurianum* was 75.57–94.21% and 44.47–98.85% for live and dead biomasses, respectively, depending on pH (Awofolu et al. 2006). The maximum $Pb^{2+}$ biosorption capacity (13.46 mg $g^{-1}$) by *Aspergillus flavus* was noticed at pH 5.0 (Akar and Tunali 2005b). Moreover, Alimohamadi et al. (2005) reported that the optimum pH for $Pb^{2+}$ biosorption by *Rhizopus arrhizus* was 5.0. Recently, Hussain et al. (2020) demonstrated that the highest removal percentages of Pb and Cr by *Penicillium digitatum* were 84% and 70%, respectively, at pH 5 within 24 h. Nevertheless, this capacity decreased from 81 to 76% at pH levels of 4 and 1.5, respectively for Pb, whereas for Cr, it was 56% at pH 4 after one hour.

Also, it was reported that biosorption of Co (II) by *Paecilomyces catenlannulatus* increased as the temperature increased from 20 to 40 °C (Cárdenas González et al. 2019). Saad et al. (2019) found that the biosorption capacity of Cu by *Aspergillus niger* was 26.2 mg/g at 31 °C where the stability of cell wall components was temperature dependent. Furthermore, Farhan and Khadome (2015) evaluated the efficiency of the yeast *Saccharomyces cerevisiae* in removing heavy metals from aqueous solutions. They investigated the effect of pH on the biosorption of Co, Pb, Cd, Zn, Cr, and Cu ions, and the results showed that fast metal removal was detected at pH, 5.0–6.0. While at an initial pH of 4.0 or less, lower biosorption was observed, moreover, at pH 2.0, the biosorption process ceased. Additionally, they highlighted that temperature affected metals biosorption, but this effect was limited under a certain temperature range, where there was an inverse relationship between temperature and biosorption capacity, and the maximum metal removal was recorded at 27 °C. Since adsorption reactions are normally exothermic, biosorption capacities increase with the decrease in temperature. Also, they mentioned that the decrease in biosorption capacity at 27 and 62 °C may be due to the damage of binding sites on the yeast cell surface.

Acosta-Rodriguez et al. (2018) reported on the optimum conditions for the fungus *Aspergillus niger* to remove heavy metals. They found the following removal percentages; 100% for Zn, 83% for F and Hg, 71.4% for Co, 48% for Ag, and 37% for Cu. The optimum biosorption conditions of 100 mg/L of the heavy metals were 100 ppm of heavy metal and one gram of fungal biomass, 28 °C and pH from 4.0 to 5.5.

Regarding the effect of initial metal concentration, this was demonstrated via the growth of *Aspergillus niger* B-77 in relation to initial Cu concentration. Tsekova and Todorova (2002) showed that as Cu concentration increased, the metal removal increased. 50, 100, and 200 mg/L Cu (II) ions resulted in biosorption capacities of 3.7, 7.2, and 19.2 mg/g, respectively. Nevertheless, as initial Cu (II) increased to 300 mg/L, neither microbial growth nor Cu biosorption was detected.

Cárdenas González et al. (2019) showed that the percentage of Co (II) adsorption by *Penicillium* spp. and *Aspergillus niger* decreased as its concentration increased from 300 to 600 mg/L. Also, it was found that with different fungal species, the removal ability differs. For example, when different fungi from soil (e.g., *Paecilomyces, Trichoderma, Aspergillus, Pythium, Penicillium, Mortierella,* and

*Rhizopus*) were investigated for biosorption of Co, the uptake of the metal increased depending on the initial Co concentration. *Mortierella* SPS 403 biomass absorbed 1036.8 mM of Co (II) (Pal et al. 2006). *Trichoderma atroviride*, in another study, was found to tolerate high (300 mg/L) concentrations of Cu (López Errasquín and Vázquez 2003). In the same vein, Romero et al. (2006) reported that metal accumulation was directly proportional to the concentration of heavy metals. The cobalt uptake by *Talaromyces helicus* also increased from 100 to 200 mg/L as its initial concentration increased from 100 to 600 ppm (Romero et al. 2006). This might be attributed to the high competition for the functional groups on the surface of the fungal biomass (Cárdenas González et al. 2019).

Other researchers investigated the effect of the initial biomass concentration, Jayaraman and Arumugam (2014) demonstrated that as the amount of fungal biomass increased, the biosorption capacity increased as well. For example, when 10 g instead of 5 g of *Paecilomyces* sp. biomass was applied, about 100% removal of the heavy metal was recorded in just 16 h. In addition, the authors demonstrated that by increasing the amount of *Aspergillus flavus* biomass from 1 to 10 mg/L, Cu removal capacity also increased (26.27 mg/g metal removal at 4 mg/mL of the biosorbent). The increase in the biosorption capacity was dependent on the amount of the applied biosorbent, the availability of free binding sites, and the surface area of the biosorbent (Cárdenas González et al. 2019).

Cárdenas González et al. (2019) studied the effect of incubation time and pH on biosorption of Co (II). They used 200 mg/L of the dried fungal biomass to determine the highest biosorption in 24 h of incubation at pH 5.0. It was found that *Aspergillus niger*, *Penicillium* spp., and *Paecilomyces* spp. showed removal percentages of 93, 77.5, and 70.4%, respectively. These results coincided with those of an experiment on *Penicillium cyclopium*, as the highest Cu and Co biosorption was observed at pH 5. When using *Paecilomyces catenlannulatus*, the removal of Co increased as pH increased from 4.5 to 7. The highest biosorption was recorded at pH 7.0. For Cu (II) removal by *Aspergillus niger*, many biomasses were incubated at separate time intervals, where the maximum biosorption capacity (25.2 mg/g) of *A. niger* was recorded for Cu within 18 h. Similarly, biosorption of Hg (II) by *Rhizopus oligosporus* showed its highest value (33.33 mg/g) at pH 6 after 6 h of incubation (Ozsoy 2010). Furthermore, Farhan and Khadome (2015) reported that biosorption of Co, Pb, Cd, Zn, Cr, and Cu ions by *Saccharomyces cerevisiae* increased as the initial concentration increased; this was attributed to the free binding sites available for metal ions (Sudhir et al. 2007). As for yeast concentration as a factor, all metal ions showed an increase in uptake capacity as the biomass increased from 0.01 to 0.1 g. in addition, more metal ions were removed at higher concentrations due to the availability of more active sites. Moreover, Saad et al. (2019) reported that the highest removal of Cu (II) and Co (II) by *Aspergillus tamarii* NRC3 biomass was 92.40 and 60%, respectively, when 5 g of *A. tamarii* biomass was used. Recently, Abdel-Wareth et al. (2023) reported that the optimum temperature for *A. niger* to remove Ni and Mn from their aqueous solutions was 50 °C. They also found that Ni was highly removed by *A. niger* at pH 6, whereas Mn preferable pH equaled 8.

# 5 New Technologies Using Fungi in Remediation of Heavy Metals

Cai et al. (2016) claimed that the use of free fungal cells can cause difficulties in the separation of biomass from the effluent. So, they immobilized the living conidia of *Penicillium janthinillum* strain GXCR, that is known to be tolerant to heavy metals, by polyvinyl alcohol (PVA)-sodium alginate (SA) beads. These PVA-SA-conidia beads were used to remove 70 mg/L of Cd, Cu, and Pb from an aqueous solution. The beads were characterized by their mechanical strength, the good settlement ability, the ease of removal from the solution, and their high biosorption rate even after four cycles of consequent sorption–desorption. These advantages make them superior to fungal biomasses when used alone for heavy metal removal. Also, Hoque and Fritscher (2019) described a unique microbial biotechnology for bioremediation of 12 metal ions. Firstly, they searched for key microorganisms in microbiomes of various sulfidic springs in Bavaria (Germany) over an area of 200 $km^2$, where they found three new strains of *Mucor hiemalis* able to remediate many metals. They combined the multimetal tolerance and hyper-accumulation of the three new strains (EH8, EH10, and EH11) to develop a novel biotechnology for simultaneous metal uptake. They showed the intracellular deposition of Hg as nanospheres in EH8's sporangiospores. Using scanning electron microscopy-energy-dispersive X-ray, binding and precipitation of other metal ions as spherical nanoparticles (~50–100 nm) at the outer cell wall-surface of EH8, EH10, and EH11 sporangiospores was confirmed. They reported that microbiomes, spores, and dead cell walls of these strains showed removal percentage up to 81–99% for Pb, Al, Cd, Hg, Co, Cr, Zn, Cu, Ni, and U simultaneously, and also restored precious Ag, Au, and Ti from water all within 48 h. These findings highlighted the potential of new biotechnologies to protect the environment from metal pollution and retrieve precious metal ions.

Recently, Park et al. (2020) developed heavy metal adsorbents from hybrids of electrospun nanofibers and Mn (II)-oxidizing fungi. They reported that the bioremediation mechanism of manganese-oxidizing fungi depended on conversion of manganese ions into manganese oxide deposits that are capable of adsorbing manganese and other heavy metals. They immobilized manganese-oxidizing fungi onto nanofiber surfaces to enhance remediation ability. Both *Coniothyrium* sp. and *Coprinellus* sp. from a superfund site water treatment system were incubated in the presence of nanofibers. As fungal hyphae showed strong affinity to bind with nanofiber surfaces, *Coniothyrium* sp. catalyzed the deposition of manganese oxide along hyphae and nanofibers, whereas *Coprinellus* sp. catalyzed its deposition only along the hyphae. Heavy metal ions were successfully adsorbed into manganese oxide crystalline structure, possibly by ion exchange with Mn within the manganese oxide.

Moreover, Ali et al. (2021) investigated the removal of Cd (II) from wastewater using fungal biomass either immobilized on loofa sponges or in Ca-alginate beads. Two fungal species, molecularly identified as *Penicillium chrysogenum* and *Cephalotheca foveolata*, were isolated from Cd contaminated wastewater pools from some Egyptian industrial plants. They used heat-inactivated mycelia of *P. chrysogenum* (PEN), heat-inactivated mycelia of *C. foveolata* (CEP), *P. chrysogenum* immobilized on loofa sponge (PEN-ILS), *C. foveolata* immobilized on loofa sponge (CEP-ILS), *P. chrysogenum* immobilized in Ca-alginate beads (PEN-IA), and *C. foveolata* immobilized in Ca-alginate beads (CEP-IA) as sorbents. They observed that the optimum conditions for the highest Cd biosorption capacity were pH 7.0, 30 min contact time and 0.5 mol/L initial Cd concentration for all tested sorbents.

## 6 Conclusion

Fungi have many advantages that made them excellent candidates that can be used for removal and sequestration of heavy metals either as live or dead biomass. Generally, it is preferred to exploit fungal spp. that are isolated from heavy metals polluted environments like wastewater, industrial effluents, contaminated soils…, etc. the process of metals absorption by fungi is controlled by many factors, such as pH, temperature, contact time, biomass dosage, and initial metal concentration. Many researchers demonstrated that various mechanisms are responsible for biosorption of metals by either live fungi or dead biomass.

## References


Abd El-Ghany TM, Abd El-Mongy M (2009) Bioremoval of heavy metals in presence of oxalic and citric acid using *Aspergillus tamarii*. Egypt Soc Exp Biol 5:53–58

Abdel-Wareth MTA, Abdel-Rahman TMA, Abdel-Ghany MNM, Hamed KA (2023) Consortium effect of *Jatropha curcas* seed husk and its endophyte *Aspergillus niger* on biosorption of manganese and nickel from wastewater. Int J Environ Stud. https://doi.org/10.1080/00207233.2022.2160591

Abdi O, Kazemi M (2015) A review study of biosorption of heavy metals and comparison between different biosorbents. J Mat Environ Sci 6(5):1386–1399

Abdullahi M, Machido DA (2017) Heavy metals resistance potential of some Aspergillus spp. isolated from Tannery wastewater. Nigerian J Basic Appl Sci 25(1):120–129

Acosta-Rodriguez I, Cardenas-González JF, de Guadalupe M-Zárate M, Perez AR, Martínez-Juárez VM (2017) Hexavalent chromium (VI) removal by *Aspergillus niger*. Handbook of metal-microbe interactions and bioremediation. CRC Press, Boca Raton, pp 673–688

Acosta-Rodríguez I, Cárdenas-González JF, Rodríguez Pérez AS, Oviedo JT, Martínez-Juárez VM (2018) Bioremoval of different heavy metals by the resistant fungal strain *Aspergillus niger*. Bioinorg Chem Appl. https://doi.org/10.1155/2018/3457196

Ahmad I, Ansari MI, Aqil F (2006) Biosorption of Ni, Cr, and Cd by metal tolerant *Aspergillus niger* and *Pencillium* sp. Using single and multimetal solution. Indian J Exp Biol 44:73–76

Ahmad I, Zafar S, Ahmad F (2005) Heavy metal biosorption potential of *Aspergillus* and *Rhizopus* sp. isolated from wastewater treated soil. J Appl Sci Environ Mgt 9:(1) 123–126

Aishwarya S, Venkateswarlu N, Chandramouli K, Vijaya T (2014) Role of endophytic fungi in the restoration of heavy metal contaminated soils. Indo Am J Pharm Sci 4:5427–5436

Akar T, Tunali S (2005a) Biosorption performance of *Botrytis cinerea* fungal by-products for removal of Cd (II) and Cu (II) ions from aqueous solutions. Minerals Eng 18(11):1099–1109

Akar T, Tunali S (2005b) Biosorption characteristics of *Aspergillus flavus* for removal of Pb (II) and Cu (II) ions from aqueous solution. Bioresour Technol 97:1780–1787

Akhtar N, Mannan MA (2020) Mycoremediation: expunging environmental pollutants. Biotechnol Rep 26:e00452

Ali EAM, Sayed MA, Abdel-Rahman TM, Hussein R (2021) Fungal remediation of Cd (ii) from wastewater using immobilization techniques. RSC Adv 11(8):4853–4863

Alimohamadi M, Abolhamd G, Keshtkar A (2005) Pb (II) and Cu (II) biosorption on *Rhizopus arrhizus* modeling mono- and multi- component systems. Minerals Eng 18:1325–1330

Alloway BJ (1995) Heavy metals in soils, 2nd edn. Blackie Academic & Professional, London, p 322

Anand P, Isar J, Saran S, Saxena RK (2006) Bioaccumulation of copper by *Trichoderma viride*. Bioresour Technol 97:1018–1025

Ashraf M (2011) Low cost biosorbent banana peel (*Musa sapientum*) for the removal of heavy metals. Sci Res Ess 6(19):4055–4064

Awofolu OR, Okonkwo JO, Merwe RRV, Badenhorst J, Jordaan E (2006) A new approach to chemical modification protocols of *Aspergillus niger* and sorption of lead ion by fungal species. Environ Biotechnol 9:340–348

Ayangbenro A, Babalola O (2017) A new strategy for heavymetal polluted environments: a review of microbial biosorbents. Int J Environ Res Public Health 14(1):94

Ayele A, Haile S, Alemu D, Kamaraj M (2021) Comparative utilization of dead and live fungal biomass for the removal of heavy metal: a concise review. Sci World J. https://doi.org/10.1155/2021/5588111

Baath E (1989) Effects of heavy metals in soil on microbial processes and populations (a review). Water Air Soil Pollut 47:335–379

Bala JD, Kuta F, Nasiru A, Adedeji AS, Al-Gheethi AAS, Fashola OH (2020) Biosorption potential of lead tolerant fungi isolated from refuse dumpsite soil in Nigeria. Acta Sci Biol Sci 42:e46753–e46753

Baldrian P (2003) Interactions of heavy metals with white-rot fungi. Enz Microb Technol 32(1):78–91

Bayramoglu G, Çelik G, Yalçin E, Yilmaz M, Arica MY (2005) Modification of surface properties of mycelia by physical and chemical methods: evaluation of their Cr (VI) removal efficiencies from aqueous medium. J Hazard Mater 119:219–229

Bilal M, Rasheed T, Sosa-Hernandez J, Raza A, Nabeel F, Iqbal H (2018) Biosorption: an interplay between marine algae and potentially toxic elements-A review. Mar Drugs 16(2):1–16

Cai CX, Xu J, Deng NF, Dong XW, Tang H, LiangY FXW, Li YZ (2016) A novel approach of utilization of the fungal conidia biomass to remove heavy metals from the aqueous solution through immobilization. Sci Rep 6(1):1–12

Cárdenas González JF, Rodríguez Pérez AS, Vargas Morales JM, Martínez JuárezV M, Rodríguez IA, Cuello CM et al (2019) Bioremoval of Cobalt(II) from aqueous solution by three different and resistant fungal biomasses. Bioinorg Chem Appl 2019:1–8

Çelebi H, GökG GökO (2020) Adsorption capability of brewed tea waste in waters containing toxic lead (II), cadmium (II), nickel (II), and zinc (II) heavy metal ions. Sci Rep 10:1–12

Chojnacka K (2010) Biosorption and bioaccumulation—The prospects for practical applications. Environ Int 36(3):299–307

Choo J, Sabri NBM, Tan D, Mujahid A, Müller M (2015) Heavy metal resistant endophytic fungi isolated from *Nypa fruticans* in Kuching Wetland National Park. Ocean Sci J 50(2):445–453

Coelho LM, Rezende HC, Coelho LM, de Sousa PA, Melo DF, Coelho NM (2015) Bioremediation of polluted waters using microorganisms. Adv Bioremed Wastewater Pollut Soil 10:60770

Congeevaram S, Dhanarani S, Park J, Dexilin M, Thamaraiselvi K (2007) Biosorption of chromium and nickel by heavy metal resistant fungal and bacterial isolates. J Hazard Mat 146(1–2):270–277

da Rocha Ferreira GL, Vendruscolo F, Antoniosi Filho NR (2019) Biosorption of hexavalent chromium by *Pleurotus ostreatus*. Heliyon 5(3):e01450

Delgado A, Novais AAM, JM (1998) Heavy metal biosorption by dried powdered mycelium of *Fusarium flocciferum*. Water Environ Res 70(3):370–375
Deng Z, Zhang R, Shi Y, Hu LA, Tan H, Cao L (2014) Characterization of Cd-, Pb-, Zn-resistant endophytic *Lasiodiplodia* sp. MXSF31 from metal accumulating *Portulaca oleracea* and its potential in promoting the growth of rape in contaminated soils. Environ Sci Pollut Res 21:2346–2357
Dhankhar R, Hooda A (2011) Fungal biosorption—An alternative to meet the challenges of heavy metal pollution in aqueous solutions. Environ Technol 32(5):467–491
Drahansky M, Paridah M, Moradbak A, Mohamed A, Owolabi F,Abdulwahab T, Asniza M, et al (2016) Application of biosorption for removal of heavy metals from wastewater. In: Derco J, Vrana B (eds) Biosorption. Intechopen, London
Dwivedi S, Mishra A, Saini D (2012) Removal of heavy metals in liquid media through fungi isolated from waste water. Int J Sci Res 1(3):181–185
El-Morsy EM (2004) *Cunninghamella echinulata* a new biosorbent of metal ions from polluted water in Egypt. Mycol 96(6):1183–1189
Ezzouhri L, Castro E, Moya M, Espinola F, Lairini K (2009) Heavy metal tolerance of filamentous fungi isolated from polluted sites in Tangier. Morocco. African J Microbiol Res 3(2):35–48
Ezzouhri L, Ruiz E, Castro E, Moya M, Espínola F, Cherrat L, Er Raioui H, Lairini K (2010) Mechanisms of lead uptake by fungal biomass isolated from heavy metals habitats. Afinidad 67(545)
Farhan SN, Khadom AA (2015) Biosorption of heavy metals from aqueous solutions by *Saccharomyces Cerevisiae*. Int J Ind Chem 6(2):119–130
Filote C, Rosca M, Hlihor RM (2020) Overview of using living and non-living microorganisms for the removal of heavy metals from wastewaters. Res J Agric Sci 52:22–31
Gadd GM (2009) Biosorption: critical review of scientific rationale, environmental importance and significance for pollution treatment. J Chem Technol Biotechnol 84(1):13–28
Gautam RK, Soni S, Chattopadhyaya, MC (2015) Functionalized magnetic nanoparticles for environmental remediation. In: Handbook of research on diverse applications of nanotechnology in biomedicine, chemistry, and engineering. IGI Global, pp 518–551
Gavrilesca M (2004) Removal of heavy metals from the environment by biosorption. Eng Life Sci 4(3):219–232
Ge W, Zamri D, Mineyama H, Valix M (2011) Bioaccumulation of heavy metals on adapted *Aspergillus foetidus*. Adsorption 17:901–910
Graz M, PawlegaB P, Jarosz-Wilkolaska A (2011) Growth inhibition and intracellular distribution of the white-rot fungus *Abortiporus biennis*. Int J Biodeter and Biodegr 65:124–129
Gupta R, Ahuja P, Khan S, Saxena RK, Mohapatra H (2000) Microbial biosorbents: meeting challenges of heavy metal pollution in aqueous solutions. Curr Sci 78(8):967–973
Gupta VG, Schmoll M, Herrera-Estrella A, Upadhyay RS, Druzhinina I, TuohyM (2014) Biotechnology and biology of *Trichoderma*. Newnes, Elsevier
Hajahmadi Z, Younesi H, Bahramifar N, Khakpour H, Pirzadeh K (2015) Multicomponent isotherm for biosorption of Zn(II), CO(II) and Cd(II) from ternary mixture onto pretreated dried *Aspergillus niger* biomass. Water Res Ind 11:71–80
Han FX, Banin A, Su Y, Monts DL, Plodinec JM, Kingery WL, Triplett GE (2002) Industrial age anthropogenic inputs of heavy metals into the pedosphere. Natur Wissen Schaften 89(11):497–504
Hansda A, Kumar VA (2016) Comparative review towards potential of microbial cells for heavy metal removal with emphasis on biosorption and bioaccumulation. World J Microbiol Biotechnol 32:170
Hassan A, Pariatamby A, Ahmed A, Auta HS, Hamid FS (2019) Enhanced bioremediation of heavy metal contaminated landfill soil using filamentous fungi consortia: a demonstration of bioaugmentation potential. Water Air Soil Pollut 230(9):1–20
Hassen A, Saidi N, Cherif M, Boudabous A (1998) Resistance of environmental bacteria to heavy metals. Bioresour Technol 64:7–15

Hemambika B, Johncy RM, Kannan VR (2011) Biosorption of heavy metals by immobilized and deadfungal cells: a comparative assessment. J Ecol Nat Environ 3(5):168–175

Hoque E, Fritscher J (2019) Multimetal bioremediation and biomining by a combination of new aquatic strains of *Mucor hiemalis*. Sci Rep 9(1):1–16

Hussain SF, Khazaal DS, Hatif ZH (2020) Biosorption of lead and chromium ions by using *Penicillium digitatum* (Pers.) Sacc. from industrial water. Iraqi J Sci:1880–1886

Idris R, Trifonova R, Puschenreiter M, Welzel WW, Seissitch A (2004) Bacterial communities associated with flowering plants of the Ni hyperaccumulator *Thlaspi goesingense*. Appl Environ Microbiol 70:2667–2677

Ifijen IH, Itua AB, MalikiM I-I, Omorogbe SO, AigbodionAI IkhuoriaEU (2020) The removal of nickel and lead ions from aqueous solutions using green synthesized silica microparticles. Heliyon 6:e04907

Igiri BE, Okoduwa S, Idoko GO, Akabuogu EP, Adeyi AO, Ejiogu IK (2018) Toxicity and bioremediation of heavy metals contaminated ecosystem from tannery wastewater: a review. J Toxicol:1–16

Imran M, Ahmad I, Barasubiye T, Abulreesh HH, Samreen S, Monjed MK, ElbannaK (2020) Heavy metal tolerance among free-living fungi isolated from soil receiving long term application of wastewater. J Pure Appl Microbiol 14:157–170

Iqbal M, Saeed A (2007) Production of an immobilized hybrid biosorbent for the sorption of Ni (II) from aqueous solution. Process Biochem 42(2):148–157

Iram S, Ahmad I, Javed B, Yaqoob S, Akhtar K, Kazmi MR, Uz-Zaman B (2009) Fungal tolerance to heavy metals. Pak J Bot 41(5):2583–2594

Iram S, Shabbir R, Zafar H, Javaid M (2015) Biosorption and bioaccumulation of copper and lead by heavy metal-resistantfungal isolates. Arab J Sci Eng 40(7):1867–1873

Iskandar NL, Zainudin NAIM, Tan SG (2011) Tolerance and biosorption of copper (Cu) and lead (Pb) by filamentous fungi isolated from a freshwater ecosystem. J Environ Sci 23:824–830

Javanbakht V, Alavi SA, Zilouei H (2014) Mechanisms of heavy metal removal using microorganisms as biosorbent. Water Sci Technol 69(9):1775–1787

Jayaraman M, Arumugam R (2014) Biosorption of Copper (II) by *Aspergillus flavus*. Int J Sci Res 3:335–340

Joshi PK, Swarup A, Maheshwari S, Kumar R, Singh N (2011) Bioremediation of heavy metals in liquid media through fungi isolated from contaminated sources. Indian J Microbiol 51(4):482–487

Kaoushik K, Mukherjee DD, Samal AC, Santra SC (2013) Isolation and characterization of Arsenic tolerant fungal strains from contaminated sites around urban environment of Kolkata. IOSR J Environ Sci Toxicol Food Technol 7:33–37

Kapahi M, Sachdeva S (2017) Mycoremediation potential of *Pleurotus* species for heavy metals: a review. Bioresour Bioprocess 4(1):32

Kapoor A (1998) Removal of heavy metal from aqueous solution by fungi *Aspergillus niger*. Doctoral dissertation, University of Regina, Regina, Canada

Kapoor A, Viraraghavan T (1995) Fungal biosorption—An alternative treatment option for heavy metal bearing wastewaters: a review. Bioresour Technol 53(3):195–206

Kapoor A, Viraraghavan T, Cullimore DR (1999) Removal of heavy metals using the fungus *Aspergillus niger*. Bioresour Technol 70(1):95–104

Khan M, Aftab S, Shakir, et al (2019) Mycoremediation of heavy metal (Cd and Cr)–polluted soil through indigenous metallotolerant fungal isolates. Environ Monitor Assess 191(9):585

Kumar A, Bisht BS, Joshi VD (2011) Zinc and cadmium removal by acclimated *Aspergilus niger*: Trained fungus for biosorption. Int J Environ Sci Res 1(1):27–30

Kurniati E, Arfarita N, Imai T, Higuchi T, Kanno A, Yamamoto K, Sekine M (2014) Potential bioremediation of mercury contaminated substrate using filamentous fungi isolated from forest soil. J Environ Sci Chin 26:1223–1231

Lakherwal D (2014) Adsorption of heavy metals: a review. Int J Environ Res Develop 4(1):41–48

Lakshmi S, Krishna R, Lakshmi S, Krishna R (2018) Application of biosorption for for removal of heavy metals from wastewater. In: Derco J, Vrana B (eds) Biosorption, Intech Open, London
Lata S, Kaur HP, Mishra T (2019) Cadmium bioremediation:a review. Int J Pharm Sci Res 10(9):4120–4128
Li H, Lin Y, Guan W et al (2010) Biosorption of Zn(II) by live and dead cells of *Streptomyces ciscaucasicus* strain CCNWHX 72–14. J Hazard Mat 179(1–3):151–159
Li HY, Wei DQ, Shen M, Zhou PZ (2012) Endophytes and their role in phytoremediation. Fungal Div 54:11–18
Lin Z, Wu J, Xue R, Yang Y (2005) Spectroscopic characterization of $Au^{3+}$ biosorption by waste biomass of *Saccharomyces cerevisiae*. Spect Acta Part a: Mol Biomol Spect 61(4):761–765
López Errasquín E, Vázquez C (2003) Tolerance and uptake of heavy metals by *Trichoderma atroviride* isolated from sludge. Chemosphere 50:137–143
Malik A (2004) Metal bioremediation through growing cells. Environ Int 30:261–278
Martino E, Turnau K, Girlanda M, Bonfante P, Perotto S (2000) Ericoid mycorrhizal fungi from heavy metal polluted soils: their identification and growth in the presence of zinc ions. Mycol Res 104(03):338–344
Massaccesi G, Romero MC, Cazau MC, Bucsinszky AM (2002) Cadmium removal capacities of filamentous soil fungi isolated from industrially polluted sediments, in La Plata (Argentina). World J Microbiol Biotechnol 18(9):817–820
Mohammadian E, Ahari AB, Arzanlou M, Oustan S, Khazaei SH (2017) Tolerance to heavy metals in filamentous fungi isolated from contaminated mining soils in the Zanjan Province. Iran. Chemosphere 185:290–296
Naveena J, Latha L (2018) Fungal cell walls as protective barriers for toxic metals. Adv Med Biol 53:19
Oladipo OG, Awotoye OO, Olayinka A, Bezuidenhout CC, Maboeta MS (2018) Heavy metal tolerance traits of filamentous fungi isolated from gold and gemstone mining sites. Brazilian J Microbiol 49:29–37
Ozsoy HD (2010) Biosorptive removal of Hg (II) ions by *Rhizopus oligosporus* produced from corn-processing wastewater. African J Biotechnol 9:8783–8790
Pal A, Ghosh S, Paul AK (2006) Biosorption of cobalt by fungi from serpentine soil of Andaman. Bioresour Technol 97:1253–1258
Pal TK, Basumajumdar A (2002) Biosorption of heavy metals by *Rhizopus arrhizus* and *Aspergillus niger*. J Ind Chem Soc 79:747–750
Panda SS, Sahoo K, Muduli SD, Sahoo G, Ahemad MDJ, Nayak BB, Dhal NK (2014) Chromium tolerant indigenous fungal strains from industrial effluents of Angul district, Odisha. India. Biolife 2(2):634–640
Park D, Yun YS, Park JM (2005) Use of dead fungal biomass for the detoxification of hexavalent chromium: screening and kinetics. Process Biochem 40(7):2559–2565
Park Y, Liu S, GardnerT JD, Keeler A, Ortiz N, Rabagh G, Ford E (2020) Biohybrid nanofibers containing manganese oxide–forming fungi for heavy metal removal from water. J Eng Fib Fab 15:1–12
Perpetuo EA, Souza CB, Nascimento CAO (2011) Engineering bacteria for bioremediation. Prog Mol Environ Bioeng: 605–632
Rao DV, Shivannavar CT, Gaddad SM (2002) Bioleaching of copper from chalcopyrite ore by fungi. Indian J Exp Biol 40:319–324
Romero MC, Reinoso HE, Urrutia MI, Moreno Kiernan A (2006) Biosorption of heavy metals by *Talaromyces helicus*: a trained fungus for copper and biphenyl detoxification. Electron J Biotechnol 9(3). https://doi.org/10.4067/S0717-34582006000300009
Rosca M, Hlihor RM, Cozma P, Comanita ED, Simion IM, Gavrilescu M (2015) Potential of biosorption and bioaccumulation processes for heavy metals removal in bioreactors. Paper presented at the 2015 E-Health and bioengineering conference (EHB), IEEE, Iasi, Romania, 19–21 November 2015

Saad AM, Saad MM, Ibrahim NA, El-Hadedy D, Ibrahim EI, El-Din AZK et al (2019) Evaluation of *Aspergillus tamarii* NRC 3 biomass as a biosorbent for removal and recovery of heavy metals from contaminated aqueous solutions. Bull Natl Res Cent 43:10

Salam KA (2019) Towards sustainable development of microalgal biosorption for treating effluents containing heavy metals. Biofuel Res J 6:948–961

Salazar-Ramírez G, Flores-Vallejo RDC, Rivera-Leyva JC, Tovar-Sánchez E, Sánchez-Reyes A, Mena-Portales J, … Folch-Mallol JL (2020) Characterization of fungal endophytes isolated from the metal hyperaccumulator plant *Vachellia farnesiana* growing in mine Tailings. Microorg 8(2):226

Salinas E, Elorzade Orellano M, Rezza I, Martinez L, Marchevsky E, Sanz deTosetti M (2000) Removal of cadmium and lead from dilute aqueous solutions by *Rhodotorula rubra*. Bioresour Technol 72:107–112

Sao K (2014) A review on heavy metals uptake by plants through biosorption. Int Proc Econom Develop Res 75(17):78–83

Say R, Yilmaz N, DenizliA (2003) Removal of heavy metal ions using the fungus *Penicillium canescens*. Ads Sci Technol 21(7):643–650

Shah SS, Palmieri MC, Sponchiado SRP, Bevilaqua D (2020) Enhanced bio-recovery of aluminum from low-grade bauxite using adapted fungal strains. Brazilian J Microbiol 51:1909–1918

Sharma S, Kour D, Rana KL, Dhiman A, Thakur S, Thakur P et al (2019) *Trichoderma*: biodiversity, ecological significances, and industrial applications. In: Yadav AN, Mishra S, Singh S, Gupta A (eds) Recent advancement in white biotechnology through fungi, diversity and enzymes perspectives. Springer International Publishing, Cham, pp 85–120

Shazia I, Uzma SG, Talat A (2013) Bioremediation of heavy metals using isolates of filamentous fungus *Aspergillus fumigatus* collected from polluted soil of Kasur. Pakistan. Int Res J Biological Sci 2(12):66–73

Siddiquee S, Rovina K, Azad SA, Naher L, Suryani S, Chaikaew P (2015) Heavy metal contaminants removal from wastewater using the potential filamentous fungi biomass: a review. J Microb Biochem Technol 7(6):384–393

Simonescu CM, Ferdeş M (2012) Fungal biomass for cu (II) uptake from aqueous systems. Pol J Environ Stud 21(6):1831–1839

Simonovicova A (2008) Use of mitosporic fungi for heavy metal removal from experimental water solutions. Czasopismo Techniczne Chemia 105(2-Ch):207–212

Srivastava S, Thakur IS (2006) Isolation and process parameter optimization of *Aspergillus* sp. for removal of chromium from tannery effluent. Bioresour Technol 97(10):1167–1173

Sudhir D, Tripathi RM, Hegde AG (2007) Biosorption of lead and copper from aqueous solutions by pretreated crab and arca shell biomass. Environmental Section, Mumbai 400085, India

Tasar S, Kaya F, Ozer A (2014) Biosorption of lead (II) ions from aqueous solution by peanut shells: Equilibrium, thermodynamic and kinetic studies. J Environ Chem Eng 2(2):1018–1026

Tewari N, Vasudevan P, Guha BK (2005) Study on biosorption of Cr (VI) by *Mucor hiemalis*. Biochem Eng J 23(2):185–192

Thakkar A, Baldi A, Murthy RS, Sharma RK (2016) Physical, chemical and phytoremediation technique for removal of heavy metals. J Heavy Metal Toxic Dis 1(2):1–5

Ting ASY, Choong CC (2009) Bioaccumulation and biosorption efficacy of *Trichoderma* isolate SP2F1 in removing copper (Cu (II)) from aqueous solutions World J Microbiol Biotechnol 25:1431–1437

Tsekova K, Todorova D (2002) Copper (II) accumulation and superoxide dismutase activity during growth of *Aspergillus niger* B-77. Z Naturforsch C J Biosci 57:319–322

Tunali S, Kiran I, Akar T(2005) Chromium (VI) biosorption characteristics of Neurospora crassa fungal biomass. Minerals Eng18 (7):681–689

Turpeinen R, Kairesalo T, Haggblom M (2002) Microbial activity community structure in arsenic, chromium and copper contaminated soils. J Environ Microbiol 35(6):998–1002

Văcar CL, Covaci E, Chakraborty S, Li B, Weindorf DC, Frenţiu T, ... Podar D (2021) Heavy metal resistant filamentous fungi as potential mercury bioremediators. J Fungi 7(5):386

Vaishaly AG, Mathew BB, Krishnamurthy NB, Krishnamurthy TP (2015) Bioaccumulation of heavy metals by fungi. Intl J Environ Chem Chromatogr 1(1):15–21

Verma S, Kuila A (2019) Bioremediation of heavy metals by microbial process. Environ Technol Innov 14:100369

Wai WL, Kyaw NAK, Nway NHN (2012) Biosorption of Lead ($Pb^{2+}$) by using *Chlorella vulgaris*. In: Proceedings of the International Conference on Chemical engineering and its applications (ICCEA), Bangkok, Thailand

Wang J, Chen C (2006) Biosorption of heavy metals by *Saccharomyces cerevisiae*: a review. Biotechnol Adv 24:427–451

Wei W, Wang Q, Li A, Yang J, Ma F, Pi S, Wu D (2016) Biosorption of Pb (II) from aqueous solution by extracellular polymeric substances extracted from *Klebsiella* sp. J1: Adsorption behavior and mechanism assessment. Sci Rep 6(1):10

Xiao X, Zeng LS, G, Wei W, Wan Y, Chen L, Guo H, Cao Z, Yang L, Chen J, Xi Q (2010) Biosorption of cadmium by endophytic fungus (EF) *Microsphaeropsis* sp. LSE10 isolated from cadmium hyperaccumulator *Solanum nigrum* L. Bioresour Technol 101:1668–1674

Xu X, Hao R, Xu H, Lu A (2020) Removal mechanism of Pb (II) by *Penicillium polonicum*: immobilization, adsorption, and bioaccumulation. Sci Rep 10(1):1–12

Yazdani M, Yap CK, Abdullah F, Tan SG (2009) *Trichoderma atroviride* as a bioremediator of Cu pollution: an in vitro study. Toxicol Environ Chem 91:1305–1314

Zafar S, Aqil F, Ahmad I (2007) Metal tolerance and biosorption potential of filamentous fungi isolated from metal contaminated agriculture soil. Bioresour Technol 98:2557–2561

Zhang D, Yin C, Abbas N, Mao Z, Zhang Y (2020) Multiple heavy metal tolerance and removal by an earthworm gut fungus *Trichoderma brevicompactum* QYCD-6. Sci Rep 10(1):1–9

# Advancements in Microbial Fuel Cell Technology

**Soumyadeep Bhaduri and Manaswini Behera**

## 1 Introduction

The world economy is shifting from fossil fuels toward renewable energy sources for its energy needs. This is evident as renewable energy production increased by 60% from 2011 to 2020 (Renewable Energy Statistics 2021). Wastewater treatment is an essential operation and wastewater treatment plants are one of the major power-consuming facilities. Although wastewater can contain as much as 9.3 times more energy than it takes to treat it, conventional treatment processes do not attempt to tap this energy (Shizas and Bagley 2004). On the other hand, MFC is a technology that generates electricity by degrading organics present in the wastewater while treating it. Thus, the treatment process becomes either a net-zero power consumer or produces electricity. MFC is one type of Bio-Electrochemical Systems (BESs) which is a technology that uses biological redox activity to get valuable products such as hydrogen, methane (Kadier et al. 2016). The major advantages of MFC compared to conventional fuel cells are (i) direct conversion of chemical to electrical energy, (ii) operation under a wide range of temperature, pH and bacterial culture, (iii) generation of lesser amount of activated sludge, and (iv) needlessness of energy for aeration in case of single chamber MFC.

MFC generally consists of two chambers one anaerobic anode and one aerobic cathode which are separated by a membrane. In this system, the bacteria are used as an electrochemical catalyst in the anode chamber where they oxidize the substrate in the wastewater producing electrons which then are transferred to an external circuit by either cellular extensions or released mediator chemicals. Proton produced in the anode pass through the PEM to the cathode to form water by combining with the electron transported through the external wire. The amount of electricity produced in

S. Bhaduri · M. Behera (✉)
School of Infrastructure, Indian Institute of Technology Bhubaneswar, Bhubaneswar, Odisha 752050, India
e-mail: manaswini@iitbbs.ac.in



M. P. Shah (ed.), *Sustainable Industrial Wastewater Treatment and Pollution Control*,
https://doi.org/10.1007/978-981-99-2560-5_11

MFC mostly depends on the internal resistance of the system which in turn depends on a number of factors, such as electrode material, membrane material, the distance between the electrodes, the type of bacterial culture used. Moreover, the overall efficiency depends on other factors like MFC configuration, number of chambers, type of wastewater, loading rate, and hydraulic retention time.

The progress in the path of MFC started when Luigi Galvani, a University professor in Bologna, in 1780 had introduced 'animal electricity' by moving muscles of a dead frog by connecting an external electric source to its nerves and putting current to its muscles (Galvani 1791). Charles Langer and Ludwig Mond in 1889 used the term 'fuel cell' while working on a practical fuel cell with air and coal gas (Santoro et al. 2017). The first reported work on MFC was in the 1900s by Michael Cresse Potter, an English botanist, who indicated that the microorganisms can be used to produce current (Potter and Waller 1911). However, the electroactive bacteria or exoelectrogens were not discovered until 1999 (Kim et al. 1999). In this chapter, we will first get to know about some key components of MFCs (Fig. 1), then we will go into the mechanism and after that the applicability of pure and hybrid MFC technologies.

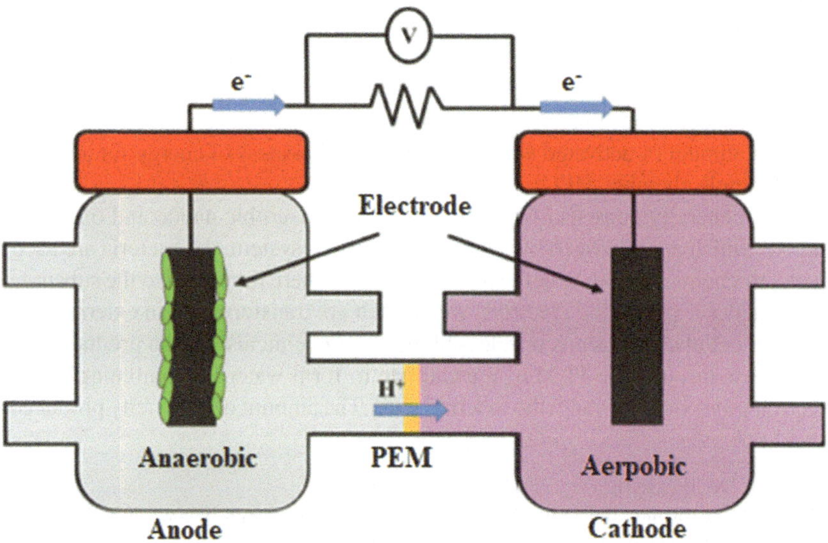


**Fig. 1** Basic components of a microbial fuel cell (Tan et al. 2021)

## 2 Components of MFC

Even though exoelectrogen is the soul of MFC reactors, other major functional components are anode, cathode, and membrane separator (PEM). The anode chamber is the reaction hub of the reactor. The wastewater is treated in the anode chamber where the substrate is degraded by the organisms, and electron and proton are generated. The membrane is used between the anode and cathode chamber to prevent any mix of catholyte and anolyte. The membrane is semipermeable and allows the proton to pass from anode to cathode. Inside the cathode chamber electron coming through an external circuit combines with the proton and turns into water with the help of oxygen supplied by aeration.

### *2.1 Anode Chamber*

The treatment of wastewater mainly takes place inside the anode chamber. In anode chamber substrate (wastewater), microorganism, electron acceptor (anode) is present. Also, electron mediator chemicals may or may not be employed. The substrate is oxidized here to produce electron that goes to the anode to be transported to the cathode through an external wire. The microorganism acts as a catalyst to lower the activation energy for the mentioned reaction (Hassan et al. 2014). The electron can be transported to the anode directly by nano-wire-like extensions of microorganism called pili or by electron mediating substances. This process of electron transfer is very important for the overall electrical efficiency of the MFC. Therefore, the interaction of anode material and microorganism becomes very important. The ideal anode material should be electroconductive, corrosion resistive, mechanically strong, economically viable, and most importantly biologically compatible (Mustakeem 2015). Carbonaceous and metallic materials are most suitable for anode as they possess all the aforesaid characteristics (Table 1).

The surface characteristics are important for the interaction of biofilm and anode material. The biofilm attachment can be increased by modifying the surface characteristics by attaching a positive functional group in the anode material (Santoro et al. 2015). A hydrophilic surface is better for biofilm attachment. Some functional groups of oxygen and nitrogen help in biofilm and anode interaction. The 3-D anode materials are preferred to 2-D surfaces as the available surface area is supposed to enhance the interaction between the bacteria and electrode (Guo et al. 2015). But in a comparison done by Blanchet et al. (2016) performance of 3-D carbon felt as an anode was reported to be at a similar level to 2-D carbon cloth despite much higher surface area. It was a result of the fact that the biofilm could not colonize into the whole structure of the 3-D carbon felt.

**Table 1** Characteristics of anode materials

| Material name | Advantages | Disadvantages | Reference |
|---|---|---|---|
| Carbon cloth | This material has a high surface area, high porosity, mechanical strength with flexibility | The cost is quite high | (Guerrini et al. 2014) |
| Carbon brush | The core titanium metal generates high conductivity while the carbon fibers twisted around it increases the surface area. The area to volume ratio is optimal here | The core titanium metal increases the overall cost of this material | (Liao et al. 2015) |
| Carbon rods | Affordable cost | Low surface area | (Jiang and Li 2009) |
| Carbon mesh | This is a low cost and flexible material | The electrical conductivity is low and possesses low mechanical strength | (Wang et al. 2009) |
| Carbon veil | This material possesses high conductivity and high porosity. It can be folded to make a 3-D electrode. The cost is low also | The single layer of this material is quite fragile | (Artyushkova et al. 2016) |
| Carbon felt | This material has high porosity, high conductivity, the pores help to colonize the biofilm internally. The cost is relatively low and mechanical strength depends on the thickness of the material | – | (Yasri and Nakhla 2017) |

(continued)

## 2.2 *Cathode Chamber*

The cathode can be of two types depending on the presence and absence of biomass inside the cathode chamber, biotic and abiotic cathode. In the cathode chamber, the oxygen is reduced by the electron produced in the anode chamber. Now this reaction can be facilitated by a catalyst that can be biological or inorganic to reduce overpotential and derive more energy (Antolini 2015). In the wastewater treatment system bacterial growth is inevitable. Therefore, the utilization of biomass as catalyst

**Table 1** (continued)

| Material name | Advantages | Disadvantages | Reference |
|---|---|---|---|
| Granular activated carbon (GAC) | This material is biocompatible and of low cost. It is used as a packing material with other anode materials. This material also helps in the adsorption of organic pollutants or heavy metals | Though the surface area is high, mostly in nanometric scale that is not useful for bio-growth. This needs another material for structural rigidity. Clogging is another disadvantage of this packing material | (Shah Maulin 2021a) |
| Granular graphite | This material possesses similar properties of GAC but much higher conductivity than GAC | The surface area is much lower due to a lack of activation | (Feng et al. 2010) |
| Carbonized cardboard | This is a 3-D structure of recycled paper that is thermally treated at 1,000 °C for 1 h in an inert ambiance. This material has very low-cost, high conductivity and high porosity | | (Kretzschmar et al. 2016) |
| Graphite plate/sheet | This also has low cost with high conductivity and possesses mechanical strength | The material has a low surface-to-volume ratio, i.e., low porosity | (Dewan et al. 2008) |
| Stainless steel plate/mesh | Highly conductive, robust, and cheap material | Lesser porosity is a major problem of this material | (Guo et al. 2016) |
| Copper | Material is highly conductive | This can be poisonous to microbes | (Baudler et al. 2017) |
| Titanium | Material is highly conductive | The cost is very high | (Zhou et al. 2016) |

may be advantageous. Biocathode has been applied to improve electricity production without increasing the expenses (Watanabe 2008). Moreover, the biofilm is self-replicate, self-repair and can be grown in the system itself. The cathode acts as an electron donor and the biofilm catalyzes the oxygen reduction reaction inside the cathode chamber. It can also be used to generate any particular product or remove a targeted pollutant. Biocathodes can be broadly classified into two types: aerobic biocathode and anaerobic biocathode. Aerobic biocathodes utilize metal ions like $Mn^{2+}$, $Fe^{2+}$ for delivering election to oxygen. The reactions are catalyzed by bacteria such as *Leptothrix discophora*, *Thiobacillus ferrooxidans* (Rhoads et al. 2005; Nemati et al. 1998). Another type of aerobic biocathode reaction is assisted by algae. This type of MFC is called photosynthetic algal microbial fuel cell. The reaction sequence in the MFC is as follows: the microorganisms in the anode utilize the organic matter and release electrons, proton, and $CO_2$, the microalgae in the cathode use these $CO_2$ and produce $O_2$ which acts as an electron acceptor for the electron generated in the anode previously. The anaerobic biocathodes are those which use other compounds like nitrate, iron, manganese, sulfate as terminal electron acceptors instead of oxygen (He 2006). Ammonia oxidation and denitrification can be done in the cathode chamber of an MFC. The disadvantages with biotic cathodes are long start-up time, process sensitivity, lesser power density than the abiotic electrode. The abiotic cathodes are mainly conductive metals, carbonaceous materials, and platinum group metal-free material on carbon support (Santoro et al. 2017). Commonly the materials used as an anode can also be employed as a cathode.

Previously, for oxygen reduction reaction the most popular catalyst was platinum-based materials. Some advanced MFC technology used platinum-based cathode (Antolini 2016). The platinum-coated cathode was reported to be toxic for the algae when it was used in an algal-assisted MFC. Nevertheless, mainly the high cost of this material prohibits the use of it as a cathode. Nowadays platinum cathodes are used for comparing the new materials for their electrical performances (Tiwari et al. 2017; Li et al. 2021). In carbonaceous materials activated carbon is mostly used (Wang et al. 2014). Apart from this, the carbonaceous materials like carbon fibers, carbon cloth, carbon felt, that were used as an anode can be used as a cathode as well. Some treatments for modification of these materials were also reported (Zhang et al. 2014). In recent years, focus has been shifted toward low-cost materials in which carbon-supported platinum group metal-free materials are mostly used. The transition metals, namely Mn, Ni, Fe, etc., have the optimal cost, high performance, and durability. In these Fe was reported to be the best performing (Santoro et al. 2020). Recently, platinum group metal-free Fe-based cathode was also used in advanced technology like direct alcohol fuel cells where alcohol is used as fuel to generate electricity (Berretti et al. 2021). It has also been reported that activated carbon embedded with metal complex, $Fe\text{-}N_4$, performed very well when used as air cathode in an MFC (Anjum et al. 2021). An anti-biofouling cathode was developed by Li et al. (2021), with carbon fiber modified with dispersing $Co_3O_4$ nanoparticles on N-doped carbon nanoflakes. Cathode plays an important role in MFC performance, and it should also be noted that choice of the cathode material should be done keeping in mind the material used for the anode and the experimental condition in the near vicinity of cathode (Berretti et al. 2021).

## 2.3 *Membrane*

Membrane in an MFC acts as separator for anaerobic anode and aerobic cathode chamber. The separator can be any proton exchange membrane (PEM), cation exchange membrane (CEM), anion exchange membrane (Rahimnejad et al. 2015). In general membranes are referred as PEM. The PEMs are compared according to their proton transport ability and ability to prevent oxygen, substrate, and mineral transport (Rahimnejad et al. 2014). Most commonly used membrane in Nafion which is a product of DuPont Inc. USA. This material has hydrophobic fluorocarbon (–CF–CF–) chain to which hydrophilic sulfonate groups ($SO_3$) are bound. The presence of the sulfonate group gives the membrane negative charge that makes it highly conductive for various cations. Power output depends on the internal resistance. Kim et al. (2007) reported that internal resistance of the MFC with the same membrane varied significantly as the electrode spacing was varied. Therefore, it indicates that membrane resistance contributes a little to the total internal resistance of the MFC. However, for overcoming problems such as pH splitting, oxygen permeation, various kind of membranes has been continuously being developed for the past several years. The most commonly used membranes are discussed in the following paragraphs.

### 2.3.1 Cation Exchange Membrane (CEM)

This type of separator allows the positive changes to be transferred to cathode chamber and also prevent substrate and oxygen transfer from anode to cathode and vice versa. Negatively charged ions such as $-PO_3^-$, $-COO^-$, $-C_6H_4O^-$. are inclusive part of this type of membrane. Therefore, it permits the cations to pass through it but prevent transfer of any negative ions. Hence, these types of membranes are also called as proton exchange membrane. Many types of materials such as Nafion, Ultrex, bipolar membrane, glass wool, polystyrene were used as CEM (Sun et al. 2009). Among these material, Nefion is the most popular as membrane in MFCs. It is not only because it has high conductivity of cations and longer life span but also because of sufficient level of hydration and thickness all of which influences the cell performance. Nevertheless, Nefion has some disadvantages as it is not suitable for neutral pH and for cations such as $Na^+$, $K^+$, $NH_4^+$ present $10^5$ times higher than $H^+$ concentrations. Moreover, the high cost of Nefion membrane is also a limiting factor for its usage. Ultrex CMI 7000 is another commonly used CEM which is relatively cheaper and has more mechanical stability. It has a polystyrene and divinylbenzene cross-link structure that contain sulfonic acid groups. Zirfon and Hyflon are other notable CEMs used in MFCs (Pant et al. 2010; Ieropoulos et al. 2010). Earthen membranes are one of the cost effective MFC separator that has oxygen permeability similar to that of polymeric membranes (Behera and Ghangrekar 2011). Some of the drawbacks of CEM are their high cost, oxygen permeability, internal resistance, and low proton transfer capacity. Besides, membrane fouling is a limiting factor for long-term use of CEMs.

### 2.3.2 Anion Exchange Membrane (AEM)

For overcoming the limitation of CEM as stated above AEM were used as an alternative. AEM contains positive charges such as $NH_3^+$, $-PR^+$, $-SR_2^+$ which help the transfer of ions through it. Moreover, AEM has better photon transfer as it uses carbonate and phosphate as pH buffer (Zuo et al. 2008). Kim et al. (2007) employing AEM achieved power density of 0.61 $W/m^2$ which was more than the power density of 0.48 $W/m^2$ got using CEM. The hydrogen fed MFC with AEM gave maximum power density (MPD) of 1 $W/cm^2$ and highest current density of 2 $A/cm^2$. This was achieved using highly active hydrogen oxidation reaction catalyst PtRu. However, the AEM membrane usage was limited by the temperature as higher (<60 °C) temperature cationic functional groups present in AEM tend to chemical degrade (Dekel 2018).

### 2.3.3 Bipolar Membrane (BPM)

This membrane is a conglomeration of both CEM and AEM. Through BPM both proton and hydroxide ions can be transferred. This type of membrane is also used for saline wastewater (Zhuang et al. 2012). The cost of these membranes is lesser. However, it possesses high conductivity. Graphite, stainless steel, and such other ergo and metallic materials are used to fabricate BPMs. Here graphite plate is used for increasing electrical conductivity and stainless steel is used for chemical stability of the BPM. Some of the exclusive benefits of using BPM are development of detachable single cells in stack MFC, distribution of fuel and oxidants uniformly inside the cell, manage heat inside the cell (Shah Maulin 2021b).

## 3 Application of MFC

The conditions and purpose of application influence the design and configuration of the MFC. Various types of MFCs, their application, and performance are discussed in this section.

One of the economic designs of MFC is single-chamber MFCs (SCMFCs) as there is no need of aeration in absence of the cathode chamber and presence of membrane is also optional. SCMFCs have been prepared in different shapes cubes (Liu et al. 2014), side-arm bottles (Logan et al. 2007), and tubular (Rabaey et al. 2005). Tubular SCMFC with continuous flow generate power as high as 90 $W/m^3$ using acetate as substrate (Rabaey et al. 2005). An air cathode membrane-less MFC used for leachate treatment reported to produce MPD of 12.8 $W/m^3$ (Zhang et al. 2008). SCMFCs become very cost effective by removing the membrane (Zhang et al. 2020).

The dual-chamber MFCs (DCMFCs) have cathode chamber which is separated by a membrane from the anode chamber. Separators such as CEM, AEM, salt bridge can be used (Logan et al. 2006). Oxidation of the organic matter takes place in anodic chamber while oxygen reduction reaction (ORR) occurs in cathode. The DCMFCs can be designed as H-type (Flimban et al. 2019), cube type (Song et al. 2015), flat-plat type (Min and Logan 2004), and tubular type (Rabaey et al. 2005). The purpose of different shapes is to maximize the electrode surfaces. H-type MFC is a favorable platform for testing membrane performance, electrode materials, microorganisms, etc. In cube and flat type MFCs, the membrane surfaces are maximized, thus the proton flux is increased. DCMFCs can be used for power production and nutrient removal that can be used in fertilizer production (Ye et al. 2020).

For higher power generation MFCs are stacked either in series of in parallel. A pilot-scale stacked MFC with 72 L volume was producing MPD of 50.9 W/$m^3$ while 97% COD removal was achieved (Wu et al. 2016). Stacking MFCs in series cause voltage reversal problem that can be eliminated by using diodes.

In up-flow MFCs the substrate flows from the bottom anode chamber to top cathode chamber before exiting. In this type of MFC, there is no separate anolyte or catholyte. This MFC was used to treat chocolate industry wastewater to recover energy. One of the major advantages of up-flow MFCs is that oxygen diffusion into anode chamber is reduced significantly. However, MFCs have high internal resistance, therefore, this type of MFCs has confined field application.

In sediment (benthic zone) MFCs, the power is produced by degrading wastewater in anode chamber while ORR occurs cathode chamber out-side the water. This type of MFCs does not have membrane as the waster between the electrodes provide resistance to oxygen diffusion. This MFCs can be used for power supply for deep sea sensors.

Photosynthetic MFCs (PMFCs) are combination of MFC with algal cultivation. Microalgae have been used for nutrients recovery while producing $O_2$. The role of algae in PMFCs is to produce $O_2$ in the cathode chamber as a terminal electron acceptor. Carbon dioxide produced in anode is supplied in cathode for growth of algae (Nayak and Ghosh 2019). It must also be noted that the algae produced can be used for biodiesel production.

Microbial electrolysis cell (MEC) is a technology that uses the electrons derived from organic matter to generate hydrogen instead of electricity. Here power input is required to initiate hydrogen evolution reaction in the cathode chamber. MECs were also used to simultaneous hydrogen and methane gas generation (Segundo-Aguilar et al. 2021).

Previously most of the work on MFC was in lab-scale as the increasing the size of individual cells tend to reduce the power density. Other several limitations of scale-up MFCs were high capital and operational cost, fouling of electrode and membrane, problem of reverse potential, etc (Tan et al. 2021). Nevertheless, through the last decade researchers developed large-scale MFCs to treat real wastewater. The first pilot-scale MFC (1,000 L) was used to treat brewery wastewater in Queensland, Australia (Logan 2010). One of the largest MFCs has volume of 1,500 L and used for human fecal treatment (Jadhav et al. 2020). The construction cost was minimized by using ceramic separator in the MFC. Various scaled-up application of MFC is given in Table 2.

A test run of MFC was done involving 250,000 people and it treated raw urine while producing power that was providing light at night time. An MFC has also been used for cellphone and lithium battery stack charging (Fischer et al. 2018). A large-scale MFC of 1,000 L containing 64 units was placed at underground gallery of a municipal wastewater treatment plant in Châteauneuf, Switzerland (Fig. 2). This

**Table 2** Application of MFC (>10 L)

| Sl No. | Reactor type | Anode (Number) | Cathode (Number), Catholyte | Membrane | Substrate | Max power ($W/m^3$) | COD removal (%) | Reference |
|---|---|---|---|---|---|---|---|---|
| 1 | DCMFC, 64 units, 1,000 L | RVC | RVC, Groundwater | VANA-Dion | Primary effluent | 0.2 | 34–95 | (Blatter et al. 2021) |
| 2 | DCMFC, 50 units, 1,000 L | GAC | GAC, Artificial catholyte | CEM | Municipal wastewater | 60 | 70–90 | (Valladares Linares et al. 2019) |
| 3 | Submergible MFC, 255 L | GFB (10) | SS-AC (16, VitoCORE®) | Glass fiber separator | Municipal wastewater | 0.3 | 41 | (Hiegemann et al. 2019) |
| 4 | SCMFC, 85 L | GFB (22) | SS-AC (15, VitoCORE®) | Glass fiber separator | Domestic wastewater | 0.605 | 80 | (Rossi et al. 2019) |
| 5 | DCMFC, 6 units, 60 L | GG | GR, Ammonium | AEM | Swine manure | 4 | 36 | (Vilajeliu-Pons et al. 2017) |

(continued)

**Table 2** (continued)

| Sl No. | Reactor type | Anode (Number) | Cathode (Number), Catholyte | Membrane | Substrate | Max power ($W/m^3$) | COD removal (%) | Reference |
|---|---|---|---|---|---|---|---|---|
| 6 | DCMFC, 2 units, 20 L | Modified carbon cloth | Modified carbon cloth | Nanofiltration membrane | Brewery wastewater | 0.44 | 95 | (Lu et al. 2017) |
| 7 | Septic tank-SCMFC (18 units, 700 L)-Disinfection | SS-GAC | Carbon cloth | Nafion 117 | Domestic wastewater | 0.00043 | 87 | (Valladares Linares et al. 2019) |
| 8 | SCMFC, 45 L | GFB (8) | Pt/C cloth/SS (2) | None | Primary effluent | 0.875 | 14–67 | (Hiegemann et al. 2016) |
| 9 | SCMFC | SS entangled wires, Synthetic medium | Air cathodes, Synthetic medium | None | Acetate | 0.811 | – | (Papillon et al. 2021) |
| 10 | Stack MFC | Activated carbon | Activated carbon | CEM | Municipal wastewater | 0.4–3.64 | 70–80 | (Liang et al. 2018) |

*Note* RVC: reticulated vitreous carbon; GAC: granular activated carbon; GG: granular graphite; GR: granular rod; SS: stainless steel; GFB: graphite fiber brush; SCMFC: single-chamber MFC; DCMFC: double-chamber MFC

MFC produce 0.2 W/m$^3$ of MPD achieving maximum 95% of COD removal. MFCs were also used as sensor for pollutant monitoring. As electrical batteries have limited lifetime the MFC can be used for powering the chemical sensor and telemetry system to transmit data to receiver. MFC technology clubbed with membrane bioreactor was used for direct water reclamation after the wastewater treatment. Air-biocathode was used in the process that showed good performance compared to cathode containing Pt. catalyst. Prefiltered (0.1 mm) domestic wastewater used as feed and MPD was 0.38 W/m$^2$ (6.8 W/m$^3$) and 0.82 W/m$^2$ (14.5 W/m$^3$) with biocathode and platinum cathode, respectively. Both the soluble COD and $NH_3$-N removals were 97% (Malaeb et al. 2013). Constructed wetland (CW) coupled with MFCs has been used to treat various types of wastewater. Application of CW-MFC reported removal up to 100% of COD while generating MPD of 0.15 mW/m$^2$ (Villaseñor et al. 2013).

## 4 Challenges on Field-Scale Application of MFC

The journey of any technology from the lab to field application mostly depends on the economic viability. The capital cost of the MFC is one of the major obstacles for practical application of the system for sustainable electricity production and wastewater treatment. The capital cost of an MFC is around 30 times higher than conventional activated sludge process. This is because of the use of expensive electrode materials and membrane materials. The cost of membrane and electrode are 62.5% and 19.5% of the total cost of the system respectively (Ge and He 2016). Low revenue of only \$0.08 per m$^3$ per day (Stoll et al. 2016) is still a challenging problem compared to other such technologies. On the other hand, MFC system has challenges like stability, lowered power generation for real wastewater and membrane fouling (He et al. 2017). Moreover, scaling up MFC results in lowing power generation levels. The distance of the positive and negative electrodes must be small for efficient transfer of ions which reduces the internal resistance of the MFC. But this can result in substrate diffusion into the cathode chamber and oxygen diffusion into the anode chamber. Biofouling is another challenge in selecting the membrane considering long-term operation of the MFC. Extensive work should be done with a view to minimizing the capital and operational cost of the MFC system to make the system commercially viable as there are thermodynamic limitations in electricity production by the system. There is also limited information available regarding the biological processes and the interaction of the bacteria with different types of substrate added in the anode chamber as fuel to the system. The electron transfer mechanism and relationship between different bacterial species need to be studied further.

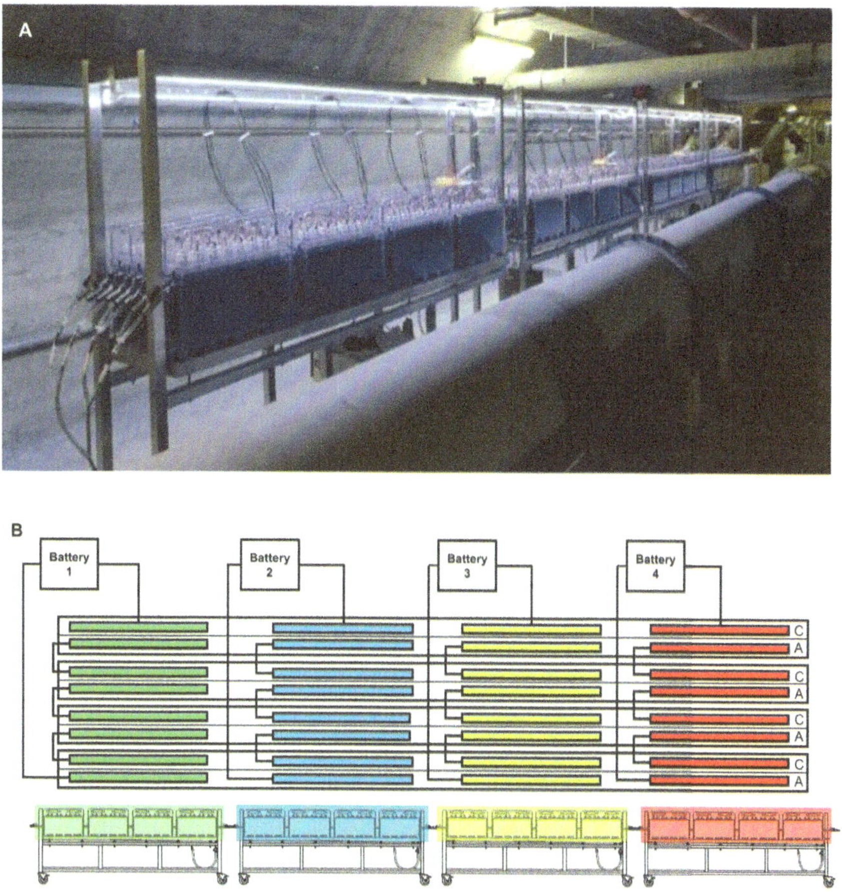


**Fig. 2** A) 1000-L MFC stack in an underground gallery, B) The 1000-L MFC was organized into four MFC-sub-stacks: green, blue, yellow and red electrodes. Each MFC-sub-stack was a serial stack of four MFCs, assembled from $4 \times 4 = 16$ MFC units (Blatter et al. 2021)

# 5 Conclusion

Microbial fuel cell is the only technology that uses wastewater as fuel to directly produce electricity. The application of MFC is broadening day by day with progress in research in lab. It is used in wastewater treatment of various sources as well as a power supply to sensors. However, MFCs cannot be operated as extreme low temperature as microbial reactions become very slow. This is a clean alternative source of power unlike the fossil fuels. The power generation by MFCs is still low, and researchers are working to improve it. The future work with MFC can be incorporating MFCs with various existing wastewater treatment technologies to harvest the energy within the wastewater.

## References


Anjum A, Mazari SA, Hashmi Z, Jatoi AS, Abro R (2021) A review of role of cathodes in the performance of microbial fuel cells. J Electroanalytical Chem 899:115673. https://www.sciencedirect.com/science/article/pii/S1572665721006998

Antolini E (2015) Composite materials for polymer electrolyte membrane microbial fuel cells. Biosens Bioelectron 69:54–70. https://doi.org/10.1016/j.bios.2015.02.013

Antolini E (2016) Structural parameters of supported fuel cell catalysts: the effect of particle size, inter-particle distance and metal loading on catalytic activity and fuel cell performance. Appl Catal B Environ 181:298–313. https://doi.org/10.1016/j.apcatb.2015.08.007

Baudler A, Langner M, Rohr C, Greiner A, Schröder U (2017) Metal–polymer hybrid architectures as novel anode platform for microbial electrochemical technologies. Chemsuschem 10(1):253–257. https://doi.org/10.1002/cssc.201600814

Behera M, Ghangrekar MM (2011) Electricity generation in low cost microbial fuel cell made up of earthenware of different thickness. Water Sci Technol 64(12):2468–2473. https://iwaponline.com/wst/article-abstract/64/12/2468/15515/Electricity-generation-in-low-cost-microbial-fuel

Berretti E, Longhi M, Atanassov P, Sebastián D, Vecchio CL, Baglio V, Serov A, Marchionni A, Vizza F, Santoro C, Lavacchi A (2016) Anodic biofilms as the interphase for electroactive bacterial growth on carbon veil. Biointerphases 11(3):031013. https://doi.org/10.1116/1.4962264

Berretti E, Longhi M, Atanassov P, Sebastián D, Vecchio CL, Baglio V, Serov A, Marchionni A, Vizza F, Santoro C, Lavacchi A (2021) Platinum group metal-free Fe-based (FeNC) oxygen reduction electrocatalysts for direct alcohol fuel cells. Curr Opinion Electrochem 29:100756. https://doi.org/10.1016/j.coelec.2021.100756

Blanchet E, Erable B, De Solan M-L, Bergel A (2016) Two-dimensional carbon cloth and three-dimensional carbon felt perform similarly to form bioanode fed with food waste. Electrochem Commun 66:38–41. https://doi.org/10.1016/j.elecom.2016.02.017

Blatter M, Delabays L, Furrer C, Huguenin G, Cachelin CP, Fischer F (2021) Stretched 1000-L microbial fuel cell. J Power Sources, 483:229130. https://www.sciencedirect.com/science/article/pii/S0378775320314245

Dekel DR (2018) Review of cell performance in anion exchange membrane fuel cells. J Power Sources 375:158–169. https://doi.org/10.1016/j.jpowsour.2017.07.117

Dewan A, Beyenal H, Lewandowski Z (2008) Scaling up microbial fuel cells. Environ sci technol 42(20):7643–7648. https://doi.org/10.1021/es800775d

Feng Y, Lee H, Wang X, Liu Y, He W (2010) Continuous electricity generation by a graphite granule baffled air–cathode microbial fuel cell. Bioresour Technol 101(2):632–638. https://doi.org/10.1016/j.biortech.2009.08.046

Fischer F, Sugnaux M, Savy C, Hugenin G (2018) Microbial fuel cell stack power to lithium battery stack: pilot concept for scale up. Appl Energy 230:1633–1644. https://doi.org/10.1016/j.apenergy.2018.09.030

Flimban SG, Ismail IM, Kim T, Oh SE (2019) Overview of recent advancements in the microbial fuel cell from fundamentals to applications: Design, major elements, and scalability. Energies, 12(17):3390. https://www.mdpi.com/1996-1073/12/17/3390

Galvani L (1791) De viribus electricitatis in motu musculari. Commentarius Bonoiensi Sci Artium Intituo Atque Acad Comment 7:363–418

Ge Z, He Z (2016) Long-term performance of a 200 liter modularized microbial fuel cell system treating municipal wastewater: treatment, energy, and cost. Environ Sci Water Res Technol 2(2):274–281. https://doi.org/10.1039/C6EW00020G

Guerrini E, Grattieri M, Trasatti SP, Bestetti M, Cristiani P (2014) Performance explorations of single chamber microbial fuel cells by using various microelectrodes applied to biocathodes. Int J hydrog energy 39(36):21837–21846. https://www.sciencedirect.com/science/article/pii/S0360319914018722

Guo K, Hidalgo D, Tommasi T, Rabaey K (2016) Pyrolytic carbon-coated stainless steel felt as a high-performance anode for bioelectrochemical systems. Bioresour Technol 211:664–668. https://doi.org/10.1016/j.biortech.2016.03.161

Guo K, Prévoteau A, Patil SA, Rabaey K (2015) Engineering electrodes for microbial electrocatalysis. Curr opinion in biotechnol 33:149–156. https://www.sciencedirect.com/science/article/pii/S0958166915000336

Hassan SHA et al (2014) Electricity generation from rice straw using a microbial fuel cell. Int J Hydrog Energy 39(17):9490–9496. https://doi.org/10.1016/j.ijhydene.2014.03.259

He Z (2006) Application of bacterial biocathodes in microbial fuel cells. Electroanalysis. https://doi.org/10.1002/elan.200603628. Accessed 28 Dec 2021

He L et al (2017) Advances in microbial fuel cells for wastewater treatment. Renew Sustain Energy Rev 71:388–403. https://doi.org/10.1016/j.rser.2016.12.069

Hiegemann H et al (2019) Performance and inorganic fouling of a submergible 255 L prototype microbial fuel cell module during continuous long-term operation with real municipal wastewater under practical conditions. Bioresour Technol 294:122227. https://doi.org/10.1016/j.biortech.2019.122227

Hiegemann H, Herzer D, Nettmann E, Lübken M, Schulte P, Schmelz KG, Gredigk-Hoffmann S, Wichern M (2016) An integrated 45 L pilot microbial fuel cell system at a full-scale wastewater treatment plant. Biores technol 218:115–122.https://www.sciencedirect.com/science/article/abs/pii/S0960852416308690?via%3Dihub

Ieropoulos I, Greenman J, Melhuish C (2010) Improved energy output levels from small-scale Microbial Fuel Cells. Bioelectrochemistry 78(1):44–50. https://doi.org/10.1016/j.bioelechem.2009.05.009

Jadhav DA, Das I, Ghangrekar MM, Pant D (2020) Moving towards practical applications of microbial fuel cells for sanitation and resource recovery. J. Water Process Eng 38:101566. https://doi.org/10.1016/j.jwpe.2020.101566

Jiang D, Li B (2009) Granular activated carbon single-chamber microbial fuel cells (GAC-SCMFCs): a design suitable for large-scale wastewater treatment processes. Biochem Eng J 47(1):31–37. https://doi.org/10.1016/j.bej.2009.06.013

Kadier A, Simayi Y, Abdeshahian P, Azman NF, Chandrasekhar K, Kalil MS (2016) A comprehensive review of microbial electrolysis cells (MEC) reactor designs and configurations for sustainable hydrogen gas production. Alex Eng J 55(1):427–443. https://doi.org/10.1016/j.aej.2015.10.008

Kim BH, Ikeda T, Park HS, Kim HJ, Hyun MS, Kano K, Takagi K, Tatsumi H (1999) Electrochemical activity of an Fe (III)-reducing bacterium, Shewanella putrefaciens IR-1, in the presence of alternative electron acceptors. Biotechnol Tech 13:475–478. https://link.springer.com/article/10.1023/A:1008993029309

Kim JR, Cheng S, Oh SE, Logan BE (2007) Power generation using different cation, anion, and ultrafiltration membranes in microbial fuel cells. Environ sci technol 41(3):1004–1009. https://pubs.acs.org/doi/abs/10.1021/es062202m?casa_token=MSkc497MWmcAAAAA:dDKWXXX_SdEA4_02zBNddLyrjp7scnOa-4aMEm0zQRwHqrlWtx--rA_62joqjriEFwVabW9NVBj5WRIcQQ

Kretzschmar J, Riedl S, Brown RK, Schröder U, Harnisch F (2016) eLatrine: lessons learned from the development of a low-tech MFC based on cardboard electrodes for the treatment of human feces. J Electrochem Soc 164(3):H3065. https://doi.org/10.1149/2.0121703jes

Li X, Lin Y, Yang Y, Zhang W, Hu M, Zhong Y, Liao Y, Li, W (2021) $Co_3O_4$ nanoparticles highly dispersed on hierarchical carbon as anti-biofouling cathode for microbial fuel cells. Electrochim Acta 391:138922. https://doi.org/10.1016/j.electacta.2021.138922

Liao Q, Zhang J, Li J, Ye D, Zhu X, Zhang B (2015) Increased performance of a tubular microbial fuel cell with a rotating carbon-brush anode. Biosens Bioelectron 63:558–561. https://doi.org/10.1016/j.bios.2014.08.014

Liang P, Duan R, Jiang Y, Zhang X, Qiu Y, Huang X (2018) One-year operation of 1000-L modularized microbial fuel cell for municipal wastewater treatment. Water Res 141:1–8. https://www.sciencedirect.com/science/article/pii/S0043135418303609?via%3Dihub

Liu B, Lei Y, Li B (2014) A batch-mode cube microbial fuel cell based 'shock' biosensor for wastewater quality monitoring. Biosens Bioelectron 62:308–314. https://doi.org/10.1016/j.bios.2014.06.051

Logan BE (2010) Scaling up microbial fuel cells and other bioelectrochemical systems. Appl Microbiol Biotechnol 85(6):1665–1671. https://doi.org/10.1007/s00253-009-2378-9

Logan B, Cheng S, Watson V, Estadt G (2007) Graphite fiber brush anodes for increased power production in air-cathode microbial fuel cells. Environ sci technol 41(9):3341–3346. https://doi.org/10.1021/es062644y

Logan, BE, Hamelers B, Rozendal R, Schröder U, Keller J, Freguia S, Aelterman P, Verstraete W, Rabaey K (2006) Microbial fuel cells: methodology and technology. Environ sci technol 40(17):5181–5192. https://doi.org/10.1021/es0605016

Lu M et al (2017) Long-term performance of a 20-L continuous flow microbial fuel cell for treatment of brewery wastewater. J Power Sources 356:274–287. https://doi.org/10.1016/j.jpowsour.2017.03.132

Malaeb L, Katuri KP, Logan BE, Maab H, Nunes SP, Saikaly PE (2013) A hybrid microbial fuel cell membrane bioreactor with a conductive ultrafiltration membrane biocathode for wastewater treatment. Environ sci technol 47(20):11821–11828. https://doi.org/10.1021/es4030113

Min B, Logan BE (2004) Continuous electricity generation from domestic wastewater and organic substrates in a flat plate microbial fuel cell. Environ Sci Technol 38(21):5809–5814. https://doi.org/10.1021/es0491026

Mustakeem (2015) Electrode materials for microbial fuel cells: nanomaterial approach. Mater Renew Sustain Energy 4(4). https://doi.org/10.1007/s40243-015-0063-8

Nayak JK, Ghosh UK (2019) Post treatment of microalgae treated pharmaceutical wastewater in photosynthetic microbial fuel cell (PMFC) and biodiesel production. Biomass Bioener 131:105415.https://www.sciencedirect.com/science/article/pii/S0961953419303642

Nemati M, Harrison, STL, Hansford GS, Webb C (1998) Biological oxidation of ferrous sulphate by Thiobacillus ferrooxidans: a review on the kinetic aspects. Biochem eng j 1(3):171–190. https://www.sciencedirect.com/science/article/pii/S1369703X98000060

Pant D, Van Bogaert G, De Smet M, Diels L, Vanbroekhoven K (2010) Use of novel permeable membrane and air cathodes in acetate microbial fuel cells. Electrochim Acta 55(26):7710–7716. https://doi.org/10.1016/j.electacta.2009.11.086

Papillon J, Ondel O, Maire É (2021) Scale up of single-chamber microbial fuel cells with stainless steel 3D anode: effect of electrode surface areas and electrode spacing. Bioresour Technol Rep 13:100632. https://doi.org/10.1016/j.biteb.2021.100632

Potter MC, Waller AD (1911) Electrical effects accompanying the decomposition of organic compounds. Proc R Soc Lond Ser B Contain Pap Biol Character 84(571):260–276. https://doi.org/10.1098/rspb.1911.0073

Rabaey K, Clauwaert P, Aelterman P, Verstraete W (2005) Tubular microbial fuel cells for efficient electricity generation. Environ Sci Technol 39(20):8077–8082. https://doi.org/10.1021/es050986i

Rahimnejad M, Adhami A, Darvari S, Zirepour A, Oh SE (2015) Microbial fuel cell as new technology for bioelectricity generation: A review. Alex Eng J 54(3):745–756. https://www.sciencedirect.com/science/article/pii/S1110016815000484

Rahimnejad M, Bakeri G, Ghasemi M, Zirepour A (2014) A review on the role of proton exchange membrane on the performance of microbial fuel cell. Polym Adv Technol 25(12):1426–1432. https://doi.org/10.1002/pat.3383

Renewable Energy Statistics (2021) p 460

Rhoads A, Beyenal H, Lewandowski Z (2005) Microbial fuel cell using anaerobic respiration as an anodic reaction and biomineralized manganese as a cathodic reactant. Environ sci technol 39(12):4666–4671. https://doi.org/10.1021/es048386r

Rossi R, Jones D, Myung J, Zikmund E, Yang W, Gallego YA, Pant D, Evans PJ, Page MA, Cropek DM, Logan BE (2019) Evaluating a multi-panel air cathode through electrochemical and biotic tests. Water res 148:51–59.https://www.sciencedirect.com/science/article/pii/S0043135418308194?via%3Dihub

Santoro C et al (2015) Influence of anode surface chemistry on microbial fuel cell operation. Bioelectrochemistry 106:141–149. https://doi.org/10.1016/j.bioelechem.2015.05.002

Santoro C, Arbizzani C, Erable B, Ieropoulos I (2017) Microbial fuel cells: from fundamentals to applications—a review. J Power Sources 356:225–244. https://doi.org/10.1016/j.jpowsour.2017.03.109

Santoro C, Serov A, Artyushkova K, Atanassov P (2020) Platinum group metal-free oxygen reduction electrocatalysts used in neutral electrolytes for bioelectrochemical reactor applications. Curr Opin Electrochem 23:106–113. https://doi.org/10.1016/j.coelec.2020.06.003

Segundo-Aguilar A, González-Gutiérrez LV, Payá VC, Feliu J, Buitrón G, Cercado B (2021) Energy and economic advantages of simultaneous hydrogen and biogas production in microbial electrolysis cells as a function of the applied voltage and biomass content. Sustain Energy Fuels 5(7):2003–2017. https://doi.org/10.1039/D0SE01797C.

Shah MP (Ed.). (2021a) Removal of emerging contaminants through microbial processes. Singapore: Springer

Shah MP (Ed.). (2021b) Removal of refractory pollutants from wastewater treatment plants. CRC Press.

Shizas I, Bagley DM (2004) Experimental determination of energy content of unknown organics in municipal wastewater streams. J ener eng 130(2):45–53.https://ascelibrary.org/doi/abs/10.1061/(ASCE)0733-9402(2004)130:2(45)?casa_token=wqKvnbBYzPYAAAAA:w1CjFBfbAH3v03kKVPCHe6pMIKz6mghXYx6GTkn3Ew2hk5onEcoJGobpo6V0WsnoSfKPWYHLdwk

Song YC, Kim DS, Woo JH, Subha B, Jang SH, Sivakumar S (2015) Effect of surface modification of anode with surfactant on the performance of microbial fuel cell. Int J Energy Res 39(6):860–868. https://doi.org/10.1002/er.3284

Stoll ZA, Ma Z, Trivedi CB, Spear JR, Xu P (2016) Sacrificing power for more cost-effective treatment: A techno-economic approach for engineering microbial fuel cells. Chemosphere 161:10–18. https://www.sciencedirect.com/science/article/pii/S0045653516308281?via%3Dihub

Sun J, Hu Y, Bi Z, Cao Y (2009) Improved performance of air-cathode single-chamber microbial fuel cell for wastewater treatment using microfiltration membranes and multiple sludge inoculation. J Power Source 187(2):471–479. https://www.sciencedirect.com/science/article/abs/pii/S0378775308021496

Tan WH, Chong S, Fang HW, Pan KL, Mohamad M, Lim JW, Tiong TJ, Chan YJ, Huang CM, Yang TCK (2021) Microbial fuel cell technology—A critical review on scale-up issues. Process 9(6):985. https://doi.org/10.3390/pr9060985

Tiwari BR, Noori MdT, Ghangrekar MM (2017) Carbon supported nickel-phthalocyanine/MnOx as novel cathode catalyst for microbial fuel cell application. Int J Hydrog Energy 42(36):23085–23094. https://doi.org/10.1016/j.ijhydene.2017.07.201

Valladares Linares R et al. (2019) Scale up of microbial fuel cell stack system for residential wastewater treatment in continuous mode operation. Water 11(2) Art. no. 2. https://doi.org/10.3390/w11020217

Vilajeliu-Pons A, Puig S, Salcedo-Dávila I, Balaguer MD, Colprim J (2017) Long-term assessment of six-stacked scaled-up MFCs treating swine manure with different electrode materials. Environ Sci: Water Res Technol 3(5):947–959.https://pubs.rsc.org/en/content/articlelanding/2017/EW/C7EW00079K

Villaseñor J, Capilla P, Rodrigo MA, Cañizares P, Fernández FJ (2013) Operation of a horizontal subsurface flow constructed wetland—microbial fuel cell treating wastewater under different organic loading rates. Water Res 47(17):6731–6738. https://doi.org/10.1016/j.watres.2013.09.005

Watanabe K (2008) Recent developments in microbial fuel cell technologies for sustainable bioenergy. J biosci bioeng 106(6):528–536. https://www.sciencedirect.com/science/article/pii/S1389172309700027

Wang Z, Cao C, Zheng Y, Chen S, Zhao F (2014) Abiotic oxygen reduction reaction catalysts used in microbial fuel cells. ChemElectroChem 1(11):1813–1821. https://doi.org/10.1002/celc.201402093

Wang X, Cheng S, Feng Y, Merrill MD, Saito T, Logan BE (2009) Use of carbon mesh anodes and the effect of different pretreatment methods on power production in microbial fuel cells. Environ sci tech 43(17):6870–6874. https://pubs.acs.org/doi/10.1021/es900997w

Wu S, Li H, Zhou X, Liang P, Zhang X, Jiang Y, Huang X (2016) A novel pilot-scale stacked microbial fuel cell for efficient electricity generation and wastewater treatment. Water res 98:396–403. https://doi.org/10.1016/j.watres.2016.04.043

Yasri, NG, and Nakhla G (2017) The performance of 3-D graphite doped anodes in microbial electrolysis cells. J Power Sources, 342:579–588. https://doi.org/10.1016/j.jpowsour.2016.12.081

Ye Y, Ngo HH, Guo W, Chang SW, Nguyen DD, Zhang X, Zhang S, Luo G, Liu Y (2020) Impacts of hydraulic retention time on a continuous flow mode dual-chamber microbial fuel cell for recovering nutrients from municipal wastewater. Sci Total Environ 734:139220. https://www.sciencedirect.com/science/article/pii/S0048969720327376

Zhang JN, Zhao QL, You SJ, Jiang JQ, Ren NQ (2008) Continuous electricity production from leachate in a novel upflow air-cathode membrane-free microbial fuel cell. Water Sci Technol 57(7):1017–1021. https://doi.org/10.2166/wst.2008.063

Zhang L, Wang J, Fu G, Zhang Z (2020) Simultaneous electricity generation and nitrogen and carbon removal in single-chamber microbial fuel cell for high-salinity wastewater treatment. J Clean Prod 276:123203. https://www.sciencedirect.com/science/article/abs/pii/S0959652620332480

Zhang X, Xia X, Ivanov I, Huang X, Logan, BE. (2014). Enhanced activated carbon cathode performance for microbial fuel cell by blending carbon black. Environ Sci Technol 48(3):2075–2081. https://doi.org/10.1021/es405029y

Zhou X, Chen X, Li H, Xiong J, Li X, Li W (2016) Surface oxygen-rich titanium as anode for high performance microbial fuel cell. Electrochim Acta 209:582–590. https://doi.org/10.1016/j.electacta.2016.05.103

Zhuang L, Zheng Y, Zhou S, Yuan Y, Yuan H, Chen Y (2012) Scalable microbial fuel cell (MFC) stack for continuous real wastewater treatment. Bioresour technol 106:82–88.https://www.sciencedirect.com/science/article/pii/S0960852411016300

Zuo Y, Cheng S, Logan BE (2008) Ion exchange membrane cathodes for scalable microbial fuel cells. Environ sci technol 42(18):6967–6972. https://doi.org/10.1021/es801055r

# Challenges of Wastewater and Wastewater Management

**Divyesh Parde and Manaswini Behera**

## 1 Introduction

For human and environmental health, the quantity of wastewater generated from households, commercial activities, hospitals, and industries is a major problem. As a result, effective wastewater management, which includes collection, conveyance, treatment, and discharge, is essential. It is vital to have the minimum possible environmental impact. Different ancient cultures had already devised wastewater management systems (Lofrano and Brown 2010). For example, in Ancient Rome about the sixth century BC, the Cloaca Maxima was constructed to serve as a massive sanitation and sewage system in the event of flooding in the metropolis (Hopkins 2012). This is a management technique that was widely used a few centuries back and is still in use in some nations, and it consists of merely transporting wastewater out from the community to preserve public health. Management of water is still a problem today, and techniques are evolving to preserve public health while minimizing the environmental effects on receiving water bodies. For example, current research adjusts to new developing pollutants produced by modern societies, with the goal of enhancing their reduction with creation of advanced treatment techniques. Furthermore, today's scenario is due to the fact that waste handling is no longer viewed as a normal issue that each municipality must address on its own, but as a global responsibility that is increasingly being addressed. Awareness about phenomena like "feminization" of fish species due to downstream of wastewater treatment plants as a result of environmental toxins released with effluents (Sumpter and Johnson 2005) has raised societal and political awareness. As a result, wastewater treatment and effluent discharge are being increasingly controlled at the national and international levels, such as in the European Union (EU), where the first unified wastewater legislation was developed

D. Parde (✉) · M. Behera
School of Infrastructure, IIT Bhubaneswar, Odisha 752050, India

M. Behera
e-mail: manaswini@iitbbs.ac.in



M. P. Shah (ed.), *Sustainable Industrial Wastewater Treatment and Pollution Control*,
https://doi.org/10.1007/978-981-99-2560-5_12

in the 1990s (EEC 1991). Some member states, however, fail to comply with EU regulations (Voulvoulis et al. 2017), which must be maintained and amended on a regular basis to maintain compatibility and environmental preservation while preventing social problems.

Even after treatment, wastewater discharges contain a substantial number of pollutants that could be harmful to humans and environment. Micropollutants, pathogens, antibiotic resistance genes, personal care products, pharmaceuticals, nanomaterials, and disinfection by-products are some of the pollutants that are increasingly being found in reclaimed wastewater. The existence of pathogens, such as viruses, bacteria, or protozoa, poses the greatest hazard to human health when it comes to water reuse because they are liable to waterborne diseases. Chemical pollutants, such as heavy metals, medications, hydrocarbons, and other substances, may represent a threat to human health. The community could be exposed to these contaminants directly by unclean water consumption, inhalation, or skin contact, or indirectly through polluted water effluents. Emerging micro-pollutants are non-regulated organic trace contaminants that have recently been introduced or found using modern analytical technologies. The phrase "emerging" refers to a contaminant that has a new source, a new route to humans, or novel treatment procedures. They are classified according to whether they pose an apprehensible, probable, or actual threat to human health and the environment. They can come from industrial wastewater, as well as municipal, agricultural, hospital, or laboratory effluent (Gogoi et al. 2018). Toxic contaminants such as heavy metals cause higher impact. The challenging task is to treat toxic wastewater with higher efficiency. Heavy metals must be removed because they are hazardous and carcinogenic compounds that should not be released back into the environment (Azimi et al. 2017).

Modern society's present needs also influence people's perceptions of wastewater and lead to paradigm shifts in wastewater study. For example, as the population of urban agglomerations grows, so does the demand for water, worsening the water footprint of metropolitan areas. This promotes a paradigm change in which wastewater is transformed from a waste product to a profitable resource (Salgot and Folch 2018). According to this paradigm, society should adopt changes to enhance the availability and quality bodies and promote resource management sustainability. The United Nations' 2030 Agenda for Sustainable Development included a particular Sustainable Development Goal (SDG) to guarantee universal access to and sustainability toward sanitation and water management.

The challenges of wastewater treatment are diverse, and they differ not only in terms of effluent control regulations, but also in terms of regional features and socioeconomic conditions. As a result, pinpointing a common obstacle that applies to all cases is challenging. Nonetheless, there is no denying the importance of implementing a cost-effective and increased wastewater treatment system (Hosomi 2016). Growing research is the outcome of the continuous progress in management of wastewater, wastewater consciousness, and new demands of modern society. The vast number of publications from many domains (engineering, analytical and experimental science, humanities and social sciences, etc.) is creating a difficulty for researchers to keep track of current developments in a particular topic.

## 2 Sources of Wastewater

Modern cultures contain an ever-increasing amount of waste and wastewater-generating activities. As a result, identifying the need of wastewater collection system, conveyance system, and treatment option is the precondition for proper management at the rural, peri-urban, and urban region. While some wastewater sources have been discovered and classified, others are either new or have been largely disregarded.

Households are the main contributors to the volume of wastewater released annually around the world. Humans, on the other hand, consume minimum of several liters of water every day to meet their needs such as drinking water, sanitation, hygiene, and for food (Gleick 1996).

Pharmaceuticals and their metabolites are detected in the wastewater via toilets, insect repellents, flames and fungicides retardants from clothes, and other personal care products, and compounds like denatonium are all contributing to the pollution load of wastewater produced (Merel and Snyder 2016).

Hospitals are another source of wastewater that is produced and handled each year. The exact generation of amount of wastewater by hospitals is not fully described because there is variation in water utilization and wastewater production that can depend on the patients utilizing water for personal hygiene and extra amount of water required to maintain the surroundings clean. Patients treated internally receive harsher therapies, such as antineoplastics, anticancer. This wastewater contains a high range of medicines that are different from the characteristic generated by households. As the advancement in medicine sector from last few years, resulting in a higher concentration of medicine found in wastewater (Mompelat et al. 2009).

Rainfall and surface runoff also contribute as source of wastewater, especially in metropolitan areas that are mostly impermeable and have a collection system. Water collected after heavy rains may contain toxins such as pesticide, diuron, carbendazim, and terbutryn that have leached from roofs and building facades (Burkhardt et al. 2011). The sources of wastewater in detail are discussed below.

Due to their distinct operations, industries are the main sources of wastewater. However, most industrial wastewater effluents are treated before being discharged, determining the exact quantity of wastewater generated and concentration of different wastewater is difficult, as it varies from one industry to another. For example, paper mills that generate a huge quantity of wastewater will focus on removing important contaminants such as BOD, whereas power plants that generate a lot of wastewater from cooling units will be concerned about the temperature of the discharged water before disposal (Ashrafi et al. 2015; Toczyłowska-Mamińska 2017). Table 1 shows the constituents present in wastewater and their effects (Ashrafi et al. 2015; Yang et al. 2020; Brausch and Rand 2011; Oviedo and Aga 2016).

**Table 1** Constituents present in wastewater

| Wastewater constituents | Effects |
|---|---|
| Suspended Solids | • Feculent condition<br>• Sludge formation |
| Biological organic matters | • Limiting oxygen<br>• Cause septicity |
| Nutrient | • Odor<br>• Eutrophication |
| Inorganics | • Foaming<br>• Plant growth |
| Metals | • Toxicity<br>• Human Health |
| Emerging Contaminants | • Prevent growth<br>• Adverse health effect |

## *2.1 Domestic Wastewater*

The wastewater discharged basically from the residential areas, communities are referred as domestic wastewater. They highly contribute to the total amount of wastewater generation. The rapid increase in the growth of population and urbanization with poor management of generated wastewater causes water pollution (Wang et al. 2014). The domestic wastewater can be classified as grey water (wastewater from kitchen sink, bathing, laundry, and other household activities, black water (wastewater from toilets), and sewage (mixture of grey and black water) (Mission 2015). The characteristics and concentration of domestic wastewater have large variation because of the heterogeneity and huge number of sources. Both characteristics and concentration of wastewater depend upon the frequency of water supply, per capita water demand, and living standards. For deciding the pollutants concentration, the level and amount of water supply play an important role. The other factors that are important for identifying the characteristic and concentration of wastewater are available initial treatment of wastewater (like septic tank, Imhoff tank, pits, etc.), weather conditions, and topography. Main analyzed parameters for wastewater are biochemical oxygen demand (BOD), chemical oxygen demand (COD), total suspended solids (TSS), nutrient mainly nitrogen and phosphorus, total coliform (TC), and fecal coliform (FC).

Domestic wastewater is a major source of nutrients in global nutrient cycling. Direct discharge of residential wastewater without treatment or after inappropriate treatment is the primary source of non-point nutrient pollution in developing nations, accounting for more than half of the overall non-point nutrient load to the aquatic environment (Shindhal et al. 2021). Domestic wastewater delivered roughly 7.7 Tg/year (Tera grams per year) of nitrogen and 1.0 Tg/year of phosphorus to the world's seas until the end of the twentieth century (Lee et al. 2019). As a result, limiting the environmental contamination concerns produced by the flow of nutrients into water bodies, particularly surface and groundwater, looks to be an essential and tough challenge for today (Luo et al. 2021). Nutrient recovery from waste streams, on the other hand, is critical for long-term wastewater treatment because it avoids depletion of scarce resources and thereby releases the circular bioeconomy. Reduce, reuse, and recycle are the simple principles of the circular bioeconomy. The recyclable nutrients recovery from a variety of wastewater sources might help fulfill the population explosion's induced rising demands. In a circular bioeconomy approach, nutrients enriched in wastewaters may, therefore, be collected and reused (Feng et al. 2021).

## 2.2 *Municipal Wastewater*

"Clean water and sanitation" is one of the 17 Sustainable Development Goals (SDGs) listed by the United Nations. Municipal wastewater management is gaining popularity around the world as a result of this broad purpose. The majority of human activities result in the production of wastewater. Municipal wastewater production now exceeds $300 \times 109\ m^3$ per year over the world. With urbanization and the resulting population growth, the wastewater production continues to rise. Increased wastewater production, along with insufficient collection and treatment facilities, has had a severe influence on the receiving aquatic ecosystem's quality of the water, biodiversity, and social well-being (Larsen et al. 2016). Due to less availability of waste, deterioration of resources, and increase in global warming, technologies for producing clean water and clean energy have gotten a lot of interest (Chung et al. 2012). Municipal wastewater can produce renewable energy, by converting the chemically bound energy content of organic pollutants to a usable energy resource. Municipal wastewater is the most common type of low-strength wastewater as compared to industrial wastewater. To tackle of the problem of global warming and wastewater production, many organizations are gradually enacting stricter pollutant discharge restrictions with a primary focus on waste reduction (Lier 2008).

The conventional biological process-based design approach is now becoming less popular because traditional biological wastewater treatment methods are encountering various hurdles. The main challenges for treating municipal wastewater are generation of huge amount of activated sludge, high consumption of energy, emission of greenhouse gases, and less energy and resource recovery (Zhang and Liu 2021). For many towns and cities, 3% of global electricity could be used if wastewater treatment plants (WWTPs) were treated with conventional technologies, while the wastewater treatment plant to treat industrial wastewater requires around 0.56–0.71 Gt $CO_2$ per year, which is similar to 4.6–5.2% of global emission of total non-$CO_2$ GHGs (Zhang and Liu 2021). Globally, production of municipal wastewater is around 360 billion cubic. Municipal wastewater consists a typical ammonium content of 40 mg/L. Therefore, the total production of ammonium is $40\ mg/L \times 360$ billion $m^3/year \times 18/\ 14 = 18.5$ billion kg ammonium/year (Shi 2011; Liu et al. 2019).

Unfortunately, in the municipal wastewater treatment plant employing biological methods, the maximum amount of ammonium is biologically converted to di-nitrogen gas. Although there are a variety of processes that can remove biological nitrogen (e.g., nitrification–denitrification process, anammox process, partial nitrification–denitrification method), all of them work on same principle that convert ammonium present in wastewater to di-nitrogen gas making nitrogen recovery nearly impossible. Clearly, there is a fundamental lack in the design process for treatment of wastewater that are addressing the municipal wastewater treatment challenges. In fact, in order to be linked with the management and reuse of wastewater must alter one's perception of municipal wastewater from an untreated waste to reuse of treated water.

## 2.3 *Industrial Wastewater*

The industrialization and urban growth have accelerated, resulting in massive amounts of wastewater being generated globally. In the context of worldwide water usage, the industrial sector uses an average of 22%. Approximately 80% of all wastewater produced is released into waterways, causing contamination and posing a risk to human health and aquatic life (The International Water Association 2018). Industrial water usage is higher in industrialized countries like North America and Europe than in developing ones, accounting up to 50% and 4–12%, respectively. Because of its advanced technologies and technical know-how, the United States of America (USA) has achieved outstanding results in wastewater treatment. According to a report done by Acumen Research and Consulting (ARC), 2019, the global industrial wastewater treatment is expected to rise to $16.5 billion by 2026, with a 4.5% compound annual growth rate. According to research issued by UN Water and UNESCO in 2017, around 70% of urban and industrial wastewater is treated in high-income nations, compared to 38% in upper-middle-income countries and 28% in lower-middle-income countries, while just 8% is treated in low-income countries (Acumen Research and Consulting [ARC] 2019; United Nations Water and UNESCO 2017). According to current trends, about 80% of worldwide wastewater will be released without treatment. India generates around 44 million $m^3$/day of industrial wastewater (Ranade and Bhandari 2014), of which nearly 6.2 billion liters are discharged into natural aquatic bodies untreated (The Trade Council 2015).

Chemical, sugar, distillery, food and dairy, textile bleaching and dyeing, paper and pulp, organic chemical, battery making industries, leather/tannery, nuclear power plants, iron and steel, electric power plant, soap and detergent, metal refining, petroleum and petrochemical pesticide and biocide, metal processing, pharmaceutical, and electroplating industries are all contain metal in their wastewater discharge. Metals present in wastewater discharge are Copper (Cu), Cadmium (Cd), Zinc (Zn), Nickel (Ni), Mercury (Hg), Chromium (Cr), Lead (Pb), Iron (Fe), and Arsenic (As). The sources of industrial wastewater contamination are shown in Table 2 (UN Environment Programme [UNEP] 2012; Bora and Dutta 2019; Shahedi et al. 2020; Rojas et al. 2013; Fu and Wang 2011). Most of the time, the untreated wastewater is released frequently into water bodies as a result of inadequate wastewater treatment facilities, such as power outages, poor maintenance, and a lack of educated and experienced

**Table 2** Sources of industrial wastewater contaminants

| Constituents | Sources |
|---|---|
| Organic chemical compounds | Brewery industries, pharmaceuticals, oil refining |
| Heavy metals | Mining, petrochemical plants, pesticide production, chemical industries |
| Acidic and alkaline | Textile industry, petroleum refineries, paper and pulp industries |
| Radioactive substances | Nuclear and thermal power plants, mining, hospitals, chemical industries |
| Emerging contaminants | Pharmaceutical, personal care product, surfactant, and chemical industries |

manpower. Contaminated water sources, especially groundwater, pose a health risk as well as pollution to the environment (UN Environment Programme [UNEP] 2012; Bora and Dutta 2019).

Increased consumption of natural resources pushed civilization to boost mining activities in order to meet metal demands. As a result, as the number of companies grows, more water and resources are consumed, resulting in the release of a large amount of metal-laden wastewater. Therefore, industrial wastewater management has become a major concern around the world. However, these processes have some drawbacks, such as limited efficiency, a high demand for solvents and reagents, and the formation of more waste products, such as waste scraps, sludge, industrial effluent, hazardous compounds, and so on (Gebretsadik et al. 2020). The economic treatment technology and sustainable wastewater treatment systems with better efficiency become a crucial necessity all over the world. The main harmful contaminants present in the wastewater (emerging contaminants and heavy metals) are described below.

### 2.3.1 Emerging Contaminants

Nowadays, emerging contaminants (ECs) are mostly found in domestic wastewater. The researchers are focusing toward the ECs due to their presence in water bodies and sewage treatment plants. No discharge standards are established for release of these contaminants into environment (Jasinska et al. 2015). Mostly their presence is found in the aquatic environments. They are found in number of varieties out of which only some are identified as toxic. The primary source of ECs in the aquatic environment is from discharge of treated sewage. The sewage treatment technologies are well developed for the treatment of organic matters, nutrients, and pathogens but they are unable to remove the organic micro-contaminants. Trace organic contaminants are widely found in the water bodies, and they are considered as emerging threat to the life of aquatic animals (Boxall et al. 2012; Brausch and Rand 2011). Various studies are done to find out the significant results for the removal of emerging pollutants from the wastewater (Gerbersdorf et al. 2015a).

ECs are categorized into wide range but some of the commonly found in the water bodies are pharmaceutical and personal care products (PCPs), polychlorinated alkanes (PCAs), perfluorochemicals (PFCs), polydimethylsiloxanes (PDMSs), polychlorinated naphthalenes (PCNs), benzotriazoles, quaternary ammonium compounds (QACs), bisphenol A (BPA), triclosan (TCS), triclocarban (TCC), hormones and steroids, benzothiazoles, synthetic musks, polar pesticides, industrial by-products, veterinary products, and food derivatives (Stuart et al. 1016).

There are various sewage treatment technologies from natural treatment to expensive treatment technology with higher degree of treatment for wastewater but they all are lacking for the treatment of micro-organic contaminants (Rojas et al. 2013). Some of the advanced wastewater treatment technologies are in the view to consider for the treatment of emerging pollutants such as membrane bio-reactor (MBR), as they are efficient for the pollutant removal from wastewater (Petrović et al. 2003). However, these technologies consume higher energy that overall increases the cost of treatment that's why implication of these technologies for treatment of each and every emerging pollutants is not a feasible option (Jones et al. 2007). Many wastewater treatment technologies are reviewed for the ECs removal, and it can be concluded that there is large variation in the different classes of ECs (Gerbersdorf et al. 2015b). The higher variation on the studies is creating difficulty to analyzing the presence of ECs in the wastewater. Meta-analysis is very much useful for such large scale of database with higher variability, and it is mainly applied in the medical, toxicological science, and ecological datasets, Hence, quantitative assessment of different treatment technology for the micro-pollutant removal is need to be done. Analogous techniques are also an option for study the comparison between experimental results (Melvin and Houlahan 2012).

The detection of emerging pollutants is creating troubles for the concerning authorities due to lack of data about the emerging contaminants in the environments. Even their future impacts to the human and environment are not certain. Some of the countries are preventing few emerging contaminants into the drinking water. A preparatory observing technique was prepared in USA to deal with the endocrine disrupters (EDCs) and to prevent its contact with the human and aquatic environment (European Commission 2011). EDCs are the chemicals that prepared artificially and if they found in drinking water, can cause harm to hormones and also effects body functioning. They reduced the concentration level of these chemicals from products. In 2013, 45 emerging contaminants are listed that can affect the aquatic life. The concentration level of pharmaceuticals and PCPs set up as not more than 10 ng/L and 10 μg/kg for soil and surface water (European commisison 2013). Such laws are also regulated by Switzerland, European Union, USA, and WHO but there are no such regulations in country India and Canada (Oviedo and Aga 2016).

Other contaminants such as hormones ethinylestrdiol and anti-inflammatory diclofenac are in the continuous observation to identify their risk and also to address them into the priority compound of emerging contaminants (Collado et al. 2014). In the Drinking Water Contamination List, various types of pharmaceutical and EDCs were enlisted. The effect of prime pharmaceutical contaminants to the water cycle caused by the naproxen, bezafibrate, carbamazepine, ibuprofen, diclofenac,

**Table 3** Effects of emerging contaminants

| Contaminants | Effects |
| --- | --- |
| Antibiotics (Penicillin, sulfonamides, Tetracyclines, Roxithromycin, clarithromycin, Tylosin) | • Consequences on food chain<br>• Algal growth inhibition |
| Endocrine disrupting compound (Bisphenol A [BPA]) | • Estrogenic effect<br>• Antiandrogen condition |
| Triclosans | • Toxicity<br>• Biocidal effect<br>• Bacterial resistance against triclosans |
| Anti-microbiological preservatives | • Estrogenic effect<br>• Carcinogenic |

sulfamethoxazole, erythromycin, gemfibrozil, and atenolol (Global Water Research Coalition 2008; US EPA 2012). The rapid increase of ECs in the water has possibilities to enter into the crops which can affect to the human life. It was reported that EDCs can cause reproductive inabilities to the human and animals. The effects of emerging contaminants are shown in Table 3 (Gerbersdorf et al. 2015a, b; Global Water Research Coalition 2008; Du et al. 2015). Emerging contaminations are critical issue for the rapid developing societies, and their effects are different in laboratory tests, due to many environmental factors, such as water level, soil type, pH, temperature, and their concentration level (Du et al. 2015).

### 2.3.2 Heavy Metals

Industries are the main sources for toxic pollutants in wastewater. Toxic pollutants are highly harmful for the human and environmental health. The toxic elements are mainly inorganic in nature. This pollutant contains heavy metal ions which are carcinogenic in nature and directly discharge of them are not recommendable. Mostly heavy metal ions found in the wastewater are chromium (Cr), mercury (Hg), nickel (Ni), cobalt (Co), zinc (Zn), copper (Cu), arsenic (As), and lead (Pb). Some of the heavy metals already exist in the environment with lower concentration but with the industrial waste discharge, increasing their concentration into the environment (Fu and Wang 2011). There is a possibility to enter heavy metals into the food chain due to their existence in soil and wastewater (Charerntanyarak 1999). However, wastewater is the major parameter to affect human that's why the concentration level of heavy metals should be within the discharge limits. They remain for higher time into the human body or in environment because they don't have biodegradable nature. The presence of heavy metal in the wastewater can cause serious health diseases, such as allergic problems to the skin dermatitis, affect to kidney, gastrointestinal issues, damage in liver or kidney skin problems, increase the risk of carcinogenic problem, high blood pressure, delays in physical or mental growth, depression, lethargy, neurological signs, and the disease of nervous system (Chen et al. 2018). The heavy metals

**Table 4** Heavy metals in industrial wastewater and their effects

| Metals | Industries | Health effect |
|---|---|---|
| Cadmium (Cd) | Chemical, distillery, sugar, food & dairy, pulp & paper, tannery, petroleum | Carcinogenic, fragile bones, lungs, and kidney damage |
| Copper (Cu) | Chemical, distillery, sugar, soap and detergents, tannery, iron & steel | Abdominal effect, diarrhea, liver issues |
| Chromium (Cr) | Textile, bleaching and dyeing, tannery, chemical | Anemia, renal failure, thrombocytopenia |
| Lead (Pb) | Chemical, sugar, pulp & paper, tannery, petroleum, soap and detergent | Nervous system issues, kidney damage |
| Zn | Chemical, pulp & paper, tannery, petroleum, soap and detergent | Reduces immunity, depression |
| Hg | Chemical, sugar, textile, plastic and resin | Respiratory, skin cancer |
| As | Textile, bleaching and dyeing, chemical | Cardiovascular problem |
| Fe | Iron & steel, chemical, textile, bleaching and dyeing | Rheumatoid arthritis, blood, and abdominal effect |
| Ni | Chemical, textile, plastic and resin, iron & steel | Respiratory, skin diseases |

present in industrial wastewater, and their effect on human body is illustrated in Table 4 (Du et al. 2015; Fu and Wang 2011; Charerntanyarak 1999; Chen et al. 2018).

Wastewater treatments are one of the important factor for conserving environment, aquatic life, and human health. For the treatment of heavy metals, technologies are based on advanced oxidation processes, electrochemical and physiochemical. The advanced oxidation processes are photocatalysis, membrane filtration, and nanotechnology. The methods of electroflotation, electrocoagulation, and electrodeposition come under electrochemical methods, and the membrane filtration, ion exchange, chemical precipitation, and adsorption are the physiochemical processes. Apart from all these methods, biological processes are one of the widely used technology for municipal and industrial wastewater. These process are considered as cost-effective technology as compared to above-mentioned methods but they are mainly designed for the biodegradable matter, not for removal of heavy metals.

## 3 Wastewater Treatment Methods

Wastewater treatment methods includes the physical process, biological and chemical process that useful for the removal of solids, organic matter, nutrients (mainly nitrogen and phosphorus), and soluble contaminants like heavy metals, emerging contamination, etc. Every treatment method has its own working performance on the basis of efficiency, cost, sludge production, reliability, feasibility, chemical operation, carbon footprint, pre-treatment required, possible formation of by-products, and operation and maintenance. The various types of wastewater treatment methods are shown in Fig. 1, and their working, advantages, and challenges are discussed below.

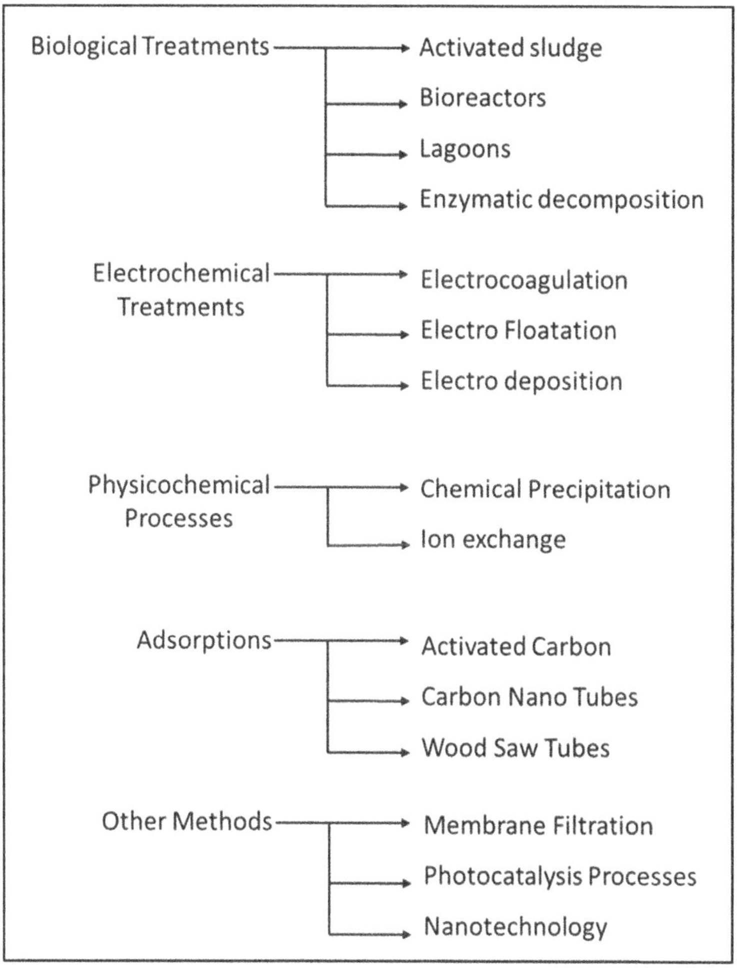


**Fig. 1** Wastewater treatment methods

## 3.1 Biological Treatment

Biological process is a widely used wastewater treatment technology. It has an ability to effectively treat municipal wastewater and various industrial wastewater. The large variety has seen on the treatment process of biological methods. The several types of biological reactors are in application, such as activated sludge, lagoons, enzymatic decomposition, bioreactors, and microbiological treatment. The activated sludge technique is used globally to treat different types of industrial and municipal wastewater. The main challenges in activated sludge process are higher energy consumption which contributes toward larger footprint, requires recirculation flow and design process is complex (Liu et al. 2018).

Over various treatment methods, biological treatment methods are highly adopted because of the following advantages:

a. Microorganism application for the degradation of organic matter is simple and easily implementable.
b. Capable to eliminate high ammonia, biodegradable organic matter and iron.
c. Efficient to reduce the color.
d. High working capacity to remove suspended solids and biochemical oxygen demand.
e. Method can be implementable for the removal of emerging contaminants and heavy metals.

Although there is a wide application of biological treatment, many questions have been raised about its operating principle, which is mostly based on biological oxidation and high energy consumption. Due to the strict discharge effluents in several countries leads to the rise in energy consumption for the treatment of high quality wastewater. The other challenges while operating on biological methods are; the requirement of controlled operation system, mainly pre-treatment of wastewater is required, management and maintenance of microorganism are necessary when toxic compounds are present, the treatment process is slow as compared to other treatment technology, bio-degradability affects when dyes are present, poor decolorization, management of foaming and sludge bulking on the reactor, production of uncontrolled degradation process and biological sludge, there is possibility to change composition and properties of cultures used, complexity in the mechanism of microbiological phenomenon (Crini and Lichtfouse 2019).

## 3.2 Electrochemical Treatments

Electrochemical treatment processes are done by the two means: direct oxidation and indirect oxidation process. Initially, pollutants are adsorbed on the surface of anode and then break down by the reaction on anodic electron, it is the direct anodic oxidation process. The strong oxidants (hydrogen peroxide, hypochlorite, and ozone) are electrochemically generated in indirect oxidation process. In indirect process, the generated oxidant reduces in bulk solution by the oxidation reaction. Electrochemical treatments are the compact and more efficient but in terms of cost they are not comparable. It requires high electricity supply and large capital investment that's why this method is not received too much attention. The main used electrochemical treatment methods are electroflotation, electrocoagulation, and electrodeposition.

Electrocoagulation process consists anode and cathode electrodes in a reactor with a low electric current. These electrodes are commonly made up of most popular material, i.e., aluminum and iron. The heavy metal ion present in the wastewater gets retained by their hydrogen bond or surface electrical charge. When the electric field is introduced, these charges neutralized with the suspended pollutants and gets coagulated. The coagulated particles bond with each other and form a sludge or floc. The produced floc can be easily removed by filtration. Electrocoagulation treatment method is a simple technology to remove wide range of pollutants by electrolysis but never be a reliable method due to its electrode reliability. This treatment method produces high sludge, and it is difficult to discharge sludge containing heavy metal ions (Il'in et al. 2002).

Electroflotation gained attraction for the wastewater treatment because of its simplicity in design and operation, adaptability, environmental compatibility, its small and compact units, and lower running cost. Electroflotation can be used for treatment of sewage, pit tank wastewater, oil–water emulsion effluent, and industrial wastewater. Electroflotation process consists of two electrodes with electric power supply, and it has three steps. Initially, reaction occurs into the cell, i.e., water electrolysis which releases hydrogen and oxygen to the solution; the present heavy metals gets absorbed or attached into the molecule of hydrogen and oxygen that makes suspended particles, and then gets destabilized and form flocs. Then, the second step is to separate the produced foam and settled flocs. Finally with the use of filtration method, pollutants can be collected (Chen and Lim 2005).

Electrodeposition process is based on converting dissolved metal ions on ionic conductors by deposition on solid particles. The process involves heavy metal reduction and oxidation in a cell consisting wastewater, electrolyte cell, anode, and cathode. The heavy metals breaks down and electroplated in cathode. The anodes are kept to be inert or insoluble, otherwise it suppress the heavy metals process recovery. The advantage of the process is that it can be applied to the solution with chelating agents or non-aqueous solution and has a better pollutant removal than aqueous solution. The challenges with the aqueous solution are; low thermal stability, hydrogen gas molecule liberation, and narrow electrochemical windows. The problems associated with the non-aqueous solutions are cell component corrosion, heat balance, and lower current efficiency (Simka et al. 2009).

## 3.3 *Physiochemical Process*

### 3.3.1 Chemical Precipitation

Chemical precipitation process is done by adding chemicals in the wastewater. The added chemical precipitant agent into the wastewater reacts with the heavy metals and forms an insoluble solid products which can be settled down in the reactor by gravity or can be removed by the filtration. The process needs proper working condition, i.e., pH should be preferred to 11 for the removal of heavy metals. The optimum condition of pH accelerates the soluble metal ion to react with solid precipitant agent and convert it into dissolved solids. For maintaining pH, alkaline agents are used which reduces the solubility of metal ion and form precipitate. The aluminum salts or iron salts are used to precipitate the phosphate and calcium chloride is used for fluorine. The dissolved precipitates can be easily separated by sedimentation (Du et al. 2015). Chemical precipitation process is easily maintained and widely used wastewater treatment technology but it requires huge amount of chemical to meet effluent as per discharge standards and can also cause for further pollution from produced precipitated sludge (Ferreira et al. 1999). Chemical precipitation can be done by hydroxide or sulfide precipitation process. In hydroxide precipitation, coagulates such as alum, iron salts, and polymers separate the heavy metal from wastewater. By sedimentation or filtration, the precipitated hydroxide soluble metals can be removed. The hydroxide precipitations are affected by the presences of organic radicals. The sulfide precipitation is also same as hydroxide precipitation, only heavy metal ion precipitates as metal sulfides. The pre-treatment and post-treatment are required for sulfide precipitation and also require controlled use of reagents due to the sulfide ion and $H_2S$ toxicity. For chemical precipitation, large variety of chemicals are used, mainly used chemicals are lime (CaO) and calcium hydroxide. The lime addition may not be considered as successful removal of heavy metals while calcium hydroxide can be effective for Cd, Mn, Cr, Pb, and Zn cations removal (Chen et al. 2018). The process forms concentrated and huge amount of sludge which increases the cost of sludge disposal, slow precipitation of metal ion, and poor precipitate settling that makes it less workable method.

### 3.3.2 Ion Exchange

In the ion exchange process, heavy metal ions are absorbed and form complex between the functional group and counterion. Then the hydration occurs at the pores of adsorbent and at the solution surface. It is based on the reversible ion interchange between liquid and solid phases. In this process, ions remove from resin in an electrolytic solution and also removes similar charged ion in an equivalent amount without change in structure of resin. To remove heavy metal ions, an insoluble resins are used. Ion exchange resins are preferable to separate or eliminate heavy metal ion from wastewater. The options for selection of resins are styrene–divinylbenzene, gel, macropore, Dowex HCR S/S, and AmberSep 132. The styrene–divinylbenzene is better resins than other. Gel resins are effective for some cases and also have low cost. Macropore are more stable resin with sponge-like structure, Dowex HCR S/S consists regeneration properties and mainly found in sodium form. Dowex HCR S/S has an ability to remove the Ni and Zn contaminated wastewater by 98%. AmberSep 132 resin consists a strong basic nature with 92.1 mg/g adsorption capacity and used to remove Cr(VI) while IRN-77 is a cation exchanger resin that absorbs Cr(III). In ion exchange process, recovery of heavy metal is also an option from inorganic effluents. Ion exchange method is always preferred over chemical precipitation because of recovery of heavy metals, production of lesser sludge volume, high efficiency, low cost, and high selectivity. It is very effective for the removal of heavy metals from wastewater. When strong cation exchangers are introduced in ion exchange cell, removal efficiency of metal increases. The factors which affect the operation of ion exchange are anions, temperature, pH, concentration of sorbate and adsorbent, and time of contact (Alyüz and Veli 2019).

### 3.3.3 Adsorption

Adsorption is a process in which sorption of contaminants occurs on the surface of adsorbent, basically it is mass transfer process between the adsorbent, i.e., solid phase and liquid phase. The process runs in reversible form as well, desorption occurs on the adsorbents. In the adsorption process, pollutants penetrate on the surface of adsorbents, then undergo adsorption on the surface of adsorbents, and penetrate into the structure of adsorbent. The suitable adsorbent provides a larger specific area and better adsorption capacity. Adsorbents can be obtained from the readily available material such as from agricultural waste, natural materials, and industrial by-products. The commonly used adsorbents are activated carbon, carbon nanotubes and sawdust. Activated carbon are widely used for the toxic metal removal. It is prepared from agricultural by product and having a 1266–3256 $m^2/g$ maximum surface area. When made at 900°C than it can remove Ni(II) more effectively. The 100% efficiency of Ni(II) was also achieved at pH range of 2–5 which is satisfactory for adsorption of cations. Carbon nanotubes are also used for adsorption having very good properties and application for the heavy metal removal but their mechanism is very complicated. It is immobilized by calcium alginate which run down the impact caused by discharging CNTs to the water bodies. CNTs show better removal efficiency for the removal of Cu(II). Wood sawdust consists lignocellulose composition with a lignin of 23–30% and cellulose of 45–50%. It is obtained as a solid product from the mechanical wood processing operation and referred as a low-cost adsorbent used for the removal of heavy metal. The cellulose and lignin structure has carboxyl, hydroxyl, and phenolic group which shows high capacity to bind metal cations. Previous studies by using wood, sawdust as adsorbents show removal efficiency of Cd(II), Cu(II), Zn(II) and were 31.9%, 76.2%, and 37.5%, respectively (Sahmoune and Yeddou 2016). The main disadvantages using adsorption process are; process is considered as a non-selective because of its non-destructive characteristics. It is not an economically viable for the industries, such as textile, pulp, and paper. It requires various types of adsorbents, chemical derivatization to improve their adsorption capacity, rapid saturation, and clogging problem occurs in the reactor, regeneration is costly.

### 3.4 Membrane Filtration Process

Membrane filtration process has a separation process of contaminants by using membranes. The treatment is not only by flowing wastewater through the membrane but in some cases also done by separation through a semi-permeable membrane with high membrane pressure; this process holds the solute one side and passes the solvent another side. Due to continuous advancement in filters or membranes, the membrane technology is expanded too much. The application of membrane filtration process is widely applied for the removal of biological solids over the conventional activated sludge process. The retention of biological solids and microorganism over the surface of membrane makes membrane filtration process an effective one. The technology is also applied as a tertiary wastewater treatment. The membrane treatment process is categorized as electro dialysis, reverse osmosis, nanofiltration, microfiltration, and ultrafiltration. Membrane filtration has an extensive application for the treatment of different types of wastewater. The advantages of treatment technology over conventional treatment technologies are that it has higher efficiency, energy consumption is very less, and there is no pollution load. The major differences on these treatment processes vary according to the pore size distribution, pore size, porosity, applied pressures, and membrane permeability. Over the various advantages, process has some limitations as well. Membrane filtration undergoes fouling which is undesirable due to disrupts in membrane selectivity, reduction of permeate flux or permeate quality. The challenging issue in the membrane filtration process is the performance of filtration because of standard blocking, complete blocking, intermediate blocking, and cake filtration. Standard blocking is the deposition of particle on the wall of pores, complete blocking occurs when the foulant particle is larger than the pore at the surface of membrane, intermediate blocking also same as complete blocking which closed the pore of the membrane surface, and cake filtration is the buildup of particles beyond intermediate and complete blocking (Murthy and Chaudhari 2009). The fouling of membranes leads to the cleaning of membrane over time periods or replacement of membrane which overall increases the operation and maintenance cost. Due to the physic-chemical interaction between the membrane and biofluid, depositions of biosolids occur which leads to the declination of flux. The solids adsorbed on the membranes which cause internal fouling and blocking of pore on the membrane are considered as irreversible, and it can be removed by use of chemicals.

## 3.5 Photocatalysis Process

Photocatalysis process method is done by using semiconductor which has no toxicity in it. The wavelength of harness light is embedded on the wastewater for the treatment instead of chemical. The process consists high stability, low-cost operation, and simple design. The application of process is for organic matter degradation, microorganism disinfection, decomposition of water for the production of hydrogen gas. It is also applicable for the removal of pollutants containing heavy metals. The method includes five steps; the contaminants retain into the surface from aqueous phase, then on the semiconductor surface they get absorbed, after that photo catalytic reaction takes place on the absorbed phase and finally contamination degraded and removed from the interface region. The process uses activator as light throughout the reaction. Chalcogenide type with oxides photocatalytic semiconductors is mostly used such as WO2, TiO2, ZnO2, CeO2, ZnO, and sulfides (CdS, WS2, ZnS). TiO2 are safest to work, chemically inert, needful for oxidizing every inorganic and organic particles, stable against photo-corrosion and also less expensive than others. The process treatment faces issues such as in absorbing visible lights, recombination of electron or hole, production of unwanted by-products (Fathinia and Khataee 2015).

## 3.6 Nanotechnology

Nanotechnology causes the structural chances of matter at their atomic and molecular level. Nanotechnology gained accomplishment as a wastewater treatment method. Nanotechnology is a science of nanomaterial. These materials are very effective for the removal of pollutants such as color, heavy metals, organic and inorganic solvents, biological toxins, and pathogens from wastewater (Kumar et al. 2014). Nanoparticles are of three types: reactive, adsorptive, and hybrid magnetic. Reactive nanomaterials like nano zero-valent iron are used to remove heavy metal from wastewater. It degrades the heavy metal into harmless products. Nano zero-valent irons are also used for in-situ wastewater treatment. It undergoes a chemical transformation with biodegradable matter present in the wastewater. Adsorptive materials are nanomagnetic oxides and are widely used due to its high stability, surface, and mesoporous structure. The efficiency depends on the nature of absorbent and physicochemical condition of system (Torgal et al. 2013). Other nanoparticles used in wastewater treatment are a hybrid-type nanoparticle. In a hybrid magnetic nanoparticle, magnetic and nanometer-scaled components are used. It has low toxicity, high surface-to-volume ratio, better adsorption capacity, and effective contaminants removal rate. Iron oxides ($Fe_3O_4$ and g-$Fe_2O_3$) are the main magnetic nano-sized material used as hybrid magnetic nanoparticles (Digigow et al. 2014).

Nanotechnology is very effective for the treatment of wastewater but the immediate consideration of technology is not recommended because to ensure the safety. As most of the nanomaterials used in the technology are toxic in nature. If they are found in effluent, so they can cause health risk even more than the contaminated wastewater. Moreover, various researches are going on for the verification of nanoparticles that ensure safety after treatment. Nanotechnology is not a cost-effective method.

## 4 Challenges of Wastewater Management

Wastewater has been a serious human health and the environment concern, and its management has been a persistent struggle. In the last few years, wastewater treatment research has gained a significant growth. Various contaminations at higher concentration have been identified in wastewater like emerging contaminants, heavy metals, etc. from the domestic wastewater as well. The main challenges while managing the wastewater are basic sanitations, social awareness and legal aspects, energy consumption, sludge production, and reuse of treated water are discussed below.

### *4.1 Sanitations*

The world population was predicted to be 7.7 billion people in the most recent census in 2018. However, basic sanitations are using only about 5.4 billion people, which includes pit latrines, amenities, and composting toilets that are not shared with other houses and are connected to sewerage system or on-site septic tank. The availability of basic sanitation to every household is human right as per the United Nations General Assembly in 2010 through Resolution 64/292, but 2.3 billion people around the world still don't have basic sanitation services (UNICEF, WHO 2017).

As per reports, there are many people lacks the toilet facilities in their household but from the last few years, there is significant growth has seen as improvisation in basic sanitation. In 2000, only 58% people have access to the water and sanitation facilities but up to 2015, around 68% people are using the facilities. However, the trajectory may vary per country. In 2015, 148 nations were capable of providing sanitation to more than 80% of their inhabitants, while only 16 countries were able to provide basic sanitation to less than 20% of their resident. Only 10% toilets are constructed in countries, such as Ethiopia, South Sudan, and Madagascar (World Bank Group 2018).

With the increase in development, the demand for access to basic sanitation poses greater threat to Africa and Asia. According to predictions, the world's population will rise in the coming decades, necessitating a huge expenditure by countries to improve and maintain sanitation systems. Economic solution needs to be identified in the countries to provide proper toilets facilities with on-site sanitation system and access to those sanitation facilities should be mandatory by keeping safe and healthy environment. Those countries that are lacking in the sanitation facilities should primarily limit their need to construct on-site sanitation facilities. This includes, among other things, water conservation (dry sanitation), feacal sludge treatment to prevent disease transmission and environmental contamination, and the recovery from nutrients.

## 4.2 Legal Aspects

The Clean Water Act, i.e., Federal Water Pollution Control Act, 1948, which was updated in 1972, was the first law adopted in the USA to manage water pollution which included financing to enhance wastewater treatment. The USA and Canada agreed in 1978 to limit certain hazardous contaminants in the Lakes. The USA's Environmental Protection Agency (EPA) issued guidelines in 1980 to help towns plan for the introduction of non-potable utilization of urban wastewater systems. Similarly, the European Union established Directive 91/271/EEC (which further updated in 2017) in 1991, which sets the standards for the collection, wastewater treatment, and release of municipal wastewater as well as wastewater from industrial sectors. (EEC 1991). This Directive also examines the possibility that a wastewater discharge would harm waterways to locally available downstream water bodies. In 2014, only 18 countries came forward in compliance with the criteria for collection of wastewater collection (article 3), for the treatment processes only 7 countries worked (article 5), and for the release of treated wastewater only 4 countries agreed (article 5). Only Germany, Austria, and the Netherlands met all of the standards. Failing to follow could result in economic fines, as seen in the north of Spain, which was ordered by the European Court to charge the European Commission 12 million euros for 17 urbanized areas with more than 15,000 residents that lacked appropriate collection of wastewater and treatment systems (European Court of Justice 2018). Parallel to this, the European Union reinforced its commitment to protect quality of the water by adopting the Water Framework Directive in 2000 (European Parliament 2000), which has now been supplemented by rules on Connecting Water Reuse into Water Management and Planning, which consider the environmental management accounting and socioeconomic advantages. Finally, a proposal for a European Union Rule on basic requirements for water resource conservation has been in the works since 2017 (Council of the European Union 2019). This proposal addresses reclaimed, risk management, water permit obligations, compliance checks, state collaboration, public information, and reclaimed quality of water, which include biochemical oxygen demand, maximum E. coli number, and turbidity values. However, a careful examination of the proposed regulation revealed that it did not adequately address some significant issues of interest, such as new pollutants and antibiotic resistance disseminated via wastewater treatment plant-treated effluent. Furthermore, it should be highlighted that the extent of this law only applies to irrigation purpose, not potable water reuse.

### 4.3 Energy

Most towns and industries utilize wastewater treatment plants (WWTPs) to limit toxic wastewater release into receiving streams. Most of the studies found out the WWTPs were mainly focused on the treatment of to their discharge limits, not on giving enough thought to the amount of energy treatment plant unit requires, and which is also an important parameter. However, because sustainable resources of both water and energy, as well as corresponding amount of carbon emissions to environment. All of them are crucial for the overall development. The energy requirement for the operation of wastewater treatment systems in WWTPs is also becoming a topic of study. The water-energy relation is now becoming a major topic in current research, motivating lots of new studies on the use of energy and water relation in WWTPs for long-term development. Water scarcity is increasing the rapid usage of energy for water transit, supply, and treatment systems across many places around the world. Furthermore, with the increased focus on climate issues, energy conservation, improved energy efficiency, and the search for alternate option for energy sources is also included as a aim in the sustainable development. Relationships among water and energy are critical in WWTPs, for example. Water quality is enhanced in most WWTPs but they increase the cost and utilization of the energy. The majority of important processes of a WWTP, such as wastewater collection units, mechanical equipment for physical treatment, working units in biological treatment, chemical treatment, pumps and motors in sludge treatment, and discharge, all necessitate a significant amount of energy. Energy consumption accounts for 25–40% of operational costs in a standard WWTP. Furthermore, the associated greenhouse gas emissions from WWTP energy use are a source of worldwide concern (Wett et al. 2007). With the increase in population, wastewater generation may rise which leads to higher degree of treatment as per their standard limits and may increase the demand of energy while treating the wastewater. According to a research conducted in southern California, WWTPs consume around 20% of the total energy consumed by the municipality. However, due to the increase in pollutant load by rapid urbanization and progressively strong regulations and related environmental standards and criteria for water quality and water reclamation in the USA, it is estimated that 20% rise in proportion of wastewater in the upcoming next 10–15% years. Due to rising energy costs and environmental concerns, the possibility for energy-independent WWTPs has now become a focus of research and development. Although carbon neutrality or zero amount of carbon discharge basically referred as energy self-sufficient WWTPs, energy neutrality, and carbon–neutral (zero GHGs emission), WWTPs are not the same. The two main are necessary to provide self-adequacy energy in treatment units are (a) savings of energy can be achieved through increasing the efficiency of process, (2) energy recovery from the alternative options such as renewable sources from naturally available resources.

## 4.4 Sludge

Excess sludge that is generated from the on-site sanitation systems and wastewater treatment technologies is still an issue. Its production rises year after year, particularly in emerging countries. While Europe's excess sludge production has doubled, in China, it has doubled in size during the previous 20 years. According to European statistics over the last five years, an individual adds 90 g of dry mass to the final product of excess sludge on a daily basis. Since the European Union's (EU) population surpassed half a billion, sewage sludge production has risen to more than 45,000 tonnes per day (Garrido-Baserba et al. 2015). As a result, a great number of EU regulations address sludge treatment and waste management issues. The directives must be implemented by all EU member states. As a result, multiple regulatory acts have been enacted that establish the maximum permitted quantities of pollutants concentration when various types of waste are released into it.

The management of sludge is a challenging task because it depends on the factor, the level of potentially toxic contaminants contained in the sludge, as well as the type and concentration of toxic substances release after the treatment of wastewater. A technique would strike a balance between feasible concentration decrease and neutralization of specific categories of toxic compounds. Due to the technologies used to neutralize groups of toxic substances, the likelihood of new, potentially hazardous compounds forming, as well as a rise in concentrations of those already present in the controlled sludge, all these issues cause challenges while sludge management. Mostly, it was found that the wastewater treatment process at the treatment plant creates an odor emission. This could have a significant impact on neighborhood people and also on the quality of the air. For the cost reasons, mostly deodorization technologies are rarely used at municipal wastewater treatment plants. The gradual but steady growth in the houses linked to sewer networks, as well as addition in the number of wastewater treatment technologies have contributed to an increase in the production of sludge which further create challenges to be manage safely. In addition, every nation has to implement their own legal actions for the sludge management. The need for raw materials restored from waste and energy recovery is expanding in together keeping in mind for providing benefits to the society and for environmental protection. The maximum permissible pollutant concentrations for sludge can't be easily determined because of huge variation in characteristics of upcoming sludge and their management methods. EU rules on waste management and the setting of maximum allowed concentrations are applied in each country that is commissioned under the European Union. It's important to note that any material released into the environment must not pose harm to animals and flora. Controlling pollutants emitted into the environment is required even if pollutant emission criteria are not set. The management of sludge without high-temperature processing could contain a wide spectrum for potentially hazardous organic chemicals, whose inspection and detection would cost a lot of money. Even 99% of potentially harmful organic compounds can be neutralized using high-temperature procedures. The concentration of heavy metal ions, on the other hand, rises during thermal processing. It does not imply that the production of ions would remain constant. Heavy metal ions should be immobilized in product or transformed to somewhere within their harmless state in both circumstances, which is attainable when using high-temperature procedures. The detection of certain species of chemicals during the execution of a management of sludge is often required by law. As a result, chemical present in the sludge should be analyzed properly and their limits should be accessed. The proper monitoring of chemical present in the sludge is necessary and it is a challenging task. Chemicals in sludge can't be avoided in any circumstances. The contaminants discovered in the sludge and other components obtained during processing are representative of the chemicals used in the wastewater treatment and sludge treatment methods, products used by the community that are coming into the wastewater and also wastewater from commercial and industrial activities. Because the present data only provide a broad picture of the characteristics of emerging pollutants in the sludge, further more research is needed to identify level of contaminants (Cieślik et al. 2018).

## *4.5 Reuse*

Due to the increase in an awareness toward water crisis and increasing the water demand, the concept of reuse of treated water stated previously is generally well recognized. However, reuse for non-potable uses such as gardening, pavement cleaning, park irrigation, etc. Reuse of treated waste for the drinking water can have low adaptation among people. The "yuck factor," a collection of negative emotional reactions that reflect the discomfort and dread associated with recycled wastewater, is a significant barrier to social acceptance. However, there are ways to reduce the yuck factor and promote acceptance of water reuse, such as by using appropriate words. Indeed, using the terms "recycled water" instead of "treated wastewater" or "purified water" instead of "effluent" resulted in a major change in public opinion.

Public support is a major challenge toward wastewater management. Many countries came with the new concepts to spread awareness among people. Similarly, Singapore uses NEWater instead of recycled waste. Similarly, in the Singapore, the authorities is don't like to use term "wastewater." They uses the term recycled water and prefer to call it as a NEWater. "NEWater," is frequently cited as an example of reuse of treated water that gains a huge public support (74%) (Timm and Deal 2018). Furthermore, the communication between the organizations and communities also involvement of public will increase acceptance of reuse of treated water to communities and also supplement availability of freshwater resources by building and maintaining confidence while giving knowledge to them about safety of available treated water.

As a result, water reuse has significant environmental and human benefits; nevertheless, management of wastewater treatment should not just restrict to increasing treatment technology, but also focus on complete treatment of wastewater up to their desired limits and toward social awareness and the involvement of public in the management. Government sector and organization, in particular, should continue

**Table 5** Challenges in reuse water

| Reuse sector | Reuse purposes | Challenges |
|---|---|---|
| Residential | Toilet flushing, gardening, washing | Plumbing arrangements, public perceptions |
| Commercial | Public parks & school yards, highway median, fire protection, nonfood crops, nurseries | Seasonal variations, sewerage arrangements, water demand |
| Groundwater/water bodies recharge | Rivers, lakes, ponds, green belt area, wetlands | Topographical changes, natural water resource quality degradation |
| Industries | Boiler water makeup, cooling tower water, process water, toilet flushings, fire protections | Constant water demand, high treatment quality |

to collaborate with stake holders and integrate public participation and societies in decision for the management of wastewater and their reuse. The challenges while reusing water are shown in Table 5 (The International Water Association 2018; United Nations Water and UNESCO 2017; UNICEF, WHO 2017; Timm and Deal 2018).

## 5 Conclusion

Several countries worldwide are experiencing rising pressure on their freshwater resources that's why water sources are running out to meet demand. Wastewaters should be treated up to their discharge limit in greater volumes and can be used for more uses around the world. As shortage for water is growing day by day in many cities. There is availability of various advancement in wastewater treatment technology. Moreover, based on their local water reuse requirements, the untreated wastewater must meet the criteria of appearance, socioeconomic acceptance, health safety, and economic feasibility. The characteristics of wastewater must be constantly reevaluated to find out the concern contaminants. The toxic and chemical element analysis remained the most difficult problems in wastewater research. The properties and concentration of collected wastewater are affected by a variety of wastewater contamination sources. These contaminants can be treated by using various wastewater treatment methods. Physical, biological, and chemical wastewater treatment methods include the removal of particles, organic debris, nutrients (mostly nitrogen and phosphorus), and soluble contaminants such as heavy metals, emergent pollution, and so on. The desired wastewater treatment process methods should be selected on the basis of efficiency, cost, sludge output, chemical operation, carbon footprint, and operation and maintenance. A brief overview of new pollutants (heavy metals and emerging contaminants) is required, including its possible environmental impact and consequence because they are detecting in a wastewater. By looking into domestic wastewater and the origins of key contaminants, treatment process methods can be selected that do not lead to their absorption into wastewater. The wastewater treatment operations are not confined for the complete removal of emerging contaminants from wastewater. They are increasing day by day into wastewater. It is necessary to get a better knowledge regarding the origins of emerging contaminants and their characteristics inside the sewerage system in order to identify more sustainable strategies to deal with wastewater management. Basic cleanliness, social awareness, and legal aspects, energy consumption, sludge generation, and reuse of treated water are the key challenges in wastewater management. The management of wastewater should be considered in the view of public's opinion. Wastewater research must also consider societal issues such as social acceptability of water recycling. Reuse of water should be increased to save freshly available wastewater.

## References

Acumen Research and Consulting (ARC) (2019) Industrial wastewater treatment market (by process: primary treatment, secondary treatment, tertiary treatment; by type: coagulants, biocides & disinfectants, flocculants; by end-use industry: mining, power generation, chemical)-global industry analysis, market size, opportunities and forecast, 2019 – 2026, 190

Alyüz B, Veli S (2019) Kinetics and equilibrium studies for the removal of nickel and zinc from aqueous solutions by ion exchange resins. J hazard Mater 167(1–3):482–488

Ashrafi O, Yerushalmi L, Haghighat F (2015) Wastewater treatment in the pulp-and-paper industry: a review of treatment processes and the associated greenhouse gas emission. J Environ Manag 158:146–157

Azimi A, Azari A, Rezakazemi M, Ansarpour M (2017) Removal of heavy metals from industrial wastewaters: a review. ChemBioEng Rev 4(1):37–59

Boxall AB, Rudd MA, Brooks BW, Caldwell DJ, Choi K, Hickmann S, Van Der Kraak G (2012) Supplemental material pharmaceuticals and personal care products in the environment: what are the big questions? Environ Health Perspect 120:9

Brausch JM, Rand GM (2011) A review of personal care products in the aquatic environment: environmental concentrations and toxicity. Chemosphere 82:1518–1532

Bora AJ, Dutta RK (2019) Removal of metals (Pb, Cd, Cu, Cr, Ni, and Co) from drinking water by oxidation-coagulation-absorption at optimized pH. J Water Process Eng 31:100839

Burkhardt M, Zuleeg S, Vonbank R, Schmid P, Hean S, Lamani X, Bester K, Boller M (2011) Leaching of additives from construction materials to urban storm water runoff. Water Sci Technol 63(9):1974–1982

Charerntanyarak L (1999) Heavy metals removal by chemical coagulation and precipitation. Water Sci Technol 39(10–11):135–138

Chen JP, Lim LL (2005) Recovery of precious metals by an electrochemical deposition method. Chemosphere 60(10):1384–1392

Chen Q, Yao Y, Li X, Lu J, Zhou J, Huang Z (2018) Comparison of heavy metal removals from aqueous solutions by chemical precipitation and characteristics of precipitates. J Water Process Eng 26:289–300

Chung TS, Li X, Ong RC, Ge Q, Wang H, Han G (2012) Emerging forward osmosis (FO) technologies and challenges ahead for clean water and clean energy applications. Curr Opin Chem Eng 1(3):246–257

Cieślik, BM, Świerczek L, Konieczka P (2018) Analytical and legislative challenges of sewage sludge processing and management. Monatshefte für Chemie Chem Monthly. 149(9):1635–1645

Collado N, Rodriguez-Mozaz S, Gros M, Rubirola A, Barcelo D, Comas J, RodriguezRoda I, Buttiglieri G (2014) Pharmaceuticals occurrence in a WWTP with significant industrial contribution and its input into the river system. Environ Pollut 185:202–212

Council of the European Union (2019) Interinstitutional File: 2018/0169(COD)

Crini G, Lichtfouse E (2019) Advantages and disadvantages of techniques used for wastewater treatment. Environ Chem Lett 17(1):145–155

Digigow RG, Dechézelles JF, Dietsch H, Geissbühler I, Vanhecke D, Geers C, Petri-Fink A (2014) Preparation and characterization of functional silica hybrid magnetic nanoparticles. J Magnet Magnetic Mater 362:72–79

Du B, Haddad SP, Luek A, Scott WC, Saari GN, Burket SR, Breed CS, Kelly M, Broach L, Rasmussen JB, Chambliss CK, Brooks BW (2015) Bioaccumulation of human pharmaceuticals in fish across habitats of tidally influenced urban bayou. Environ Toxicol Chem

European commisison (2013) Directive 2013/39/EU of the European Parliament amending Directives 2000/60/EC and 2008/105/EC as regards priority substances in the field of water policy JO-EU L226/1.2013

EEC (1991) Council Directive 91 /271 /EEC concerning urban waste water treatment, p. 13

European Commission (2011) Screening report Iceland: Chapter 27-Environment, European Commission, Brussels. http://ec.europa.eu/enlargement/pdf/iceland/keydocuments/screening_report_27_is_internet_en.pdf

European Court of Justice (2018) European Commission V Kingdom of Spain, Case C-205/17

Fathinia M, Khataee A (2015) Photocatalytic ozonation of phenazopyridine using TiO2 nanoparticles coated on ceramic plates: mechanistic studies, degradation intermediates and ecotoxicological assessments. Appl Catal A 491:136–154

Feng S, Ngo HH, Guo W, Chang SW, Nguyen DD, Cheng D, Varjani S, Lei Z, Liu Y (2021) Roles and applications of enzymes for resistant pollutants removal in wastewater treatment. Biores Technol 335:125278

Ferreira SL, de Brito CF, Dantas AF, de Araujo NML, Costa AS (1999) Nickel determination in saline matrices by ICP-AES after sorption on Amberlite XAD-2 loaded with PAN. Talanta 48(5):1173–1177

Fu F, Wang Q (2011) Removal of heavy metal ions from wastewaters: a review. J Environ Manag 92(3):407–418

Garrido-Baserba MMSPG, Molinos-Senante M, Abelleira-Pereira JM, Fdez-Güelfo LA, Poch M, Hernández-Sancho F (2015) Selecting sewage sludge treatment alternatives in modern wastewater treatment plants using environmental decision support systems. J Cleaner Prod 107:410–419

Gebretsadik H, Gebrekidan A, Demlie L (2020) Removal of heavy metals from aqueous solutions using Eucalyptus Camaldulensis: An alternate low cost adsorbent. Cogent Chem 6(1):1720892

Gerbersdorf SU, Cimatoribus C, Class H, Engesser K-H, Helbich S, Hollert H, Lange C, Kranert M, Metzger J, Nowak W, Seiler T-B, Steger K, Steinmetz H, Wieprecht S (2015a) Anthropogenic Trace Compounds (ATCs) in aquatic habitats—research needs on sources, fate, detection and toxicity to ensure timely elimination strategies and risk management. Environ Int 79:85–105. https://doi.org/10.1016/j.envint.2015.03.011

Gerbersdorf SU, Cimatoribus C, Class H, Engesser K-H, Helbich S, Hollert H, Lange C, Kranert M, Metzger J, Nowak W, Seiler T-B, Steger K, Steinmetz H, Wieprecht S (2015b) Anthropogenic Trace Compounds (ATCs) in aquatic habitats —research needs on sources, fate, detection and toxicity to ensure timely elimination strategies and risk management. Environ Int 79:85–105

Gleick PH (1996) Basic water requirements for human activities: meeting basic needs. Water Int 21(2):83–92

Global Water Research Coalition (2008) Tool for risk management of water utility assess Report Ref. No. 08/RG/05/25. London: UK water Industry Research

Gogoi A, Mazumder P, Tyagi VK, Chaminda GT, An AK, Kumar M (2018) Occurrence and fate of emerging contaminants in water environment: a review. Groundw Sustain Dev 6:169–180

Hopkins J (2012) The 'sacred sewer': tradition and religion in the Cloaca Maxima. Rome, Pollution, and Propriety. Dirt, Disease, and Hygiene in the Eternal City from Antiquity to Modernity, British School at Rome studies (Cambridge), 81–102

Hosomi M (2016) New challenges on wastewater treatment. Clean Technol Environ Policy 18(3):627

Il'in VI, Kolesnikov VA, Parshina YI (2002) Purification of highly concentrated industrial sewage from the porcelain and faience industry by the electric flotation method. Glass Ceramics 59(7):242–244

Jasinska EJ, Goss GG, Gillis PL, Van Der Kraak GJ, Matsumoto J, de Souza Machado AA, Giacomin M, Moon TW, Massarsky A, Gagné F, Servos MR, Wilson J, Sultana T, Metcalfe CD (2015) Assessment of biomarkers for contaminants of emerging concern on aquatic organisms downstream of a municipal wastewater discharge. Sci Total Environ 530–531:140–153. https://doi.org/10.1016/j.scitotenv.2015.05.080

Jones OAH, Green PG, Voulvoulis N, Lester JN (2007) Questioning the excessive use of advanced treatment to remove organic micropollutants from wastewater. Environ Sci Technol 41:5085–5089. https://doi.org/10.1021/es0628248

Kumar S, Ahlawat W, Bhanjana G, Heydarifard S, Nazhad MM, Dilbaghi N (2014) Nanotechnology-based water treatment strategies. J Nanosci Nanotechnol 14(2):1838–1858

Lee E, Rout PR, Shin C, Bae J (2019) Effects of sodium hypochlorite concentration on the methanogenic activity in an anaerobic fluidized membrane bioreactor. Sci Total Environ 678:85–93

Larsen TA, Hoffmann S, Lüthi C, Truffer B, Maurer M (2016) Emerging solutions to the water challenges of an urbanizing world. Science 352(6288):928–933

Liu YJ, Gu J, Liu Y (2018) Energy self-sufficient biological municipal wastewater reclamation: present status, challenges and solutions forward. Biores Technol 269:513–519

Liu Y, Gu J, Zhang M (2019) AB processes: towards energy self-sufficient municipal wastewater treatment. IWA Publishing

Lofrano G, Brown J (2010) Wastewater management through the ages: A history of mankind. Sci Total Environ 408:5254–5264

Luo L, Kaur G, Zhao J, Zhou J, Xu S, Varjani S, Wong JW (2021) Optimization of water replacement during leachate recirculation for two-phase food waste anaerobic digestion system with off-gas diversion. Biores Technol 335:125234

Melvin SD, Houlahan JE (2012) Tadpole mortality varies across experimental venues: do laboratory populations predict responses in nature? Oecologia 169:861–868

Merel S, Snyder SA (2016) Critical assessment of the ubiquitous occurrence and fate of the insect repellent N,N-diethyl-m-toluamide in Water. Environ Int 96:98–117

Mission SB (2015) Solid and liquid waste management in rural areas. Report by Ministry of Drinking Water & Sanitation. Govt. of India

Mompelat S, Le Bot B, Thomas O (2009) Occurrence and fate of pharmaceutical products and by-products, from resource to drinking water. Environ Int 35(5):803–814

Murthy ZVP, Chaudhari LB (2009) Separation of binary heavy metals from aqueous solutions by nanofiltration and characterization of the membrane using Spiegler-Kedem model. Chem Eng J 150(1):181–187

Oviedo KN, Aga DS (2016) Lessons learned from more than two decades of research on emerging contaminants in environment. J Hazard Mater 316:242–251

Petrović M, Gonzalez S, Barceló D (2003) Analysis and removal of emerging contaminants in wastewater and drinking water. Trends Anal Chem 22:685–696. https://doi.org/10.1016/j.ejca.2003.09.021

Ranade VV, Bhandari VM (2014) Industrial wastewater treatment, recycling and reuse. Butterworth-Heinemann

Rojas MR, Leung C, Bonk F, Zhu Y, Edwards L, Arnold RG, Sáez AE, Klecka G (2013) Assessment of the effectiveness of secondary wastewater treatment technologies to remove trace chemicals of emerging concerns. Crit Rev Environ Sci Technol 43:1–36

Salgot M, Folch M (2018) Wastewater treatment and water reuse. Curr Opin Environ Sci Health 2:64–74
Sahmoune MN, Yeddou AR (2016) Potential of sawdust materials for the removal of dyes and heavy metals: examination of isotherms and kinetics. Desalin Water Treat 57(50):24019–24034
Shahedi A, Darban AK, Taghipour F, Jamshidi-Zanjani A (2020) A review on industrial wastewater treatment via electrocoagulation processes. Curr Opin Electrochem 22:154–169
Shi CY (2011) Mass flow and energy efficiency of municipal wastewater treatment plants. IWA Publishing
Shindhal T, Rakholiya P, Varjani S, Pandey A, Ngo HH, Guo W, Ng HY, Taherzadeh MJ (2021) A critical review on advances in the practices and perspectives for the treatment of dye industry wastewater. Bioengineered 12(1):70–87
Simka W, Puszczyk D, Nawrat G (2009) Electrodeposition of metals from non-aqueous solutions. Electrochim Acta 54(23):5307–5319
Stuart M, Lapworth D, Crane E, Hart A (2012) Review of risk from potential emerging contaminants in UK groundwater. Sci Total Environ 416:1–21. https://doi.org/10.1016/j.scitotenv.2011.11.072
Sumpter JP, Johnson AC (2005) Lessons from endocrine disruption and their application to other issues concerning trace organics in the aquatic environment. Environ Sci Technol 39(12):4321–4332
The International Water Association (2018) Wastewater report. The Reuse Opportunity. https://www.iwa-network.org/wp-content/uploads/2018/02/OFID-Wastewater-report2018.pdf
The Trade Council, Embassy of Demark, India, New Delhi. Sector Analysis. Water industry, India. 2015
Timm SN, Deal BM (2018) Understanding the behavioral influences behind Singapore's water management strategies. J Environ Planning Manage 61(10):1654–1673
Toczyłowska-Mamińska R (2017) Limits and perspectives of pulp and paper industry wastewater treatment–A review. Renew Sustain Energy Rev 78:764–772
Torgal FP, Diamanti MV, Granqvist CG (2013) Nanotechnology in eco-efficient construction. Woodhead Publishing. EDS
UNICEF, WHO (2017) Progress on drinking water, sanitation and hygiene: 2017 update and SDG baselines, Geneva
UN Environment Programme (UNEP) (2012) Geo 5, Global Environmental Outlook. Environment for the Future we want
United Nations Water and UNESCO (2017) UN world waster development report wastewater: the untapped resource
US EPA (2012) Water: contaminant candidate list 3. US. Environmental Protection Agency, Washington, DC (http://water.epa.gov/scitech/drinkingwater/dws/ccl/ccl3.cfm)
Van Lier JB (2008) High-rate anaerobic wastewater treatment: diversifying from end-of-the-pipe treatment to resource-oriented conversion techniques. Water Sci Technol 57(8):1137–1148
Voulvoulis N, Arpon KD, Giakoumis T (2017) The EU Water Framework Directive: From great expectations to problems with implementation. Sci Total Environ 575:358–366
Wang H, Wang T, Zhang B, Li F, Toure B, Omosa IB, Chiramba T, Abdel-Monem M, Pradhan M (2014) Water and wastewater treatment in Africa—current practices and challenges. Clean Soil Air Water 42(8):1029–1035
Wett B, Buchauer K, Fimml C (2007) Energy self-sufficiency as a feasible concept for wastewater treatment systems. InIWA Leading Edge Technol Conf 21:21–24
World Bank Group (2018a) People using at least basic sanitation services. World Bank Group
Yang J, Monnot M, Ercolei L, Moulin P (2020) Membrane-based processes used in municipal wastewater treatment for water reuse: State-of-the-art and performance analysis. Membranes 10(6):131
Zhang X, Liu Y (2021) Reverse osmosis concentrate: an essential link for closing loop of municipal wastewater reclamation towards urban sustainability. Chem Eng J 421:127773

The manufacturer's authorised representative in the EU is Springer Nature Customer Service Centre GmbH, Europaplatz 3, 69115 Heidelberg, Germany. If you have any concerns regarding our products, please contact ProductSafety@springernature.com

Printed and bound by CPI Group (UK) Ltd, Croydon, CR0 4YY
30/09/2026
02225318-0008